The November 23rd, 1980 Irpinia-Lucania, Southern Italy Earthquake: Insights and Reviews 40 Years Later

The November 23rd, 1980 Irpinia-Lucania, Southern Italy Earthquake: Insights and Reviews 40 Years Later

Editors

Sabina Porfido
Giuliana Alessio
Germana Gaudiosi
Rosa Nappi
Alessandro Maria Michetti

MDPI • Basel • Beijing • Wuhan • Barcelona • Belgrade • Manchester • Tokyo • Cluj • Tianjin

Editors
Sabina Porfido
Istituto di Scienze
Dell'alimentazione del
Consiglio Nazionale Ricerche
(ISA-CNR), Avellino, Italy
and Istituto Nazionale di
Geofisica e Vulcanologia
Sezione di Napoli
Osservatorio Vesuviano,
(INGV-OV)
Napoli, Italy

Giuliana Alessio
Istituto Nazionale di
Geofisica e Vulcanologia,
Sezione di Napoli
Osservatorio Vesuviano,
(INGV-OV)
Napoli, Italy

Germana Gaudiosi
Istituto Nazionale di
Geofisica e Vulcanologia
Sezione di Napoli
Osservatorio Vesuviano,
(INGV-OV)
Napoli, Italy

Rosa Nappi
Istituto Nazionale di
Geofisica e Vulcanologia
Sezione di Napoli
Osservatorio Vesuviano,
(INGV-OV)
Napoli, Italy

Alessandro Maria Michetti
DISAT Università degli Studi
dell'Insubria Como, Italy and
Istituto Nazionale di
Geofisica e Vulcanologia
Sezione di Napoli
Osservatorio Vesuviano
Napoli, Italy

Editorial Office
MDPI
St. Alban-Anlage 66
4052 Basel, Switzerland

This is a reprint of articles from the Special Issue published online in the open access journal *Geosciences* (ISSN 2076-3263) (available at: https://www.mdpi.com/journal/geosciences/special_issues/earthquake_1980).

For citation purposes, cite each article independently as indicated on the article page online and as indicated below:

LastName, A.A.; LastName, B.B.; LastName, C.C. Article Title. *Journal Name* **Year**, *Volume Number*, Page Range.

ISBN 978-3-0365-6816-4 (Hbk)
ISBN 978-3-0365-6817-1 (PDF)

Cover image courtesy of Albert Pissart
The photo, taken on 4-4-1981, shows the primary surface faulting on the bedrock fault scarp near Senerchia town: It is visible the free face on the limestone fault plain

Contents

About the Editors

Sabina Porfido

Dr. Sabina Porfido is engaged in research on the evaluation and reduction of natural hazards. The research activities mainly address the study of recent and historical seismicity, macroseismic studies, seismic microzonation, seismotectonic processes and the definition of seismogenic areas, as well as the study of induced seismic effects: primary and secondary ground effects, and a new macroseismic scale for the evaluation of the effects induced by the environment (INQUA scale, 2003–2004; ESI scale, 2007). She also researches natural hazards connected to hydrogeological events: floods in the Campania region, coastal hazards; the environment, land use and Resilience. She has participated in numerous national and international research projects: INQUA-TERPRO, EEE Catalogue ISPRA; Cooperazione Bilaterale Italia-Ungheria (CNR-MTA); IODP-Campi Flegrei; study of the surface faulting accompanying the 2017 Casamicciola earthquake (Ischia Island); Proyecto de red interuniversitaria regional Italo-Centroamericana en análisis de los fenómenos naturales para la evaluación de la peligrosidad en Centro América (Guatemala, Nicaragua and El Salvador-Ministero Affari Esteri-Cooperazione allo sviluppo); Accordo bilaterale Italia-Moldavia.

Giuliana Alessio

Dr. Giuliana Alessio mainly deals with the seismicity and seismotectonics of active faults in the Central-Southern Apennines (Irpinia; Sannio-Beneventano; Matese chain; Salento peninsula) and in the volcanic areas of Vesuvius, Campi Flegrei and Ischia island, studying their geometry and activity according to the space–time correlation of local seismicity, field data recognition, recent deformation data, and historical seismicity. She participated in field surveys for the seismic emergencies of L'Aquila, 2009; Emilia-Romagna, 2012; Central Italy (Amatrice and Norcia) 2016; the island of Ischia, 2017; as well as for the Stromboli volcanic eruptions of 2002/2003 and 2019. She also has applied the multiparameter approach for hydrogeologic hazard assessment in the Somma-Vesuvius and Sorrento peninsula areas. She operates for the mitigation of seismic and volcanic risk with scientific dissemination activities for various national and international Institutions and geoarcheological campaigns.

Germana Gaudiosi

Dr. Germana Gaudiosi studies the geodynamical models of Western Mediterranean and shallow structure of volcanic areas of southern Italy from joined analysis of active and passive seismic data, deep crustal structure of Italian peninsula from active seismological techniques (Vesuvio, Campi Flegrei, Ischia, Campanian Plain, Eolie, Etna, southern-central Appennines, Eastern Sicily, Ionian sea). Moreover, through the integrated analysis of tectonic, seismic and gravimetric data, she found the geometry of active/silent outcropping/buried faults of Central-Southern Italy (campanian Plain, Mt. Matese, central Apennines). Additionally, she re-evaluated the macroseismic effects at the MCS scale and ground effects in natural environment according to the ESI 2007 scale. She has participated in numerous national and international research projects and collaborated with Universities of Naples Federico II, University of Cosenza, University of Palermo, University of Milano, University of Trieste, University of Paris, CNR-ISA.

Rosa Nappi

Dr. Rosa Nappi studies the fault process in tectonic and volcanic area in the seismotectonic framework, earthquake geology and induced primary and secondary seismic ground effects. Through integrated geological (surface coseismic deformation) and geophysical data, processed in GIS environment, she performed a quantitative geomorphic analysis and ground field deformation to identify the seismogenetic source, with the aim to contribute to the evaluation and reduction of seismic hazards. Additionally, she re-evaluated the macroseismic effects in MCS scale and ground effects in a natural environment according to the ESI 2007 scale. The recent study areas are as follows: Salento Peninsula (Apulia Region, Southern Italy); Central Apennines (l'Aquila earthquake and Amatrice-Visso-Norcia 2016/17 seismic sequences, Ischia island (August 21, 2017, Casamicciola earthquake; Campanian Plain. She has participated in numerous national and international research projects (INQUA project; METIQ Project; INGV Ricerca Libera; FAC Project) and collaborated with the Universities of Naples Federico II, University of Insubria and CNR-ISA.

Alessandro Maria Michetti

Prof. Alessandro Maria Michetti. Author of more than 94 papers, with 2809 citations (Scopus 11 September 2022). Most of my research is focused on the reconstruction of the recent landscape evolution to define geological criteria for assessing natural hazards, particularly seismicity. This includes the integration of historical and instrumental seismicity data with observations from Quaternary, predominantly Holocene, stratigraphy, geomorphology and tectonics, to understand the styles and rates of ongoing crustal deformation, and their interaction with climatically controlled surface processes and, with that information, define the seismic potential at both local and regional scales. This research ranges from laboratory analysis of vintage airphoto coverages and topographic imagery, to the field mapping of recent landforms and deposits, to geomorphic reconstruction of slip rates and paleoearthquake histories of capable faults, as well as detailed studies of trench exposures showing evidence for Holocene earthquake surface faulting. Of particular interest for my research is the study of earthquake ground effects and their use in macroseismic intensity scales. Recent and ongoing projects in which I am involved include: a study of the surface faulting accompanying the August 23 and October 30, 2016, Amatrice and Norcia earthquakes in Central Italy; Surface (INQUA Project 1620R) SURface FAUlting Catalogue, worldwide dataset of Surface Ruptures generated during earthquakes to build reliable empirical relationships (http://www.earthquakegeology.org/index.php?page=projects&s=4).

Acknowledgments: The Editors kindly acknowledge the invaluable contribution of all the referees who assisted in the scientific revision; special thanks to Kathleen Nicoll for help and advice during the final editing of this volume.

Preface to "The November 23rd, 1980 Irpinia-Lucania, Southern Italy Earthquake: Insights and Reviews 40 Years Later"

A little less than two months after 23 November 1980. While on 15 and 16 January 1981, the "Antonio Gramsci" Institute gathered together intellectuals from every sector in Avellino to define "Politics and culture for the reconstruction of the Mezzogiorno", the editor Einaudi published the report of the study group directed by Manlio Rossi-Doria on the Situazione, problemi e prospettive dell'area più colpita dal terremoto del 23 Novembre 1980- Situation, problems and prospects of the area most affected by the earthquake of 23 November 1980'.

At the Avellino conference, Alberto Samonà, Professor at the Faculty of Architecture of the Federico II University in Naples, gave a clear, precise presentation calling for analysis and accountancy of the dramatic events that occurred in the Belice area (Sicily), when it was hit by the earthquake on 15 January 1968; he urged that this case inform responsibility in urban planning and architecture; on the one hand, removing the tendency to be alienated by localised differences and, on the other, implementing a genuine form of participation in the choices to be made. In other words, not locking decisions into academic or professional alternatives between rebuilding towns and cities *where they were* and *as they were*, or building them from scratch, which is, after all, the forbidden dream of every architect.

"I believe that town planning and architecture would be completely powerless if they did not change their attitude", he warned, to invoke "completely different ways of expressing all the capacity for renewal that the situation requires in both practice and theory".

If I look up today and see the outline of the houses built in Bisaccia or Conza della Campania, in Teora or San Mango, and in the many other places in the epicentral area between Irpinia, north of Salerno and Basilicata at that time, I realize that Samonà's appeal has gone tragically unheeded.

Not only has respect for the spirit of the places been betrayed and the lesson of Christian Norberg-Schulz mortified, but what Vezio De Lucia described in a 30 September 1990 article for the Pci newspaper 'l'Unità': "I have seen architectural havoc, they have called it reconstruction". The heart of the southern Apennines had become a marginalized area, like a suburb without a metropolis.

The analysis by Rossi-Doria and his Center for Specialisation and Economic-Agricultural Research for the Southern Italy in Portici, instead, contained special indications of merit for planning development of the earthquake-affected territory, in timely accordance with vocations and within the framework of a more general economic development plan. "This requirement corresponds not only to a sort of moral commitment that the Nation has contracted with itself over the last month, but also to an objective situation", said Manlio Rossi-Doria, strongly emphasizing the work. The area had been divided into zones, and for each one there was a detailed description of the socio-economic situation, land structures, natural features, atmospheric and climatic conditions, and the extent of damages caused by the earthquake. I always go back and reread these reflections whenever there is a discussion about an agri-food prospect for Irpinia, about plans to revive a productive role for "the province of bone", a landscape enhancement to accommodate tourists: in comparison with what Manlio Rossi-Doria foreshadowed, with the meticulous care and passion in the approach by the Center's group, so much debate now seems to me an exercise in unbearable rhetoric, the tired and downright ritualistic ceremony of desires that inconsistency causes to degrade into wishful thinking and of plans that fade into utter inadequacy.

One should assign the reasons for the current state of Irpinia in the lack of response to that call for ideas in January 1981. The reasons for the state of Irpinia today should be sought in the failure to respond to the suggestions offered in January 1981. Four decades after the earthquake that destroyed it, progressively emptied of population and hopes, nailed to a marginality that preludes to irrelevance, I wonder why the content of those documents was not adequately taken into account when defining strategies and choices: why the suggestion of establishing the laboratory of a new and better South on the rubble of the earthquake soon gave way to the to the old politics of welfare and exploitation and the priority of a long vision that reconsidered the mistakes of the past was not affirmed and amended to gain a perspective for the future.

We should start from here and retrace the various events that have taken place, from judicial enquiries to political complaints, from scandals to controversies, from exploitation to convenience. Remembering in order to understand, in an in-depth examination—sincere and merciless—an examination of the reasons for the failure to implement the proposals presented in the conference of the "Gramsci" Institute and in the survey of the Manlio Rossi-Doria Center would lead to identifying the truths.

Forty years constitute the epochal threshold that Ian Assmann describes as "the moment when living memory is threatened with decline and the forms of cultural remembrance become problematic". For another, there is no event such as an earthquake that one wants to forget so quickly: it may be that this is the only way that one is able to calm the awareness that the ground on which one lives will inevitably suddenly begin to shake again, and then setting aside this certainty constitutes the only possibility of continuing to inhabit the places of catastrophes and tragedies. But it also happens, however, that earthquakes never come to pass, and that the aftermath, the consequences and reconstructions go hopelessly into a kind of temporal loop, almost as if they were on a Moebius strip that twists onto itself endlessly toward eternity.

113 years since Messina's earthquake on 28 December 1908, there was a measure to dismantle the shacks erected during the emergency and evidently inhabited by the successors of that enduring temporary status. And it comes as no surprise that there are still some funds being allocated for the reconstruction of private buildings destroyed or damaged in Campania by the 23 November 1980 earthquake, a residual expense 41 years after the enactment of Law 219, which has overcome the barriers of the euro, centuries, millennia, and modesty, entrusted by the Region to a paradoxical "Committee for the simplification and speeding up of bureaucratic procedures".

The past and present chase each other, the clock hands stop, and the course of history grinds to a halt.

In a paper written in 1992, "Sulle rovine", Ettore Sottsass states that "the present always has its tail in some form of the past".

According to Sottsass, the present and the past end up narrowing the perspective, taking away space from the future, and in this way, he adds, "there are few possibilities left. All we are left with is the present as a place to meet with existence; an ambiguous present, a curious present, an uncertain present, both for the projects of the future and for what we find when we enter the desert—or cemetery—of the ruins of the past". The Irpinia of the villages emptied by the new emigration and the towns of lunar architecture, the dismay of the generation of emptiness and the anger of those who have the stubbornness to want to remain, the nostalgia for an invented past and the struggle to reflect themselves in a sincere memory.

The text is collected in a valuable posthumous book with a title that seems to refer precisely to the inland areas of this time: "Di chi sono le case vuote? Whose empty houses are they?".

Generoso Picone
Journalist, Writer

geosciences

MDPI

Editorial

40 Years Later: New Perspectives on the 23 November 1980, Ms 6.9, Irpinia-Lucania Earthquake

Sabina Porfido [1,2], Giuliana Alessio [2], Germana Gaudiosi [2], Rosa Nappi [2,*] and Alessandro Maria Michetti [2,3]

1 Consiglio Nazionale delle Ricerche-ISA, Via Roma 64, 80100 Avellino, Italy; sabina.porfido@isa.cnr.it
2 Istituto Nazionale di Geofisica e Vulcanologia, Sezione di Napoli Osservatorio Vesuviano, Via Diocleziano 328, 80124 Napoli, Italy; giuliana.alessio@ingv.it (G.A.); germana.gaudiosi@ingv.it (G.G.); alessandro.michetti@uninsubria.it (A.M.M.)
3 Dipartimento di Scienza e Alta Tecnologia, Università degli Studi dell'Insubria, Via Valleggio 11, 22100 Como, Italy
* Correspondence: rosa.nappi@ingv.it; Tel.: +39-0816108324-80124

Citation: Porfido, S.; Alessio, G.; Gaudiosi, G.; Nappi, R.; Michetti, A.M. 40 Years Later: New Perspectives on the 23 November 1980, Ms 6.9, Irpinia-Lucania Earthquake. *Geosciences* **2022**, *12*, 173. https://doi.org/10.3390/geosciences12040173

Received: 9 March 2022
Accepted: 31 March 2022
Published: 15 April 2022

After more than forty years since the 1980 Irpinia-Lucania earthquake, with this Special Issue "The 23 November 1980 Irpinia-Lucania, Southern Italy Earthquake: Insights and Reviews 40 Years Later" we revisit this milestone geological and seismological event, bringing together the latest views and news on this earthquake, with the aim of improving the dissemination of wide-ranging information on this remarkable case history.

This earthquake struck Irpinia-Lucania region (Lucania is also called Basilicata; Southern Italy) on 23 November 1980 (Ms 6.9, Io X MCS) [1,2], and it is remembered in Italy not only for being the strongest earthquake recorded in the last 100 years causing devastation of entire regions and severe loss of human life, but also for the destruction of the cultural heritage in the epicentral area.

It was felt throughout Italy, from Sicily in the South, to Emilia Romagna and Liguria in the North (Figure 1) causing damage in over 800 localities spread in the regions of Campania and Basilicata with a total of 75,000 houses destroyed and 275,000 seriously damaged. The number of victims was about 3000, with 10,000 injured people [1–8].

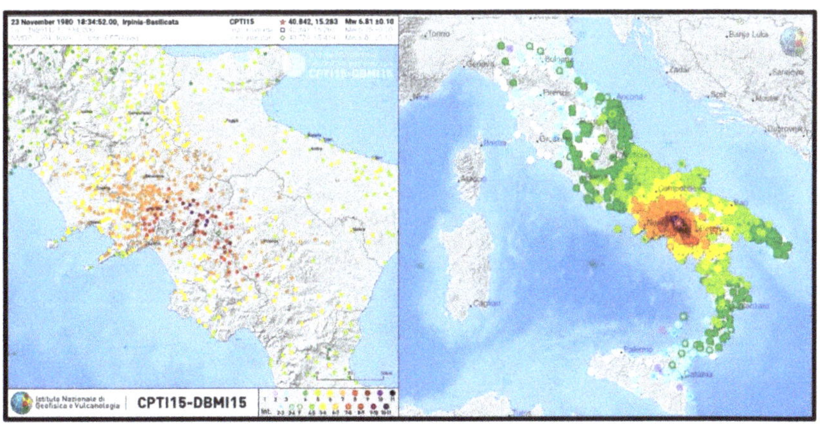

Figure 1. Intensity felt reports of the Irpinia-Basilicata 23 November 1980 earthquake (modified by CPTI15 [2]).

The earthquake also caused several striking effects on the natural environment, including extensive coseismic surface faulting which is still visible today (Figures 2 and 3), and was mapped in the following years for a total length of about 40 km [9–19].

Figure 2. The photos (**a**,**b**), taken in 2004, show the coseismic 1980 fault scarps along M.te Carpineta; the photo (**c**) shows a paleoseismological trench wall along the Piano di Pecore plain (Photos by Rosa Nappi).

Figure 3. The coseismic 1980 fault scarp along M.te Carpineta, 40 years later (Photos by Giuliana Alessio, October 2020).

Moreover, over 200 landslides occurred [20–24]; also, widespread soil fracturing was observed, and minor liquefaction effects [25,26]. Wide changes in water flow rate in some regional karst springs [27,28] were reported. In that extremely dramatic context for Italy, the national scientific community played an important role, through the Geodynamic Finalized Project of the National Research Council ("Progetto Finalizzato Geodinamica", PFG-CNR), that involved in the field many researchers, from different universities and research institutes, contributing to collect instrumental data, macroseismic surveys (Figures 4 and 5), geological fieldwork and mapping (Figures 6 and 7), and seismic engineering analyses, indispensable for the knowledge of the 1980 earthquake and for implementing a proper risk mitigation strategy [1,28–36].

CONSIGLIO NAZIONALE DELLE RICERCHE
PROGETTO FINALIZZATO GEODINAMICA

IL TERREMOTO DEL 23-11-1980
RILIEVO MACROSISMICO
STATO DI AVANZAMENTO AL 27-1-1981

CONVEGNO STRAORDINARIO
DEL PROGETTO FINALIZZATO GEODINAMICA

Roma 27-28 Gennaio 1981

Figure 4. First macroseismic study published about two months after the 23 November 1980, earthquake (January 1981), showing preliminary intensity assessment by PFG researchers [29].

In particular, during the emergency activity for urban reconstruction, preliminary seismic microzonation studies were carried out in collaboration with the Tuscany and Emilia-Romagna regions, involving 39 towns in the epicentral area of Campania and Basilicata, with the objective of providing technical maps, on a scale of 1:5000, with indications of areas with different geological characteristics and the most suitable areas for reconstruction [35]. Other two important publications edited by the CNR- PFG in the field of seismology were certainly the Atlas of Isoseismal maps of Italian Earthquakes, and the first modern Italian Seismic Catalogue (Figure 8) [1,36].

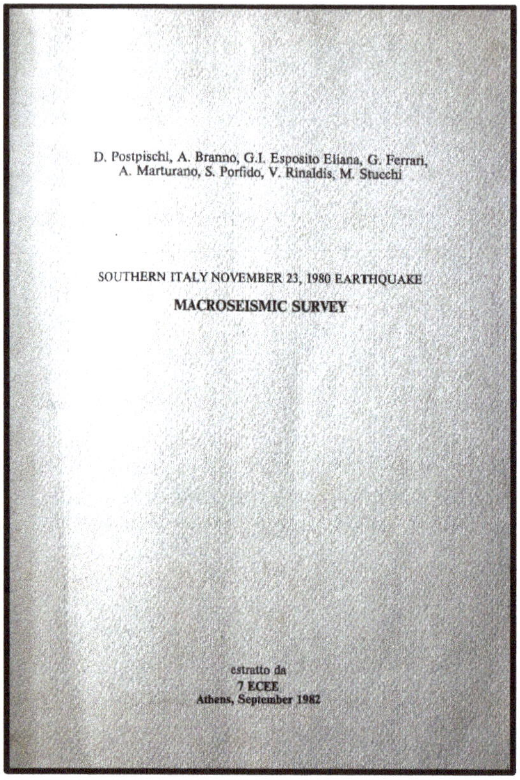

D. Postpischl, A. Branno, G.I. Esposito Eliana, G. Ferrari, A. Marturano, S. Porfido, V. Rinaldis, M. Stucchi

SOUTHERN ITALY NOVEMBER 23, 1980 EARTHQUAKE

MACROSEISMIC SURVEY

estratto da
7 ECEE
Athens, September 1982

Figure 5. Macroseismic survey of the 23 November 1980 earthquake presented at the 7ECEE conference in Athens in 1982 [3].

The Irpinia-Lucania 1980 earthquake is still considered a crucial seismic event for the study of seismicity in Italy and abroad, and for the development of modern seismology, Quaternary geology and active tectonic studies, also including the emerging methodology of paleoseismology in Europe (Figure 2).

The recognition of the primary surface faulting due to the 1980 seismic event, and of its extent, was not obvious at all, for the limited technological tools, difficult environmental conditions and doubts of the scientific community; it required several months of field work by various researchers, and was firstly properly interpreted in 1984 by Rob Westaway and James Jackson [12] near Piano di Pecore, where a ca. 1 m high fault scarp was observed and mapped [11], then detected in the following months and in the neighbouring areas for over 40 km [12–17].

Paleoseismological studies, based on the excavation of trenches across the coseismic fault scarps detected on the surface, began in Italy only ten years later [15]. Thanks to these studies it has been possible to reconstruct the seismic history of the 1980 seismogenic fault, through the recognition of past earthquakes (historic and prehistoric) detectable by the faulted geological strata and the relative dating.

Forty years after the earthquake and with the introduction of modern scientific knowledge, it becomes important and fundamental to reconsider the many relevant and still open research lines that have been triggered by the Irpinia-Lucania earthquake.

Figure 6. Detail of the original 1:25,000 scale topographical map with the most important geological effects surveyed by Italian geologists immediately after the earthquake of 23 November 1980 (fractures, landslides, faults by Carmignani et al., 1981 [11], see also Supplementary Materials S1) right top corner, trace of the most relevant primary earthquake ruptures located at Piano di Pecore on Monte Marzano (see also Figure 2c; courtesy of Paolo Scandone, PFG).

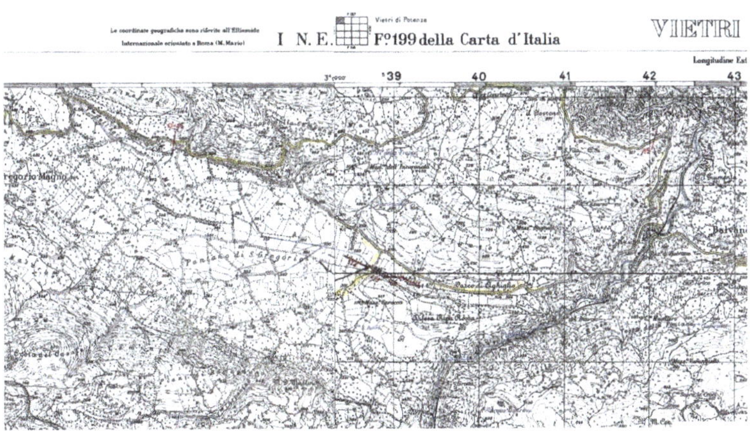

Figure 7. Detail of the original 1:25,000 scale topographical map with the most important geological effects surveyed by Italian geologists immediately after the earthquake of 23 November 1980 (fractures, landslides, faults by Carmignani et al., 1981 [11], see also Supplementary Materials S1); note the trace of the most relevant primary earthquake ruptures located at Pantano di San Gregorio (courtesy of Paolo Scandone, PFG).

Figure 8. Some of the most important CNR-PFG publications; left, the volume of preliminary seismic microzonation studies for 39 locations severely affected by the 1980 earthquake [35]; center, the atlas collecting the isoseismal maps of the most important Italian earthquakes (1985, [36]); right, the catalogue of Italian earthquake from 1000 to 1980 of the CNR-PFG [37].

The Special Issue "The 23 November 1980 Irpinia-Lucania, Southern Italy Earthquake: Insights and Reviews 40 Years Later" contains 13 articles proposed by 44 researchers with different expertise, with a multidisciplinary approach that highlights the most important aspects of the earthquake from a seismological and geological point of view, without neglecting the reconstruction of cultural heritage, the resilience of the population, and the socioeconomic development of the internal areas of the Southern Apennines after the earthquake. No doubt, lessons learned from the Irpinia-Lucania event are relevant at the local level, for the whole Mediterranean region, and in similar seismotectonic and cultural environments around the world.

The volume is organised in five virtual sections in which the authors deal different features about the 1980 Irpinia earthquake: the historical-scientific framework, the geological and seismotectonic setting, the seismological framework, examples of applied geophysics, post-earthquake and resilience aspects.

The historical-scientific framework is represented by the papers of Lombardi [38] and Gizzi and Potenza [8]. The first paper introduces the "earthquake" topic by highlighting the far-reaching significance of this calamitous event not only in the social history of our country, but also and above all in the context of the various disciplines, from the scientific to the humanistic ones, giving a critical review of the academic debate that still exists today. According to the Author, this socio-historical vision can also support the demand for knowledge and risk mitigation coming from citizens and communities living in seismic areas, focusing on attention to social memory and awareness of seismic risk and natural risks.

The second paper by Gizzi and Potenza [8] can be rightly considered a milestone on the topic because it analyses about 640 papers with a tailored methodological approach, international and national (Italian) studies initiated and advanced since the earthquake occurred. They built and analyzed statistically two bibliographic databanks regarding the earthquake studies: (a) the international version of Irpinia Bibliographic database (IR_BASE_ENG), selecting and standardizing the pertinent scientific documents extracted from Scopus, Web of Science, and other databases and (b) the national version of the database (IR_BASE_IT) using the Google Scholar search engine to search for the most relevant papers in Italian. The review provides a rich and useful bibliography (123 papers) that includes studies on seismic source, environmental effects, seismic damage, seismic microzonation, disaster response and recovery, disaster epidemiology, ground motion estimates, and other research. The results of their research confirm that the 1980 Irpinia-Lucania earthquake was a significant occasion for the scientific community to expand the knowledge on the seismic phenomena, as well as to learn lessons in view of setting up preventive actions to mitigate the seismic risk.

The section about geological and seismotectonic setting includes papers by Ascione et al. [39], Matano et al. [19], and Galli [18]. Three papers that provide a broad overview of new insights not only into the detailed geology and geomorphology of the area but also into the seismotectonic interpretation and post seismic deformation of the epicentral area through different methodologies: the analysis of PS-InSAR data, new stratigraphic, petrographic, structural data, paleoseismological and archaeoseismological evidence. In detail, Ascione et al. [39], analyzed eighteen years of PS-InSAR data after the earthquake, showing that in the past decades soil deformation has affected the 1980 earthquake epicentral area. The analysis showed that cumulative deformation is consistent with coseismic deformation inferred from both seismological data (rupture mechanisms of the three main shocks which occurred in a 40 s timespan), levelling data and coseismic surface faulting. It is also consistent with evidence of Late Quaternary active faults at the surface. The Authors identify continuing uplift of the footwall and subsidence of the hanging wall blocks of the two major faults activated by the 23 November earthquake; they also show that the region in the mid part between the main seismic structures is currently affected by slow uplift. Moreover, the results of PS-InSAR data show that postseismic deformation is still occurring 30 years after the earthquake.

Matano et al. [19] collected geological data from the studies for the excavation works of the Pavoncelli bis hydraulic tunnel, developing between Caposele and Conza della Campania, highly damaged during the 1980 earthquake and described the geology of the epicentral area of the 1980 earthquake with new stratigraphic, petrographic and structural data. Through a multi-disciplinary and updated datasets the Authors have achieved (1) new insights on the tectono-stratigraphic evolution and stratigraphic architecture of the southern Apennines foreland and basin system, as well as on the structural and stratigraphic relations of Apennines tectonic units (2) the timing of their kinematic evolution, (3) a better understanding of the relationships between internal and external basin units within the Apennine thrust belt and its tectonic evolution.

Galli [18] in his paper deals with both paleoseismological clues from the Monte Marzano Fault System (the structure responsible for the catastrophic, Ms 6.9, 1980 earthquake) and archaeoseismological evidence of settlements founded in its surroundings, in order to cast light on two poorly known earthquakes that occurred at the onset and at the end of the first millennium CE, likely in 62 and in 989 CE. The Author tried to demonstrate that both earthquakes should share the same seismogenic structure and the same power as the 1980 seismic event.

With regard to the seismological section, the papers presented by Festa et al. [40] and Piombino et al. [41] starting from the 1980 event, give precise information on the current seismicity. Festa et al. [40], provided detailed location and characterization of events of the 3–7 July 2020 Irpinia sequence (southern Italy) that occurred at the northern tip of the main segment that ruptured during the 1980 Irpinia earthquake. Using an autocorrelation technique, they detected more than 340 events within the sequence, with local magnitude ranging between 0.5 and 3.0. The Authors provided double difference locations, source parameter estimation, and focal mechanisms determination for the best quality events and found that the sequence ruptured an asperity with a size of about 800 m, along a fault structure having a strike compatible with one of the main segments of the 1980 Irpinia earthquake fault system, and a dip of 50–55 at depth of 10.5–12 km and 60–65 at shallower depths (7.5–9 km).

Piombino et al. [41] merged historical records of seismicity with new satellite techniques to allow for the precise determination of ground movements, and then derived physical dimensions, such as strain rate. In this way, the Authors verified that in Irpinia, the occurrence of new strong shocks forty years after the 1980 event (one of the strongest known seismic events in the district) with Mw 6.8 is still a realistic possibility. They hypothesized that the reason for this is due to the fact that, since 1990, only areas characterized by high rates of deformation have hosted significant earthquakes, also confirmed by analyzing the historical catalog of events with seismic completeness for magnitude M 6 over the last four centuries. Moreover, they show that strong seismic events with magnitude M 6 generally occurred at a relatively short time distance between one another, with a period of 200 years without strong earthquakes between the years 1732 and 1930.

Nardone et al. [42], Forcellini [43] and Mina and Forcellini [44] provide a useful overview of applied geophysics investigating some specific cases, both real and theoretical on the seismic assessment, the site effects, the soil–structure interactions, the response of the cultural heritage, applying various numerical simulations. Nardone et al. [42], analyze the ground response in the Avellino town (Campania, Southern Italy) and its correlation with the effects caused by the 23 November 1980 Irpinia earthquake on the historical buildings with the aim to get some clues about the earthquake damage and the cause-effect relationship. They use the seismic hazard disaggregation for estimating the ground motion response for Avellino, where strong motion recordings are not available. For assessing the seismic ground motion, the authors use borehole data to build the lithological model. The results indicate that the complex subsoil layers influence the ground motion, particularly in the lowest period (0.1–0.5 s). Moreover, the comparison with the observed damage of the selected historical buildings and the maximum acceleration expected indicates that the damage distribution cannot be explained by the surface geology effects alone.

Forcellini [43] deals with the role of the water level, closely related to changes in the degree of soil saturation, in the assessment of seismic vulnerability for the 1980 Irpinia–Basilicata earthquake, performing several 3D numerical finite element modeling in order to consider the effects of soil–structure interaction (SSI) on a representative benchmark structure. The results show the importance of considering the water level for buildings on shallow foundations in terms of settlements, base shear forces and floor displacements.

Mina and Forcellini [44] present a systematic study of the effects caused by the strong earthquake that struck southern Italy on 23 November 1980 (Ms = 6.9) and affected the Campania and Basilicata regions. In detail the Authors study the response site effects by considering several soil free-field conditions and the assessment of the role of the soil–structure interaction (SSI) on a representative benchmark structure through the numerical simulations performed with the advanced platform Open Sees, which can consider non-linear models for both the structure and the soil. The results show the importance of considering the SSI in the seismic assessment of soil amplifications and its consequences on the structural performance.

Last, but certainly not least, are the papers in the section relating generically to post-earthquake, but including papers on seismic risk and resilience of heavily earthquake-damaged villages. The paper of Wyss and Rosset [45], using the computer code QLARM, calculates the number of dead and injured in near-real time taking into account data from the 1980 earthquake. The results show that the number of casualties and injuries in large and major earthquakes in Italy can be estimated correctly within less than an hour by using QLARM, very important for definition of the seismic risk and for the civil protection actions to be prepared.

The paper by Moscaritolo [46] approaches the post-1980 earthquake reconstruction problem as a complex social process in which cultural backgrounds, expectations, and ideas of the future come into play, without neglecting geological, historical, legislative, economic, and political factors. Combining oral historical sources and archival records, the article shows the paths taken by two small towns among the most affected by the 1980 earthquake.

The last paper by Porfido et al. [7], aims to present, through a photographic reportage, the current state of rebuilding of the most devastated villages by the 1980 earthquake. Forty years after the seismic event, the photographs show villages almost completely rebuilt with modern techniques where reinforced concrete prevails. Only in few instances, the reconstruction was carried out trying to recover the pre-existing building heritage, without changing the original urban planning, or modifying it. Even more than this, documenting the rebuilding process in a large epicentral area reveals the human legacy to the natural landscape, and our ability, or failure, to properly interpret the environmental fate of a site.

This volume, far from being exhaustive, is nevertheless intended to be an important point of reference for the new generations, who will be able to have a historical as well as a multidisciplinary vision a seismic event that some of the researchers involved in the drafting of the volume have experienced personally [7,47–55]. An event that, due to its catastrophic consequences, not only modified and conditioned the lives of many people, but whose effects are still felt today.

Supplementary Materials: The following are available online at https://www.mdpi.com/article/10.3390/geosciences12040173/s1, File S1: Supplementary_material_Editorial_Porfido et al_2022.

Author Contributions: For this editorial all authors have contributed to the work reported. All authors have read and agreed to the published version of the manuscript.

Funding: This research received no external funding.

Acknowledgments: The Guest Editors thank all the Authors, the "Geosciences" Editors, and the reviewers for their appreciable contributions and commitment to this Special Issue. Many thanks go "Geosciences" Assistant Editors, for their dedication to this project and their valuable collaboration in the setup, promotion, and management of the Special Issue. Special thanks to Paolo Scandone (1939–2016) who made available the topographic maps with the seismic-induced effects survey carried out immediately after the earthquake. We would like to take this opportunity to thank once again all the researchers who, despite the many difficulties of the moment, took part in the post-earthquake investigations organised by the PFG of the CNR.

Conflicts of Interest: The authors declare no conflict of interest.

References

1. Postpischl, D.; Branno, A.; Esposito, E.; Ferrari, G.; Marturano, A.; Porfido, S.; Rinaldis, V.; Stucchi, M. The Irpinia earthquake of November 23, 1980. In *Atlas of Isoseismal Maps of Italian Earthquakes*; CNR-PFG: Bologna, Italy, 1985; Volume 114, pp. 152–157.
2. Rovida, A.; Locati, M.; Camassi, R.; Lolli, B.; Gasperini, P. *CPTI15, the 2015 Version of the Parametric Catalogue of Italian Earthquakes*; Istituto Nazionale di Geofisica e Vulcanologia (INGV): Roma, Italy, 2016. [CrossRef]
3. Postpischl, D.; Branno, A.; Esposito, E.; Ferrari, G.; Marturano, A.; Porfido, S.; Rinaldis, V.; Stucchi, M. Southern Italy November 23, 1980 earthquake: Macroseismic Survey. In Proceedings of the Seventh European Conference on Earthquake Engineering, Athens, Greece, 20–25 September 1982.
4. Commissione Parlamentare d'Inchiesta Sulla Attuazione Degli Interventi per la Ricostruzione e lo Sviluppo dei Territori della Basilicata e Della Campania Colpiti dai Terremoti del Novembre 1980 e Febbraio 1981. Relazione Conclusiva e Propositiva. 1991, Volume I. Available online: https://storia.camera.it/lavori/sedute/27-novembre-1990-bf-10-19901127-00-01 (accessed on 24 December 2020).
5. Proietti, G. *Dopo la Polvere: Rilevazione Degli Interventi di Recupero Post-Sismico del Patrimonio Archeologico, Architettonico ed Artistico delle Regioni Campania e Basilicata Danneggiato dal Terremoto del 23 Novembre 1980 e del 14 Febbraio 1981*; Ministero per i Beni Culturali e Ambientali: Rome, Italy, 1994; Volume 5.
6. Mazzoleni, D.; Sepe, M. (Eds.) *Rischio Sismico, Paesaggio, Architettura: L'Irpinia Contributi per un Progetto*; CRdC, AMRA: Naples, Italy, 2005.
7. Porfido, S.; Alessio, G.; Gaudiosi, G.; Nappi, R.; Michetti, A.M.; Spiga, E. Photographic Reportage on the Rebuilding after the Irpinia-Basilicata 1980 Earthquake (Southern Italy). *Geosciences* **2021**, *11*, 6. [CrossRef]
8. Gizzi, F.T.; Potenza, M.R. The Scientific Landscape of November 23rd, 1980 Irpinia-Basilicata Earthquake: Taking Stock of (Almost) 40 Years of Studies. *Geosciences* **2020**, *10*, 482. [CrossRef]
9. Bollettinari, G.; Panizza, M. Una faglia di superficie presso San Gregorio Magno in occasione del sisma del 23/11/1980 in Irpinia. *Rend. Soc. Geol. Ital.* **1981**, *4*, 135–136.
10. Cinque, A.; Lambiase, S.; Sgrosso, I. Su due faglie nell'alta valle del Sele legate al terremoto del 23.11.1980. *Rend. Soc. Geol. Ital.* **1981**, *4*, 127–129.
11. Carmignani, L.; Cello, G.; Cerrina Feroni, A.; Funiciello, R.; Klin, O.; Meccheri, M.; Patacca, E.; Pertusati, P.; Plesi, G.; Salvini, F.; et al. Analisi del campo di fratturazione superficiale indotto dal terremoto Campano-Lucano del 23/11/1980. *Rend. Soc. Geol. Ital.* **1981**, *4*, 451–465.
12. Westaway, R.; Jackson, J. Surface faulting in the southern Italian Campania-Basilicata earthquake of 23 November 1980. *Nature* **1984**, *312*, 436–438. [CrossRef]

13. Cotecchia, V. Ground deformations and slope instability produced by the earthquake of 23 November 1980 in Campania and Basilicata. *Geol. Appl. Idrogeol.* **1986**, *21*, 31–100.
14. Bernard, P.; Zollo, A. The Irpinia (Italy) 1980 earthquake: Detailed analysis of a complex normal faulting. *J. Geophys. Res.* **1989**, *94*, 1631–1648. [CrossRef]
15. Pantosti, D.; Valensise, G. Faulting mechanism and complexity of the November 23, 1980, Campania-Lucania earthquake, inferred from surface observations. *J. Geophys. Res.* **1990**, *95*, 319–341. [CrossRef]
16. Blumetti, A.M.; Esposito, E.; Ferreli, L.; Michetti, A.M.; Porfido, S.; Serva, L.; Vittori, E. Ground effects of the 1980 Irpinia earthquake revisited: Evidence for surface faulting near Muro Lucano. In *Large-Scale Vertical Movements and Related Gravitational Processes*; Dramis, F., Farabollini, P., Molin, P., Eds.; Studi Geol. Camerti: Camerino, Italy, 2002; pp. 19–27.
17. Ascione, A.; Mazzoli, S.; Petrosino, P.; Valente, E. A decoupled kinematic model for active normal faults: Insights from the 1980, MS = 6.9 Irpinia earthquake, southern Italy. *Geol. Soc. Am. Bull.* **2013**, *125*, 1239–1259. [CrossRef]
18. Galli, P. Roman to Middle Age Earthquakes Sourced by the 1980 Irpinia Fault: Historical, Archaeoseismological, and Paleoseismological Hints. *Geosciences* **2020**, *10*, 286. [CrossRef]
19. Matano, F.; Di Nocera, S.; Criniti, S.; Critelli, S. Geology of the Epicentral Area of the November 23, 1980 Earthquake (Irpinia, Italy): New Stratigraphical, Structural and Petrological Constrains. *Geosciences* **2020**, *10*, 247. [CrossRef]
20. Del Prete, M.; Trisorio Liuzzi, G. Risultati dello studio preliminare della frana di Calitri (AV) mobilitata dal terremoto del 23/11/1980. *Geol. Appl. Idrogeol.* **1981**, *16*, 153–165.
21. Esposito, E.; Gargiulo, A.; Iaccarino, G.M.; Porfido, S. Distribuzione dei fenomeni franosi riattivati dai terremoti dell'Appennino meridionale. Censimento delle frane del terremoto del 1980. In *Proceeding Internal Convention on Prevention of Hydrogeological Hazards*; CNR(IRPI): Torino, Italy, 1998; pp. 409–429.
22. Porfido, S.; Esposito, E.; Michetti, A.M.; Blumetti, A.M.; Vittori, E.; Tranfaglia, G.; Guerrieri, L.; Ferreli, L.; Serva, L. Areal distribution of ground effects induced by strong earthquakes in the Southern Apennines (Italy). *Surv. Geophys.* **2002**, *23*, 529–562. [CrossRef]
23. Porfido, S.; Esposito, E.; Guerrieri, L.; Vittori, E.; Tranfaglia, G.; Pece, R. Seismically induced ground effects of the 1805, 1930 and 1980 earthquakes in the Southern Apennines. *Boll. Soc. Geol. Ital.* **2007**, *126*, 333–346.
24. Wasowski, J.; Pierri, V.; Pierri, P.; Capolongo, D. Factors controlling seismic susceptibility of the Sele valley slopes: The Case of the 1980 Irpinia Earthquake Re-examined. *Surv. Geophys.* **2002**, *23*, 563–593. [CrossRef]
25. Carulli, G.B.; Migliacci, A.; Onofri, R.; Porfido, S. Indagini geologiche ed in prospettiva sismica a S. Michele di Serino. *Rend. Soc. Geol. Ital.* **1981**, 161–164.
26. Galli, P. New empirical relationships between magnitude and distance for liquefaction. *Tectonophysics* **2000**, *324*, 169–187. [CrossRef]
27. Cotecchia, V.; Salvemini, A. Correlation between seismic events and discharge variations at Caposele and Cassano Irpino springs, with particular reference to the 23 November 1980 earthquake. [Correlazione fra eventi sismici e variazione di portata alle sorgenti di Caposele e Cassano Irpino, con particolare riferimento al sisma del 23 novembre 1980]. *Geol. Appl. Idrogeol.* **1981**, *16*, 167–192.
28. Esposito, E.; Pece, R.; Porfido, S.; Tranfaglia, G. Hydrological anomalies connected to earthquakes in Southern Apennines, Italy. *Nat. Hazards Earth Syst. Sci. EGS* **2001**, *1*, 137–144. [CrossRef]
29. Postpischl, D.; Mercu, G.; Pergalani, F.; Puglielli, C.; Signanini, P.; La Bagnara, R.; Bonardi, G.; Carannante, G.; Ortolani, F.; Orsi, G.; et al. Il terremoto del 23-11-1980. Rilievo Macrosismico Stato di Avanzamento al 27-1-1981. In Proceedings of the Convegno straordinario del progetto Finalizzato Geodinamica, Roma, Italy, 27–28 January 1981.
30. Berardi, R.; Berenzi, A.; Capozza, F. *Terremoto Campano-Lucano del 23 Novembre 1980. Registrazioni Accelerometriche Della Scossa Principale e Loro Elaborazioni*; Tech. Rept.; CNEN-ENEL: Rome, Italy, 1981.
31. Braga, F.; Dolce, M.; Liberatore, D. A statistical study on damaged buildings and an ensuing review of the MSK-76 scale. In Proceedings of the Seventh European Conference on Earthquake Engineering, Athens, Greece, 20–25 September 1982; pp. 65–84.
32. Del Pezzo, E.; Iannaccone, G.; Martini, M.; Scarpa, R. The 23 November 1980 Southern Italy earthquake. *Bull. Seism. Soc. Am.* **1983**, *73*, 187–200. [CrossRef]
33. Faccioli, E. Engineering Seismology Aspects of the M-6.5, Southern Italy Earthquake of November 23, 1980: A Preliminary Review, 1981. In Proceedings of the International Conferences on Recent Advances in Geotechnical Earthquake Engineering and Soil Dynamics, Bengaluru, India, 13–16 July 2020; pp. 1215–1222.
34. Gruppo di Lavoro Sismometria terremoto del 23.11.1980. Il terremoto campano-lucano del 23.11.1980: Elaborazione dei dati sismometrici. *Rend. Soc. Geol. Ital.* **1981**, *4*, 427–450.
35. Siro, L.; Bigi, G.; Testoni, P.; Luongo, G. Indagini di microzonazione sismica. In *Intervento Urgente in 39 Comuni Della Campania e Della Basilicata Colpiti dal Terremoto del 23 Novembre, 1980*; Pubbl. n. 492; CNR-PFG: Bologna, Italy, 1983; 221p.
36. Postpischl, D. *Atlas of Isoseismal Maps of Italian Earthquakes*; CNR-PFG: Bologna, Italy, 1985; p. 114.
37. Postpischl, D. *Catalogo dei Terremoti Italiani Dall'anno 1000 al 1980*; Consiglio Nazionale Delle Ricerche, Progetto Finalizzato Geodinamica, Sottoprogetto Rischio Sismico e Ingegneria Sismica: Bologna, Italy, 1985.
38. Lombardi, G. Irpinia Earthquake and History: A Nexus as a Problem. *Geosciences* **2021**, *11*, 50. [CrossRef]
39. Ascione, A.; Nardò, S.; Mazzoli, S. The MS 6.9, 1980 Irpinia Earthquake from the Basement to the Surface: A Review of Tectonic Geomorphology and Geophysical Constraints, and New Data on Postseismic Deformation. *Geosciences* **2020**, *10*, 493. [CrossRef]

40. Festa, G.; Adinolfi, G.M.; Caruso, A.; Colombelli, S.; De Landro, G.; Elia, L.; Emolo, A.; Picozzi, M.; Scala, A.; Carotenuto, F.; et al. Insights into Mechanical Properties of the 1980 Irpinia Fault System from the Analysis of a Seismic Sequence. *Geosciences* **2021**, *11*, 28. [CrossRef]
41. Piombino, A.; Bernardini, F.; Farolfi, G. Assessing Current Seismic Hazards in Irpinia Forty Years after the 1980 Earthquake: Merging Historical Seismicity and Satellite Data about Recent Ground Movements. *Geosciences* **2021**, *11*, 168. [CrossRef]
42. Nardone, L.; Gizzi, F.T.; Maresca, R. Ground Response and Historical Buildings in Avellino (Campania, Southern Italy): Clues from a Retrospective View Concerning the 1980 Irpinia-Basilicata Earthquake. *Geosciences* **2020**, *10*, 503. [CrossRef]
43. Forcellini, D. The Role of the Water Level in the Assessment of Seismic Vulnerability for the 23 November 1980 Irpinia–Basilicata Earthquake. *Geosciences* **2020**, *10*, 229. [CrossRef]
44. Mina, D.; Forcellini, D. Soil–Structure Interaction Assessment of the 23 November 1980 Irpinia-Basilicata Earthquake. *Geosciences* **2020**, *10*, 152. [CrossRef]
45. Wyss, M.; Rosset, P. Near-Real-Time Loss Estimates for Future Italian Earthquakes Based on the M6.9 Irpinia Example. *Geosciences* **2020**, *10*, 165. [CrossRef]
46. Moscaritolo, G.I. Reconstruction as a Long-Term Process. Memory, Experiences and Cultural Heritage in the Irpinia Post-Earthquake (November 23, 1980). *Geosciences* **2020**, *10*, 316. [CrossRef]
47. Le macerie invisibili. In *Rapporto 2010 Osservatorio Permanente sul Doposisma*; MidA (Ed.) 2010; ISBN 97890-51-48-1488. Available online: http://www.osservatoriosuldoposisma.com/ricerche/ricerche/le-macerie-invisibili-il-rapporto-2010 (accessed on 1 January 2022).
48. Picone, G. *Paesaggio Con Rovine*; Ed. Mondadori: Milano, Italy, 2020; p. 226. ISBN 978-88-04-72480-3.
49. Festa, G.; Iuliano, G.; Saggese, P. *Irpinia-1980–2020*; Delta 3Edizioni; p. 312. ISBN 978-88-6436-860-3.
50. Porfido, S. *Temporary Housing in San Potito Ultra*; Blurb: San Francisco, CA, USA, 2020; ISBN 978-1-71-476363-4.
51. Porfido, S.; Spiga, E. *Ricostruzione 1980–2020 Ed*; Blurb: San Francisco, CA, USA, 2020; Volume I, ISBN 978-1-71-571504-5.
52. Porfido, S.; Spiga, E. *Ricostruzione 1980–2020 Ed*; Blurb: San Francisco, CA, USA, 2020; Volume II, ISBN 978-1-71-572142-8.
53. Spiga, E.; Porfido, S. *Bisaccia Piano di Zona Ed*; Blurb: San Francisco, CA, USA, 2020; ISBN 978-1-71-555296-1.
54. Available online: https://ingvterremoti.com/category/terremoti_storia/terremoti-del-900/terremoto80/ (accessed on 1 January 2022).
55. Available online: https://www.facebook.com/groups/83021542900/?multi_permalinks=10158673022082901&notif_id=1643320398042162&notif_t=group_highlights&ref=notif (accessed on 1 January 2022).

geosciences

Article

Irpinia Earthquake and History: A Nexus as a Problem

Giovanni Lombardi

CNR—Institute for Studies on the Mediterranean (IsMed), 80134 Naples, Italy; giovanni.lombardi@ismed.cnr.it

Abstract: Forty years from the 23 November 1980, Irpinia-Basilicata earthquake date represents much more than a commemoration. It has been a fracture for the history of Italy. Important for many reasons, this earthquake has been a watershed for the studies and the public role of research. Historians have been solicited to work on the topic by scholars of the geological and seismological sciences: in the face of the repetition of disastrous seismic events in Italy, earthquakes remained 'outside the history'. However, the real difficulty of socio-historical science is not neglecting seismic events and their consequences, but rather the reluctance to think of 'earthquake' as a specific interpretative context. This means to deal with the discipline 'statute' as well as the public commitment of scholars. In this way, the circle earthquake-history-memory requires broad interdisciplinarity, which offers insights to work on historical consciousness and cultural memory: important aspects to understand the past as well as to favour a seismic risk awareness.

Keywords: earthquake history; memory

Citation: Lombardi, G. Irpinia Earthquake and History: A Nexus as a Problem. *Geosciences* **2021**, *11*, 50. https://doi.org/10.3390/geosciences11020050

Received: 1 December 2020
Accepted: 21 January 2021
Published: 27 January 2021

1. Introduction

Forty years from the 23 November 1980, much has been said about the Irpinia earthquake, and any socio-historical reading can be submerged by such vast records. Indeed, the event immediately focused the world's gaze on that little-known and remote land of Southern Italy; the concern was also shown by the generous chain of international solidarity. After all, the memory of this broad mobilization is alive and still visible in the place names, as in the case of the "Villaggio Italo-Canadese" of San Mango sul Calore—built with the help of Canadians—or the "Bergamo condominium" of Lioni, funded by Bergamo's citizens (Northern Italy) [1]. Furthermore, the rescues on such a large-scale showed different visions of emergency management, exposing cultural barriers other than solidarity between peoples [2]. Several generations have been affected by the events or wrapped by their consequences over time. Volunteers who flocked as rescuers remained bonded to that memory. In a land branded by emigration, some came back to help; many others left, while the remittances from abroad supported the reconstruction. Some enriched themselves; others simply rolled up their sleeves to go on. The lobbies looked for business. Meanwhile, political struggle enhanced civil protection, seismic risk mapping and safe practices. Ultimately, interests of all kinds and new settlements upset the anthropic impact in the area, polarizing the national choices. In other words, territories and communities visibly changed along with institutions, which were urged to rethink themselves. The events involved scientists of every order and degree. First of all, who rushed to help and then to study: working on the field as well as on theoretical elaboration, they rewrote the history of earthquakes in Italy, generating tools and knowledge, nowadays patrimony of all. The Irpinia-Basilicata earthquake appeared, from the start, a fracture in the history of Italy.

2. About History and the Earthquake: An Interesting Relation

A careful scholar noted that earthquake disasters happened in Italy on average every 4–5 years in the last 150 years, urging more attention from historians, regretting historiography with earthquakes "outside the history" and low memory of the risk [3]. The reflection is reasonable. Nevertheless, the question invests the role of the humanities and, specifically, of history as a discipline. Science has gained sound knowledge, offering ideas, proposals and hard data on the "Irpinia earthquake". From the first days, the scientific community arose with an unprecedented mobilization, setting up coordination around the Vesuvius Observatory—the oldest in the world, founded in 1841. The observatory was historically an attractor for the scholars and was close to the disaster sites. Hence, it immediately became a reference point, hosting, just in the aftermath of the earthquake, the earlier operative meetings. [4–6] Certainly, historical disciplines' perspective is not quite the same as other sciences: events gain sense within the "historical" narration, and this process needs "time". Moreover, it depends on the contemporary viewpoint. That not for neglecting "objectivity"—destroyed towns are destroyed towns—but for how semiotic contexts shape the representation of the past [7]. It is interesting, for example, how the controversial concept of resilience—adaptive resilience, post-disaster resilience and so forth—can influence the historical interpretation in the light of present-time mechanisms—may be showing the limits of the authoritarian/military emergency management and the marginalization of the communities from post-disaster choices [8]. Moreover even, it is enlightening how historical analysis can decode an interview, showing meanings materialized only with time. However, the "free will" of the scholars is both the pivot and matter for the discipline, judging what to highlight, transmit or leave out. Very roughly, the contemporary perspective shapes the past as history. These issues would lead us off-topic, but here it is just important to remember that these matters exist. Despite the instances of objectivation and measurement influence now strongly the social sciences, historians remain essentially anchored to holistic views.

Earthquakes come with Italian history; just to think of great twentieth-century disasters like those of Messina (1908), Marsica (1915), Belice (1968), Friuli (1976), Irpinia (1980); or the more recent of L'Aquila (2009) and the others happened all along the Appennini Mountain chain in the twentieth-first. The infinite seconds of any seism have marked the "land of the thousand bell towers". In addition to the pars destruens, all along the centuries, the earthquake has molded the forma urbis, as the recent case of L'Aquila has brought back to general attention [9]. Moreover, seismic events shaped the morphology of vast territories, scattered settlements, updated governing praxis, as has been shown by recent natural hazard studies [10]. Then the crucial issue remains how and if "earthquake" dwells in the memory of the country. No understanding, no safety plan, zoning, prevention, recovery or omission make sense without reference to "memory" [11]. Many events were "unexpected" due to a lacking narrative and also by psychological and emotional removal of disaster memory. Hence, the nexus earthquake-history-memory emerges as both a slippery and focal point.

The issue about the historians" attention does not concern the omission of disasters and their consequences. It rather concerns—looking at Italian historiography—a reluctance to think of "earthquake" as a special interpretative habitat in the critical context of general history. About the Irpinia earthquake, the historians had reflected on the social problems arisen due to "the worst Italian disaster from the Second World War"—as it was immediately recognized by international public opinion. The State institutions have been shaken, while the inadequacy and anachronism of the Italian ruling classes have been brought to light [12,13]. From the beginning, aid and reconstruction also meant conflict, criminality, material and moral ravage, and bloated investment for this endless "emergency". In this scenario, the historians adopted traditional approaches to studying the events. Ultimately, the related geological and environmental studies—and in general of hard sciences—were not really taken into account when elaborating historical analysis. Nowadays, after investigations and with the new generations, the scholars began to sense the interpretative limits of the "Irpiniagate"—to use a journalistic term—putting the earthquake of 1980 in the view of a slightly more comprehensive version of the general Italian history [14]. Moreover, recently, the earthquake–history connection as "disaster narrative" seems to revive the historiography, while new studies widen the discipline on the socio-anthropological side and towards a sociology of disaster [15]. It sounds like an "earthquake" can shake even academic "resilience".

3. Shaking Land, Shake the State

The date of 23 November 1980 has been fixed in the collective memory by a jarred audiotape aired months later on radio, in which earthquake disturbingly rumbles over a folk music track. In 1970, a reference law had set an overall groundwork about "Rules on relief and assistance to disaster victims—Civil Protection" [16]: it concerned the emergency, the definition of "disasters", and the commitment of the state. A subsequent regulation should have ensured the law's application, but it was not made. However, that evening of November 1980 showed the absolute unpreparedness of the state, although only four years had passed since another terrible earthquake, namely that of Friuli (1976). The command chain was unclear. No information, no electricity, no telephone line, no coordination. After the one of Rome, the Turin fire department sent right away its firemen: many men came from the affected areas of South, and that helped the rescues. They camped in Avellino, the main city of the area, in harsh conditions and without food: all its 119 district municipalities were hit. Firemen most readily dealt with the difficulties of mountain localities, often unreachable, while the Italian army—at the time, a military force of conscription—took longer to arrive. Similar initiatives started everywhere, among the confusion and the incompetence of territorial authority. In the villages, local "carabinieri" and volunteers dig right away. Rescue vehicles, equipment and expertise, coffins were all missing. Food was scarce, and squads gave up their rations to the victims. Once again, the political agenda and social behavior had ignored the past experiences and environmental risk: the consequences were clear. On average, the earliest significant help arrived after more or less five days. On 23 November, the President of the Republic, Sandro Pertini, had just returned to Rome after an unofficial state visit to the Hellenic Republic. On the 24th, after having welcomed the UK PM Margaret Thatcher, he left for the disaster region. On 25th, Pertini reached the earthquake sites by helicopter, despite the contrariety of the PM Arnoldo Forlani, ministers and advisers. The next day, in a discourse to the nation, the President reported the desperate situation and denounced the political responsibilities; the coordination centers, planned in 1970, had never been realized [17]. That day, the historical newspaper of South Italy—"Il Mattino"—suggested 10.000 deaths: the front page had an international echo and soon became a symbol [1,18]. Everything happened under the sight of the world.

Much has been written on the political struggle, the interests, the dreams and the betrayal, the outputs due to the Irpinia 1980 [14]. To have an idea, the dossier of Deputy House of 2009 reported, for the quake of 23–24 November 1980, around 200 municipalities affected in Campania and Basilicata regions, about 60 of that severely damaged, 2.914 deaths, 8800 injured, 280,000 displaced/homeless, 150,000 buildings to rebuild, 47.5 billion euro (currency evaluation at 2008) for the reconstruction of disaster areas, excluding other costs like mortgages and tax benefits, and 17.5 billion for Naples—the last funds were related to the earthquake, but linked to specific law 219/1981, urged by the social tension of the city [19]. Still, the so-called "Milleproroghe Act" (2017, art. 9) included the renewal of the commissioner for ongoing issues related to the Irpinia earthquake (considering the period 1980–1981), pointing to long-run implications. After all, several municipalities still report sites for prefab housing that survived here and there—as in Avellino province—sometimes used, mostly abandoned. In some localities, "temporary" housings still exist [1,20].

Different parameters about damages may touch up some estimates, but the key figures are clear. The Irpinia-Basilicata earthquake was the strongest seismic event (Mw 6,9) in Italy in the last 100 years. Felt throughout the peninsula, particularly affected the regions of Campania and Basilicata—the last also named Lucania—harming more than 800 locations. Fifteen municipalities in the Avellino, Salerno and Potenza districts were almost destroyed. Even damages to the natural environment and the hydrogeological instability were huge, although the emergency needs of the populations focused for a long time on the urban settlements [21].

On the historical horizon, the dynamics between disaster and institutions—or emergency and institutions—can conceal recurrent patterns. The symbolic universe of the dominion over nature/territory concerns the bases itself of the institutional power: not by chance, in Ancien Régime, royal gardens ruled the reluctant nature showing, metaphorically, the sovereign order over chaos. In a long-run historical perspective, symbolic and social schemes can be repeated. One of these concerns the need to reaffirm the government authority—shaken by the crisis and the impotence in the face of nature—and to offer a narrative of the events. In addition, the naming of governing commissioners to deal with an emergency, as an adaptive response, is a recurrency in history; in the past, as a step of the state-building, in 1980 as a lack of consideration of the seismic risk in an ordinary state commitment.

Overall, the new perception of the risk improved institutions and approaches. One of the most visible examples was the enhancement of the civil protection department. However, above all, there was impressive enforcement of the public role of science. A grassroots mobilization involved scholars of all kinds, from students to academics. Practically, a generation of young graduates—among the first rescuers or subsequently involved in assistance, studies and reconstruction—participated in the events and remained tied to this experience also as experts and professionals. This social adherence gathered around a scientific network in full reorganization. Just before the first Irpinia-Basilicata shock, earthquake studies were a task of various institutions, including the Istituto Nazionale di Geofisica (ING, now known as INGV). At the time, the National Research Council (CNR) had a section for the study on the territory, involving seismic and volcanic risk [22] and, from 1976 (year of Friuli earthquake), it was carrying out the "Progetto Finalizzato Geodinamica" (PFG). Many scholars—often students and young researchers—had known the Irpinia area through this project as well as the Progetto Finalizzato Energetica, both promoted by CNR. Immediately after the earthquake, the Vesuvian Observatory took on the coordination of the "geodynamics" network and the scientific mobilization: the historical volcanology observatory became a leading reference [6,23]. In the following years, the scientific network was renewed; expertise and practices were improved, thanks to an extraordinary collaboration among such research institutions, universities and other organizations. It is noteworthy the strengthen of seismic mapping and historical seismology, as shown by the catalog of strong earthquakes in Italy [24]; inquiring geological settings linked to ground deformation and infrastructure matters [25]; zoning and building laws; the long-run studies on seismic fragility for Italian RC buildings [26]; the evolution of paradigms and incubators involving academy and public administration [27]; the update of financial strategies. In short, the Irpinia earthquake prompted processes that have influenced and daily influence the lives of millions of people beyond their awareness of these processes.

4. Memory: A Complicated Matter

Conza della Campania suffered seismic disasters in 1466, 1517, 1694 and 1732, yet the old country was still there when, in 1980, 90% of the houses collapsed; only then people decided to abandon the ancient settlement. It was 9 km away from the epicenter, made by buildings with often no foundations and leaning against each other on two unstable hills. San Mango del Calore suffered a destructive earthquake in 1732, but it was there that night of 23 November to be destroyed, with its inadequate housing and its landslides. The nearby Calitri was struck by the shock and by following incidents including a slump-earth flow that moved 23 million cubic meters of land, but similar things had already happened with the earthquakes of 1694 (X MCS), 1805, 1910 (IX MCS) and 1930 [28]. "Histoire événementielle" is not enough to explain such resilience or to avoid such recurrent exposition to the natural hazard.

Nonetheless, the nexus history-memory came out of the Irpinia earthquake with force, and it is crucial to understand events and behaviors. Beyond the scientific and technical assumption for the reconstruction, the communities had to look back to their own historical identity. That followed many ways: the "narration" went from a reconstruction as a continuum with the origin up to breaking the link with the past in the name of brave choices. In this range, terms such as *Genius loci*, memory/oblivion, tradition/shock acquired real meaning. Among the most known cases, San Mango sul Calore and Calitri were rebuilt on-site, Conza della Campania was relocated. San Mango, reusing emergency wooden buildings as touristic resources, embraced the grief shaping within a more truthful and accepting narrative that goes beyond the practical outcomes. Conza—where the temporary village has been quite abandoned—focalized its origin on the archaeological site discovered on the ancient settlement—*Compsa*, ancient Roman city—nowadays elevated to a symbol of city and tourist resource. While the old village stays similar to the immediate after quake time, the new town shows contradictory urban and architectural signs. Moreover, the cemetery stays as a barycenter between the old hit hamlet and new town, like a memorial with the graves of the victims. The new industrial facility is near, displaying the territorial policies that occurred with the reconstruction. This view shows the fragmentation of historical memory and how the meaning of the places can change [29].

Visual cesura labeled some localities [1]. Bisaccia, in the province of Avellino, has a background of earthquakes and landslides: destroyed by the 1694 seism, devastated by those of 1732, 1930, 1980. After the last one, the municipality rebuilt the town on a safer site—so-called "piano"—identified in a master plan related to the 1930 event. Nowadays, two Bisaccia exist, the medieval hamlet that clings to the castle and the new city. The last is mostly rebuilt according to the urban plan of a renowned architect, Aldo Loris Rossi. Many people criticize the new urban spaces far from local uses, the otherness of its architectonic-urban codecs, the harsh contrast with a landscape dotted by wind turbines [30]. Nonetheless, some appreciate the radical changes based on anti-seismic criteria and unordinary ideas. On everything, it hovers a sense of suspended lives, perceivable at the sight of uncompleted and run-down public housings. The last scene is common in several post-quake localities. Moreover, beyond the cases, it coincides with a trend of the Apennine mountains range: between 1946 and 2000, in a context of depopulation of mountains, 3.5 new buildings were erected for every lost inhabitant, with an impressive rate of unoccupied dwellings [14,31]. Contrasting hydrogeological instability and seismic risk with non-essential buildings is quite a contradiction. Moreover, if the "semantization" of the sites depends on the community, which community we are talking about is a real question referring to holiday presences, migration and many more "absent" actors. The "quality" of the community influenced the decision-making process as well as the leading values of the reconstruction. These are matters that historians generally leave out, especially in the light of globalization: when, after all, there are the turbines to characterize the new skyline of Bisaccia.

Indeed, identity memorials are never just tangible but often related to a system of territorial relationships. In the case of Sant'Angelo dei Lombardi, the self-perception of its role in the region—having, before the events, a Hospital, a Court and other public services for the zone—worked as an ideal starting point, likewise to the wrecked medieval core. Hence, people considered its lost role as a reference and recalled this feature by setting up hosting camps, reconstruction commissions and offices. Social studies took on board these aspects, checking local archives and recovering interesting accounts [32]. Moreover, here and there appeared a "resilience" to change the balance between towns, an aspect that is both related to practical interests and the need to hold onto their identity.

People affected by the Irpinia earthquake had shaped the agrarian and mountain landscape as well as the urban scenery. Their way of life improved biodiversity, gave semantic sense to the environment and played a part in the dynamics of the natural hazard. This puts delicate questions upon subjects, languages and contexts of the "memory". Melito Irpino had suffered earthquakes and landslides during the twentieth century and before, and it had been already relocated in 1980: the earthquake just accelerated the demolition of the ancient site. An elderly inhabitant—a "privileged eye-witness" of the urban changes and keeper of important pictures and cinematographic patrimony—said: "In my heart, there is that beautiful country that you do not see, but I feel" [33]. This speech reflected the fracture among generations that did not share the same experience/imagination of the place: the bricks crumbled, but their meaning too; material marks go together with the gazes that give sense to them.

Very roughly, the bonds commonly collected under the concept of "memory" show how the earthquake–history nexus could be problematic, not because simply unclear, but because it is fundamentally open, dynamic and debatable. Yet, the semantization of the events, the explanation of settlements and landscapes, the self-understanding of a community, all these depend on it.

We can sum up the issues on two levels. The first concerns the commitment of historians on the study on the earthquakes, and this case of the Irpinia earthquake. As mentioned, we begin to include earthquakes in a more general Italian history [14]. However, the difficulty of considering earthquakes as interpretative spaces with characteristics, and developing approaches and languages for them, remains; not simply to add suggestions and records. In this regard, it is interesting how a great part of the literature about the Irpinia earthquake is fundamentally linked to memorials, experienced journalism, literary production. In a way, it could not be otherwise for the impact of this seismic event: this narrative expresses a collective need. On the other side, this production has not been accompanied by an equivalent historical reflection. The wide use of the term "crater" to indicate the area of the 1980 earthquake, as well as the reception of simple interviews as *tout court* expression of oral history, are just two examples of semantic slip from a calibrated lexicon; these simplifications introduce confusing elements, but also testify the disengage of the historians from a disciplinary narrative. The second—but not secondary—level concerns the contribution of the historical studies to the disaster memory and then to the cultural memory [34]. Without dwelling on the classification details, that "memory" is an expression of the communities/groups/individuals and how they process traumatic events. It can keep knowledge about natural hazards and inspire adaptive behaviors regarding risk. As long-term and future-oriented memory, it needs to be fed or/and reinvented, using every kind of tool, such as museums, acting and narration, festivals and commemorations. In this perspective, they also become basic communication, exercises, education and games—especially involving young people—to raise awareness of the risk and consciousness of history/fragility of the communities [35]. That implies working for an intergenerational memory too and, indirectly, for the social cohesion around shared values. Ultimately, the cultural memory appears at the same time prelusive, inclusive and consequent of any educational process to promote a seismic risk awareness.

5. Sweeping the Past, Reshaping the Past

Forty years means a generation shift. Moreover, in the areas affected by the earthquake of 1980, that change went along with an astonishing modification of the landscapes too. In visual culture terms, it is not just an esthetic matter but rather a semiotic issue. The Irpinia earthquake hit mostly mountains and rural areas with little towns, low urbanization and people of reserved customs. Their "well-known landscape" was not a frame around the communities but an expression of the human presence. Its springs have been poles of life for centuries; its country lanes age-old routes; its square were crossroads of active awaiting and encounters; its biodiversity came from centuries of hard-working lives. Briefly, the landscape expressed ways of life. After the earthquake, the changes were deep and fast. The urban and the rural landscape mirrored the historical fracture caused by the quake, widespread anomie and the revolution of the social and economic structures. In this context, the self-understanding of the communities concerning the environment, and the meaning of the places themselves, were fragmented. The migration, which already was part of a longer-term trend [14], favored the alienation from the settlements.

Pressed by harsh social issues, the major cities used their political weight to get as much as possible by the emergency funds; mainly Naples and its hinterland, where the earthquake had struck, but, above all, inflamed housing problems [36]. A trivialization of the core-periphery paradigms as a spot-refrain accompanied the investments. Hence, demographic and political poles worked as spending outposts, fostering top-down economic models on those interior and "periphery" areas which had been the main victims of the disaster. The "emergency" catalyzed political-economic interests and changed the spending centers. The involvement of crime upset the life of the whole country. It became necessary a parliamentary anti-mafia committee face and stopped corruption and territorial mafia organizations from an extraordinary social penetration, the main one being "Camorra" [14,37]. Among the consequences, they influenced regional planning. The law 219/1981 was about rebuilding, but it favored demolition and urban reconstruction rather than the recovery of boroughs of historical and architectural value. A sort of general amnesia facilitated the loss of cultural heritage, compounded by expropriations pursued by foreigners economically stronger than residents or/and in agreement with local ruling classes. The same law made room for infrastructures and industrial facilities—many short-lasting while some became important for the territory—that branched out in a rural world. Industrialization involved areas like San Mango sul Calore, Guardia dei Lombardi, Nusco, Morra de Sanctis as well as the highway Ofantina-bis.

After forty years, some settlements are dependent on migrant remittances, exposing the inadequacy of development models. In particular, "High Irpinia" seems a marginal system for the general socioeconomic process; some localities come across as out of general time. Nonplace, the disappearance of historical landmarks without a coherent rewriting of the rural and urban landscape, poor territorial planning and urban sprawl, all that are now debated among scholars [29]. Useless to list the cases with pros and cons. Our focus is that the earthquake of 23 November 1980 unleashed a powerful change, but that was accompanied by the disorientation of the memory, a disorder of the community references. Top-down decisions, experiments, speculations and successful choices concerned a land shocked in both seismic and metaphorical sense. Induced from the outside, some models of economic growth seem currently outdated. Therefore, there is a need for a new participatory regional planning that would take into account comparative studies [29].

There is undergoing a recomposition of the cultural memory driven by wider civil participation, reinventing traditions and reshaping the past. High-quality agro-food production linked to ancient anthropic presence—for example, Irpinia has got fine grapes varieties—rural tourism, festivals, biodiversity promotion, reflect the effort to recompose its history, processing the trauma. These trends related to the territories hit by the Irpinia earthquake can be transposed in an international context of studies on heritage and identity [38].

Just like the current pandemic shakes the use/meaning of the spaces, the connection earthquake-history-memory stimulates a reflection on those used, forgotten or provisory places that marked the post-quake Irpinia territory. Between the disorder and bureaucratic reconstruction, they are interstices of the disaster memory. As previously mentioned, in San Mango, the provisional recoveries have been converted in a touristic hypothesis. Use and reuse of these cabins, beyond the practical implications, recall the wish of the community to metabolize the past and, eventually, to produce storytelling. The case of San Mango is not isolated [1]. Romagnano al Monte—gold medal for civil merits: an honor earned for its reconstruction effort—has been rebuilt far from its former millennial settlement. Metaphorically, the wooden provisory houses signed the passage to the new village from the old one; the last, it is now a closed ghost town. Except for some houses used by the local families, they are now available for tourists or migrants returning to the country for a while. Beyond the logistic solution and the economic aspects, the safekeeping of this temporary landscape is a meaningful symbol of a shared past. There is extensive iconological documentation on the days of the earthquake and those of the reconstruction [1]. We would like to dwell on some aspects of a temporary nature that deal with the disaster memory. For example, in San Potito Ultra, the local administration set thermo-igloos provided by the Sicily Region—i.e., dome houses in expanded polystyrene. This solidarity pattern is recurrent in

Italian history, a sign of national unity. The igloos were inhabited for over twenty years and dismantled in 2015 [39]. There was a consistent flow of families that occupied them one after the other. This was a common phenomenon in the first decades after the Irpinia earthquake and reached challenging moments in the not far city of Naples. The footprint of past lives dots the "residual" spaces of the rural and urban landscape, as well as the material ruins of abandoned workplaces. Just recently, scholars have tried to read the landscape of the earthquake areas as a semiotic and anthropological scenery, with an eye to the symbolization of the space evoked by Marc Augé and to the so-called *délaissé* places by Gilles Clément [40]. These sites mirrored the new hierarchies of poverty, the contrasts and the solidarities between inhabitants, the succession of the generations. They were part of the social recomposition. In this sense, reused or not, restored or degraded, these sites are symptoms of a complicated memory and pieces of the post-earthquake dynamic landscape. Tangible and intangible signs have been swept away or reinterpreted by the consequence of the disaster. Moreover, within all these things, we can see the processing of the trauma and a reshaping of the cultural memory. The Irpinia-Basilicata earthquake was not just a fracture in a linear story—of course, it was too—but the beginning of a more overall rethink of the local and national history. In this sense, it involves the historians' role and their interpretative approaches.

6. Conclusions

When the Irpinia-Basilicata earthquake struck, the trauma of the disaster was immediately associated with the vivid memory of Belice (1968) and Friuli earthquakes (1976). The two events had deeply affected the country. Furthermore, the last one had coincided with the renewal and reorganization of the studies on earthquakes and their impact [6,23,41]. Nevertheless, because of the energy released, the highest number of victims, the responsibilities and the consequences, the Irpinia earthquake immediately appeared in the international press as the worst Italian disaster from the Second World War [12,13], grabbing more attention due to the deaths, the slowness of rescue, the villages erased by maps [42]. There is a before and an after the earthquake of 23 November 1980. Italy changed, painfully, and the disaster acted as the detonator, starting processes whose consequences are still visible. The commemoration of the 40 years since this event may be an opportunity to reflect, not only on the history of the earthquake itself but—it is the focus of this paper—on the problematic connection between the Irpinia earthquake and historical discipline. The challenge for historians is to overcome disciplinary fences, considering the "earthquake" as an interpretive space, both specific and complex. In this view, the reference earthquake-history-memory became, essentially, a hermeneutical circle. This attitude can shed light on the Irpinia earthquake as well as on aspects closely connected to historical understanding. Memory, landscape, identity, participation, processing trauma, disaster narrative are some of many interpretative focuses. It is not just about approaching earthquakes as part of the general history. It is about grabbing the insights that come from the study of the earthquake to refine the disciplinary tools. This also questions the role of history and historians in this regard.

In this view, social-historical studies can also support the demand for knowledge and mitigation of the risk that comes from citizens and communities living in seismic areas. This means paying attention to social memory, the consciousness of the seismic risk and natural hazard, without forcing the historical discipline towards distant topics or teleological instances. We have touched just some points to reflect.

"Don't worry. She's the Earth cradling you" is a common saying to calm the children in some South American countries, a tender metaphor that makes "comprehensible" the events. Pacha Mama Andine culture is far away from postmodern society yet, the need to accept the earthquake as a possibility of life, perhaps through an inclusive narration, is an instance sharable everywhere, in any culture.

Funding: This research was funded by CNR—Institute for studies on the Mediterranean (IsMed).

Institutional Review Board Statement: Not applicable.

Informed Consent Statement: Not applicable.

Data Availability Statement: Not applicable.

Acknowledgments: I'd like to thank the anonymous reviewers. Their suggestions accompanied me in improving the writing and clarifying the topics. Their questions led me to deepen several aspects, learning a lot from their advice and this interdisciplinary reflection. All that has opened new horizons for me; one more reason to thank.

Conflicts of Interest: The author declares no conflict of interest.

References

1. Porfido, S.; Alessio, G.; Gaudiosi, G.; Nappi, R.; Michetti, A.M.; Spiga, E. Photographic Reportage on the Rebuilding after the Irpinia-Basilicata 1980 Earthquake (Southern Italy). *Geosciences* **2021**, *11*, 6. [CrossRef]
2. Hamburger Abendblatt, Historisches Archiv Nr. 283 vom 04.12.1980 Seite 32. 'Dank für die Deutsche Hilfe', Materdomini, 4. Dezember. Available online: https://web.archive.org/web/20140728192238/http://www.abendblatt.de/archiv/article.php?xmlurl=%2Fha%2F1980%2Fxml%2F19801204xml%2Fhabxml801012_7712.xml (accessed on 30 December 2020).
3. Guidoboni, E. Il Valore Della Memoria. Terremoti e Ricostruzioni in Italia nel lungo periodo. *Quellen Forsch. Ital. Arch. Bibl.* **2017**, *96*. [CrossRef]
4. Gasparini, P.; Musella, S.; Pierattini, D. Macedonio Melloni and the foundation of the Vesuvius Observatory. *J. Volcanol. Geotherm. Res.* **1992**, *53*, 1–10. [CrossRef]
5. Cubellis, E.; Luongo, G.; Obrizzo, F. Cultural climate in Naples between the birth and development of volcanology. *Soc. Geol. Ital.* **2017**, *43*, 64–78. [CrossRef]
6. INGV Terremoti, Terremoto80—Il terremoto del 1980 e le Iniziative del Progetto Finalizzato Geodinamica (Parte Prima). Available online: https://ingvterremoti.com/2020/11/13/terremoto80-il-terremoto-del-1980-e-le-iniziative-del-progetto-finalizzato-geodinamica-parte-prima/ (accessed on 28 December 2020).
7. Roth, P.A. Whistling History: Ankersmit's Neo-Tractarian Theory of Historical Representation, Rethinking History. *J. Theory Pract.* **2013**, *17*, 548–569. [CrossRef]
8. Imperiale, A.J.; Vanclay, F. Top-down Reconstruction and the Failure to "Build Back Better" Resilient Communities after Disaster: Lessons from the 2009 L'Aquila Italy Earthquake. *Disaster Prev. Manag.* **2020**. [CrossRef]
9. Properzi, P. *L'Urbanistica e i Terremoti nella Costruzione della forma Urbana*; "Città & Storia", VI/1; Università Roma Tre-CROMA: Rome, Italy, 2011; pp. 189–206. [CrossRef]
10. Blumetti, A.M.; Guerrieri, L.; Porfido, S. Cataloguing the EEEs Induced by the 1783 5th Calabrian earthquakes: Implications for an Improved Seismic Hazard. In *Earthquake Environmental Effects for Seismic Hazard Assessment: The ESI Intensity Scale and the EEE Catalogue*; Edition: Mem. Descr. Carta Geol. D'It. XCVII; Guerrieri, L., Ed.; A.T.I.-SYSTEMCART srl-S.EL.CA. srl: Rome, Italy, 2015; pp. 153–164. [CrossRef]
11. Parrinello, G. To Whom Does the Story Belong? Earthquake Memories, Narratives, and Policy in Italy. *RCC Perspect.* **2018**, *3*, 69–77. [CrossRef]
12. Der Spiegel. Erdbeben: "Wir brauchen Zelte, nicht Politiker". 1 December 1980. Available online: https://magazin.spiegel.de/EpubDelivery/spiegel/pdf/14332601 (accessed on 31 December 2020).
13. Der Spiegel. Italien: "Armes, unglückseliges Vaterland". 8 December 1980. Available online: https://magazin.spiegel.de/EpubDelivery/spiegel/pdf/14334606 (accessed on 31 December 2020).
14. Ricciardi, P.; Picone, G.; Fiorentino, L. *Il Terremoto dell'Irpinia. Cronaca, Storia e Memoria dell'Evento più Catastrofico dell'Italia Repubblicana*; Donzelli Editore: Roma, Italy, 2020.
15. Cecere, D.; De Caprio, C.; Gianfrancesco, L.; Palmieri, P. (Eds.) *Disaster Narratives in Early Modern Naples Politics, Communication and Culture*; Viella: Roma, Italy, 2018.
16. Law 996. 8 December 1970. Available online: https://www.gazzettaufficiale.it/atto/serie_generale/caricaDettaglioAtto/originario?atto.dataPubblicazioneGazzetta=1970-12-16&atto.codiceRedazionale=070U0996&elenco30giorni=false#:~{}:text=1.,fronteggiate%20con%20interventi%20tecnici%20straordinari (accessed on 30 December 2020).
17. Institutional Agenda. Available online: https://archivio.quirinale.it/aspr/presidente/alessandro-pertini (accessed on 31 December 2020).
18. Il Mattino, Wednesday 26 November 1980, LXXXIX, n. 301, Naples, Italy. Front page title: "Fate Presto". Available online: https://ambasciator.it/carlo-franco-fate-presto/ (accessed on 1 December 2020).
19. Camera dei Deputati. XVI Legislatura. Documentazione e Ricerche. I Principali Eventi Sismici a Partire dal 1968. Schede di Lettura n. 67. 14 Maggio 2009. Available online: https://documenti.camera.it/Leg16/Dossier/Testi/AM0065.htm (accessed on 25 January 2021).

20. Comune di Montella. Provincia di Avellino. Piano Urbanistico Comunale. Legge Regionale n. 16/2004 "Norme sul Governo del Territorio". Regolamento n. 5/2011 di Attuazione Delle Norme per il Governo del Territorio. Relazione Illustrativa. Available online: https://www.comune.montella.av.it/doc/puc/01-%20Relazione%20illustrativa%20%20-%20adeguato.pdf (accessed on 25 January 2021).
21. Porfido, S.; Alessio, G.; Gaudiosi, G.; Nappi, R.; Spiga, E. Effetti Ambientali Indotti dai Terremoti: Il caso di Studio di Alcune Località Colpite dal Sisma del 1980.Conferenza Nazionale ASITA, Salerno, Italy, 21–23 November 2017, 899–906. Available online: http://atti.asita.it/ASITA2017/Pdf/182.pdf (accessed on 25 January 2021).
22. Valensise, M.R. *I Settanta anni del Consiglio Nazionale delle Ricerche (1923–1993)*; CNR: Roma, Italy, 1993; (reprint 1994).
23. INGV Terremoti, Terremoto80—Il Terremoto del 1980 e le Iniziative del Progetto Finalizzato Geodinamica (Parte Seconda). Available online: https://ingvterremoti.com/2020/11/20/terremoto80-il-terremoto-del-1980-e-le-iniziative-del-progetto-finalizzato-geodinamica-parte-seconda/ (accessed on 28 December 2020).
24. Catalogo dei forti terremoti, WEB GIS, release 4 (2007): E. Guidoboni/G. Ferrari/D. Mariotti/ A. Comastri/G. Tarabusi/G. Valensise, CFTI4Med, Catalogue of Strong Earthquakes in Italy from 461 B.C. to 2000 and in the Mediterranean area, from 760 B.C. to 1500. An Advanced Laboratory of Historical Seismology 2007. Available online: http://storing.ingv.it/cfti/cfti5/ (accessed on 1 December 2020).
25. Matano, F.; Di Nocera, S.; Criniti, S.; Critelli, S. Geology of the Epicentral Area of the November 23, 1980 Earthquake (Irpinia, Italy): New Stratigraphical, Structural and Petrological Constrains. *Geosciences* **2020**, *10*, 247. [CrossRef]
26. Del Gaudio, C.; Di Ludovico, M.; Polese, M.; Manfredi, G.; Prota, A.; Ricci, P.; Verderame, G.M. Seismic Fragility for Italian RC Buildings Based on Damage Data of the Last 50 Years. *Bull. Earthq. Eng.* **2020**, *18*, 2023–2059. [CrossRef]
27. Mazzoleni, D.; Sepe, M. (Eds.) *Rischio Sismico, Paesaggio, Architettura: L'Irpinia, Contributi per un Progetto*; Università degli Studi di Napoli Federico II—CRdC-AMRA: Naples, Italy, 2005.
28. Porfido, S.; Alessio, G.; Gaudiosi, G.; Nappi, R.; Spiga, E. The Resilience of Some Villages 36 Years after the Irpinia-Basilicata (Southern Italy) 1980 Earthquake. In *Advancing Culture of Living with Landslides, Volume 5 Landslides in Different Environments*; Mikoš, M., Vilímek, V., Yin, Y., Sassa, K., Eds.; Springer: Dordrecht, The Netherlands, 2017; pp. 121–134. [CrossRef]
29. Mazzeo, G. *Città a Meno del Piano. L'Indifferenza delle Strutture Urbane alla Pianificazione*; FrancoAngeli: Milano, Italy, 2011.
30. Spiga, E.; Porfido, S. *Bisaccia Piano di Zona*; Blurb: San Francisco, CA, USA, 2020; ISBN 978-1-71-555296-1.
31. Zullo, F.; Marucci, A.; Fiorini, L.; Romano, B. The Italian Apennines between Earthquakes, High Naturalness and Urban Growth. *Environ. Plan. B Urban Anal. City Sci.* **2020**, *47*, 716–731. [CrossRef]
32. Moscaritolo, G.I. Reconstruction as a Long-Term Process. Memory, Experiences and Cultural Heritage in the Irpinia Post-Earthquake (November 23, 1980). *Geosciences* **2020**, *10*, 316. [CrossRef]
33. Porfido, S.; Spiga, E. Prima Che Si Perda La Memoria: Viaggio Iconografico in Irpinia Tra Dissesti e Terremoti. Before Memory Is Gone: An Iconographic Journey among the Landslides and Earthquakes of Irpinia, Italy. Available online: http://eprints.bice.rm.cnr.it/15294/ (accessed on 1 December 2020).
34. Pfister, C. 'The Monster Swallows You': Disaster Memory and Risk Culture in Western Europe, 1500–2000. *RCC Perspect.* **2011**. [CrossRef]
35. Solarino, S.; Amaral Ferreira, M.; Musacchio, G.; Rupakhety, R.; O'Neill, H.; Falsaperla, S.; Vicente, M.; Lopes, M.; Sousa Oliveira, C. What Scientific Information on the Seismic Risk to Non-structural Elements do People Need to Know? Part 2: Tools for Risk Communication. Annals of Geophysics, [S.l.], Volume 63, p. AC05. October 2020. Available online: https://www.annalsofgeophysics.eu/index.php/annals/article/view/8439/7266 (accessed on 21 December 2020).
36. Camera dei Deputati. *X Legislatura. Doc. XXIII, n. 27. Commissione Parlamentare di Inchiesta Sulla Attuazione degli Interventi per la Ricostruzione e lo Sviluppo dei Territori della Basilicata e della Campania Colpiti dai Terremoti del Novembre 1980 e Febbraio 1981, etc. Data di Presentazione 25 Gennaio 1991. Progetto di Relazione Premessa alla Relazione Conclusiva, in Part. p. 9*. Available online: http://www.senato.it/leg/10/BGT/Schede/docnonleg/30412.htm (accessed on 4 January 2021).
37. Barbato, E. Qualcuno Fece Presto. E fu la Camorra. Terre di Frontiera. 23 December 2019. Available online: https://www.terredifrontiera.info/camorra-e-terremoto-1980/ (accessed on 1 December 2020).
38. Macdonald, S. *Memorylands. Heritage and Identity in Europe Today*; Routledge: London, UK, 2013.
39. Porfido, S. *Temporary Housing in San Potito Ultra*; Blurb: San Francisco, CA, USA, 2020; ISBN 978-1-71-476363-4.
40. Piermattei, S. Le "case al mare" a Civita di Cascia. Estetiche e pratiche di una ricostruzione post-terremoto tra liminalità, non luoghi e terzo paesaggio. In *Antropologia del "Terzo Paesaggio"*; Lai, F., Breda, N., Eds.; CISU: Roma, Italy, 2011; pp. 149–171.
41. Peruzza, L.; Sarao, A.; Barnaba, C.; Massolino, G. What do young people think, 40 years after the earthquake in Friuli? *Boll. Geofis. Teor. Appl.* **2019**, *60* (Suppl. 1). [CrossRef]
42. Plusieurs Milliers de Morts. La lenteur des secours suscite de nombreuses protestations "Faites vite". Le Monde, 27 Novembre 1980. Available online: https://www.lemonde.fr/archives/article/1980/11/27/bull-plusieurs-milliers-de-morts-bull-la-lenteur-des-secours-suscite-de-nombreuses-protestations-faites-vite_3075432_1819218.html (accessed on 1 December 2020).

geosciences

MDPI

Review

The Scientific Landscape of November 23rd, 1980 Irpinia-Basilicata Earthquake: Taking Stock of (Almost) 40 Years of Studies

Fabrizio Terenzio Gizzi * and Maria Rosaria Potenza

Institute of Heritage Science, National (Italian) Research Council (ISPC-CNR), 00010 Rome, Italy; mariarosaria.potenza@cnr.it

* Correspondence: fabrizioterenzio.gizzi@cnr.it

Received: 29 October 2020; Accepted: 23 November 2020; Published: 28 November 2020

Abstract: The November 23rd, 1980 Irpinia-Basilicata (Southern Italy) earthquake is one of the strongest earthquakes ever occurred in Italy. The earthquake was a natural laboratory for the scientific community, which was engaged highly and promptly in investigating the event, thus publishing a flood of papers in different research areas over time. Just these research outputs are the focus of the article, which examines, with a tailored methodological approach, the international and national (Italian) studies started and advanced since the occurrence of the earthquake. First, we built and analyzed statistically two bibliographic databanks regarding the earthquake studies: (a) the international version of IRpinia Bibliographic databASE (IR_BASE_ENG), selecting and standardizing the pertinent scientific documents extracted from Scopus, Web of Science, and other databases and (b) the national version of the database (IR_BASE_IT) using the Google Scholar search engine to search for the most relevant papers in Italian. Second, IR_BASE_ENG was analyzed in a bibliometric perspective through the data mining VOSviewer software (Waltman et al., 2010) that builds co-occurrence term maps useful in perspective of investigating the wide-ranging studies on the earthquake. Third, taking a cue from this network analysis, we recognized the main research topics and performed a minireview of the related international studies, integrating in it a quick reference to the literature in Italian. Finally, we associated the scientific outputs to each cluster/topic, also performing the frequency analysis of the published documents for each subject, thus gaining information on the temporal trends of studies and getting a more exhaustive evidence of the scientific landscape on the earthquake over the last 40 years.

Keywords: 1980 Irpinia-Basilicata earthquake; earthquake effects; environmental effects; disaster epidemiology; disaster response

1. Introduction

The November 23rd, 1980 Irpinia-Basilicata (Southern Italy) earthquake is one of the strongest earthquakes ever occurred in Italy (Me 6.7, [1]). The event caused over 2700 casualties while felt effects occurred in a very wide area, from south to north, from Sicily to Liguria. Among the 687 municipalities affected, 37 were declared devastated, 314 seriously damaged, and 336 damaged [2]. Out of around 1,850,000 houses, 75,000 were destroyed, 275,000 seriously damaged, and 480,000 slightly damaged. In particular, the villages of Castelnuovo di Conza, Conza della Campania, Laviano, Lioni, Sant'Angelo dei Lombardi, and Santomenna, in the Avellino province, were almost razed to the ground. The damage to cultural heritage was also significant, with several hundreds of churches, bell towers, castles, palaces, and archaeological remains damaged with different degrees of severity. Furthermore, the event caused primary effects consisting of surface faulting extended for about 40 km in length [3] and secondary effects with the triggering or the reactivation of numerous landslides such as those in Senerchia, Caposele, and Calitri in the Avellino province, San Giorgio La Molara in the Benevento province, and Grassano in the Matera province [4–6].

Due to the broad-spectrum of effects, the area most involved in the earthquake was considered a sort of natural and open-air laboratory to perform in-depth studies covering multidisciplinary and interdisciplinary areas. Research activities by the scientific community promptly started in the aftermath of the earthquake so that the first papers were published in the weeks and months immediately following the earthquake.

That being stated, in order to shed light on these studies in qualitative and quantitative way, the article aims to illustrate and discuss the international and national scientific publications. To do that, we built two databanks: (1) the international version of the IRpinia Bibliographic databASE (IR_BASE_ENG), making use mainly of Scopus and Web of Science citation indexes. Furthermore, BIOSIS, Data Citation Index, and MEDLINE were also considered to extract and analyze the pertinent papers published over a period of about 40 years, since 1980. Additionally, in order to increase the database completeness, we made some manual additions of bibliographic records, (2) the national version of the database (IR_BASE_IT) using the Google Scholar search engine to search for papers in Italian. The combined use of the databases allowed to widely tracing the typology, amount, and evolution of the most typologies of studies regarding the earthquake.

Once IR_BASE_ENG was built, we examined it to focus particularly on the hot issues and research lines covered by the papers. To do that, we made use of the VOSviewer data mining bibliometric software that builds distance-based maps (www.vosviewer.com [7]). The software is widely used to perform analysis in different fields such as information and communication technologies, medicine, agriculture, earth sciences, and applied geophysics [8–12].

Taking a cue from the network analysis, we identified the main research branches. We also carried out a minireview of some relevant international studies, also including a quick look at the 1980 literature in Italian.

The article is divided into four main parts: (1) the criteria followed to build the two 1980 IR_BASE and analyze the IR_BASE_ENG through the bibliometric software, (2) the statistical result analysis related to the two databases, (3) the VOSviewer map analysis by identifying the main clusters/topics and performing a minireview about each of them, and (4) the statistical analysis of the research outputs for each topic over time.

2. Materials and Methods

As Figure 1 shows, the approaches followed to build the two 1980 literature databases were quite different.

As regards to IR_BASE_ENG, we considered four main steps: (1) selection of the international bibliographic databases from which the pertinent data is extracted; (2) identification of the proper queries to interrogate the archives and extract the data; (3) checking, homogenization, and fusion of the bibliographic data extracted by the different digital repositories; and (4) manual adding of both missing records and incomplete bibliographic information.

Concerning the first point, we considered Scopus and Web of Science, which are the most considered databases commonly used to search for the literature [13]. In addition, in order to expand our searches to as many research branches as possible, these two databases were supplemented by BIOSIS, Data Citation Index, and MEDLINE.

Scopus was launched in November 2004, and it includes over 25,000 active serial titles from more than 5000 international publishers. Scopus delivers the most comprehensive overview of the world's research output in the field of science, technology, medicine, social science, and arts and humanities. Scopus also includes "Secondary Documents" (SDs) that are documents, which have been extracted from a Scopus, document reference list, but are not available directly in the Scopus database since they are not indexed. In order to increase our findings for the 1980s and 1990s especially, SDs were also considered in building the 1980 international database. This choice implied to manually adding some missing information to the SD records, as they are not indexed by Scopus. This aspect will be discussed in a deeper way later.

The Web of Science Core Collection (1965 to present) consists of 10 indexes containing information gathered from thousands of scholarly journals, books, book series, reports, conferences, and more. Citation databases are Science Citation Index Expanded (SCI-EXPANDED), Social Sciences Citation Index (SSCI), Arts & Humanities Citation Index (A&HCI), Conference Proceedings Citation Index-Science (CPCI-S), Conference Proceedings Citation Index-Social Science & Humanities (CPCI-SSH), Book Citation Index–Science (BKCI-S), Book Citation Index–Social Sciences & Humanities (BKCI-SSH), Emerging Sources Citation Index (ESCI), Current Chemical Reactions (CCR-EXPANDED), and Index Chemicus (IC). Web of Science includes over 12,000 highly acclaimed impact journal worldwide.

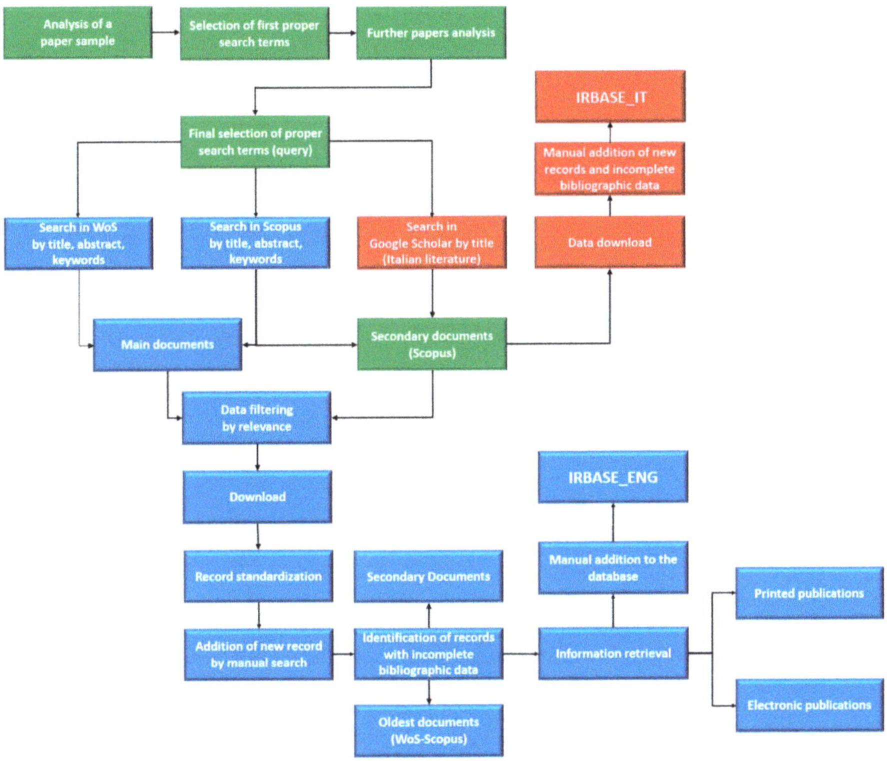

Figure 1. Flow chart summarizing the methodology used to build the two databases. In green, the methodological path common for the two databases. In blue, the path followed only for assembling IR_BASE_ENG. In red, the path for the building of IR_BASE_IT.

As said before, other databases were used to search for the 1980 earthquake documents: BIOSIS, Data Citation Index, and MEDLINE. BIOSIS Citation Index, with coverage from 1926 to present, is the world's most comprehensive reference database for life science research. It also includes cited references to primary journal literature on medical research findings. In addition, it covers original research reports and reviews in traditional biological and biomedical areas. The Data Citation Index contains source records relating to research data available in Web-based data repositories and covers all scholarly disciplines. MEDLINE, with coverage from 1950 to present, is the premier database of the US National Library of Medicine (NLM). It contains over 12 million records of journal articles in all areas of the life sciences, with particular emphasis on biomedicine. Web of Science Core Collection,

BIOSIS, Data Citation Index, and MEDLINE databases were accessed by the option "All Databases" of the Web of Science platform.

Looking at the second point of the methodology, in order to identify the search terms and queries useful to extract as many records as possible by the above mentioned international databases indexes, we performed a preliminary quick reading of a few dozen of papers, taking into account the overall time period under investigation here (1980–2020). After that, we fixed the first query to be used to pull records from the citation indexes. From these preliminary records, we identified and selected further search terms. This approach was followed until no new search terms were found (for the search terms used see Appendix A). Joining all the key terms, we execute the final query to interrogate Scopus and Web of Science by title, abstract, and keywords, both author and indexed ones. The choice to include all keywords implies that selected documents include both direct studies on 1980 earthquake that research where the 1980 events are not the main target of the investigations.

We executed the query considering all document typologies coming from the repositories taken as a reference. Furthermore, we considered the search results without restriction for specific research areas. Once executed the query, the results were saved for the two searches (Scopus and Web of Science) separately.

The third point of the approach refers to the preprocessing of data. Due to their differences in the format, we homogenized the data deriving from Scopus and Web of Science and fused them in an Excel sheet, deleting duplicate records and manually checked the pertinence of each item with our search aim, analyzing the abstract and/or the full texts.

The fourth step concerns the manual adding of information, in both Secondary Documents of Scopus and oldest documents retrieved in Web of Science as well as in Scopus. As said before, Secondary Documents are not indexed. Therefore, the bibliographic records are only partial and they do not include, e.g., the abstracts and source information as they are frequently incomplete, at least for the items published in the 1980s and early 1990s. Furthermore, such records are frequently duplicate reflecting the different way in which the documents were cited. These limitations required a deep and time-consuming manual check of the records, with corrections and/or adding of missing information (authors, year of publication, abstracts, and publication source) by consulting the original scientific works retrieved on the web and libraries. However, also the indexed documents required to be made complete in the bibliographic data. Indeed, the documents of Web of Science and Scopus published in the 1980s frequently did not have abstracts. Therefore, we performed the manual addition of them, in perspective to be analyzed by the VOSviewer software (version 1.6.8).

However, the use of Scopus as well as Web of Science alone did not allow finding some documents. We refer, e.g., to some papers published in the Special Issue of *Annali di Geofisica* edited in 1993 dedicated to the Irpinia-Basilicata earthquake [14]. To overcome this drawback and increase the completeness of the IR_BASE_ENG database as much as possible, we added manually in it some missing records and the related abstracts, by analyzing the references of some relevant papers on the 1980 earthquake published in the 1980s and 1990s, with particular attention to the works reported in Valensise (1993) [15]. After these steps, we built the final version of the IRpinia Bibliographic databASE (IR_BASE_EN). Later on, the data were read in and explored by the VOSviewer software (see next sections).

As regards the construction of IR_BASE_IT, we used Google Scholar (GS). GS is not a bibliographic database, but a specialized search engine that looks for scholarly materials that may be available online. GS was accessed by the Publish & Perish software (https://harzing.com/resources/publish-or-perish). However, unlike of Scopus and Web of Science, in GS, it is not possible to perform search also in abstracts and keywords [16]. In addition, the search analysis in GS performed through the "Keywords" field of Publish & Perish provided too much irrelevant data. Therefore, we oriented our analysis of Italian literature to find only the most relevant documents, interrogating GS only by the title field. Anyway, with the aim to increase the corpus of IR_BASE_IT records, the GS data were supplemented by searches performed in the Secondary Documents of Scopus. Furthermore, also for IR_BASE_IT, manual additions of records coming from Valensise (1993) [15] were performed. We used the following search terms: *terremoto/i 1980* (earthquake/s 1980), *sisma 1980* (earthquake/s 1980), *terremoto/i irpinia* (earthquake/s irpinia), and *faglia irpina* (fault irpinia). Then, we extracted the bibliographic data and saved them as *.csv file to be analyzed statistically joined with the other databank (IR_BASE_ENG). The analysis of IR_BASE_IT by the VOSviewer software was not done, but a quick discussion of the related research items was fused within the minireview based on the international Scopus-Web of Science derived papers, as we have discussed better later. Data for both IR_BASE_ENG and IR_BASE_IT databases were searched and downloaded between 25th July 2020 and 3rd August 2020.

As said before, IR_BASE_ENG was analyzed by the VOSviewer software. This builds co-occurrence networks, suitable to give an idea of relationships between words. Indeed, word co-occurrence networks, among the most common linguistic networks studied in the past due to their topological features [17], are used to detect semantic similarity between terms [18]. The software uses the text mining technique to identify the noun phrases from titles or title and abstracts. The noun phrases are classified based on a relevance score: high relevance score is assigned when terms co-occur mainly with a limited set of other noun phrases so showing a more precise connotation in the field considered. This means that noun phrases with low relevance score are those that tend to be too general and meaningless for the domain of interest: they are omitted from the data processing. The software grouped the high relevance noun phrases (referred as terms) together into clusters to identify possible subfield or research topics. The default option of the software is to select the 60% most relevant terms among the noun phrase that occurred 10 times at least [19,20].

In our analysis on the 1980 earthquake, we selected the titles and abstracts option to extract noun phrases. Furthermore, in order to have a picture as more representative as possible of the research activity, we lowered considerably the threshold values of noun phrase, fixing the occurrence at 3 times at least. Conversely, regarding the percentage of most relevant terms, we considered a percentage of 70%. Furthermore, the option to insert a thesaurus text file was considered. This option is helpful in order to merge different spellings of the same term (e.g., *hypocenter* and *hypocentre*; *isnet* and *irpinia seismic network*). In addition, the option was considered also for merging different terms referring to the same concept (e.g., *earthquake sequence* and *seismic sequence*). After that, the resulting terms were cleaned by deleting the irrelevant words (e.g., *March*, *introduction*, *paper*, *review*).

The software builds three typologies of maps: network visualization, overlay visualization, and density visualization. The first map shows the items by their label and by a circle. For each term, the size of the term's label and the size of the term's circle depends on the weight of the term. Furthermore, the color of an item is determined by the cluster to which the item belongs. The overlay visualization is the same as the network visualization except that items are colored in a different way depending on the user choice. Lastly, density visualization shows the density of an item at a certain point. In our analysis, we built and analyzed comparatively the network and the density visualizations.

3. Results

The Two Databases

Overall, the two databases, made up mainly of journal articles and conference proceedings, include 636 bibliographic records, 512 of which related to IR_BASE_ENG and 124 to IR_BASE_IT.

Figure 2 shows the yearly pattern of number of bibliographic records for both the databanks over the period of almost 40 years. On the one hand, IR_BASE_ENG covers all the time window; on the other hand, IR_BASE_IT shows an intermittent lacking of data for 11 years, since the early 1990s. However, over the entire period, at least one international or national document was published yearly, including the remaining period of just over a month after the earthquake occurrence. The average annual number of documents is equal to 16, the maximum number of documents published in one year (1981) is 64.

Analyzing Figure 3, which shows the cumulative yearly percentage of documents published in the two databases taken alone and as a whole, we can see that in the 1980s, about 50% of the total documents was published. Considering the two databases separately, we can also realize that approximately the same percentage (50%) of documents in Italian was reached only after 6 years since the event occurrence, while the equivalent percentage for the Scopus and Web of Science derived documents was reached in a double period.

Figure 4 shows the number of total authors involved in the authoring the documents yearly (researchers that authored more papers were counted more times according to the number of papers they signed or cosigned). Overall, 2035 authors were involved in the studies, with a yearly mean of about 51. The maximum yearly of 197 was in 1981 and the minimum of 16 in 2001 and 2006. However, for the most of the period, the number of authors was less than 60.

The term map obtained by analyzing the IR_BASE_ENG database with the VOSviewer software shows three main clusters made up of about 330 words (Figure 5). However, the green and red clusters, in turn, can be divided into a series of subclusters that identify different research areas. We will discuss of these in the following sections. An interesting data about the mutual relationship between the three clusters can be provided by Figure 6, where we can see the density view map. In the visualization, the connection between the red and green clusters (171 and 131 items, respectively) appears to be slightly stronger than of either these subfields with the blue cluster (27 terms). What is more, we can also see that relationship within the red and green clusters are quite relevant, thus showing that subclusters/research topics cannot be considered as independent to each other, with special attention to the closest group of terms.

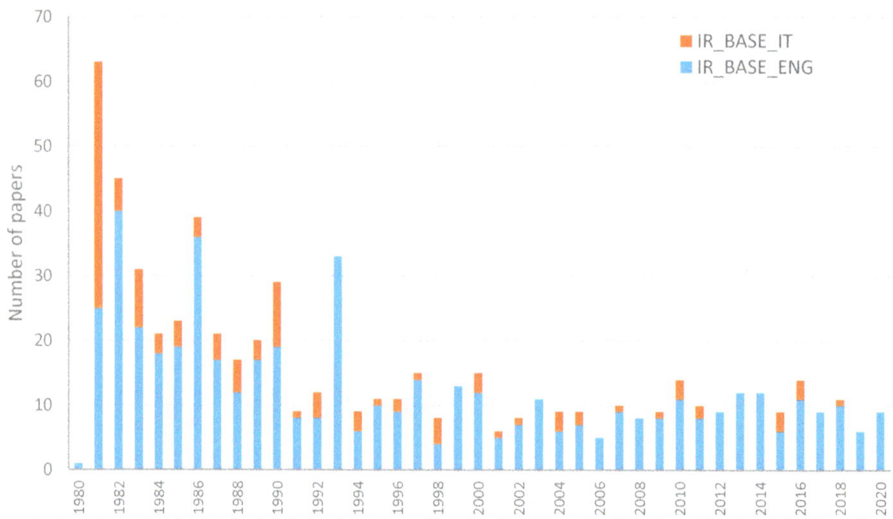

Figure 2. Yearly number of documents in the two 1980 Irpinia earthquake databases.

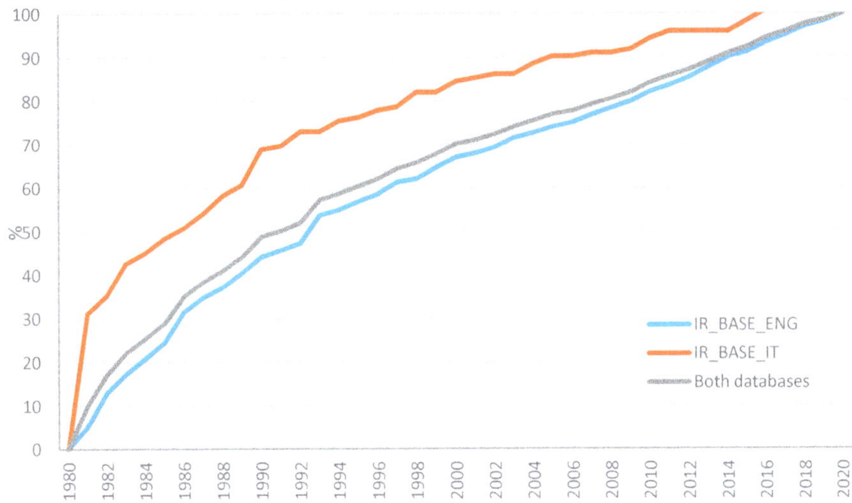

Figure 3. Cumulative percentage of yearly documents in the two databases taken both separately and as a whole.

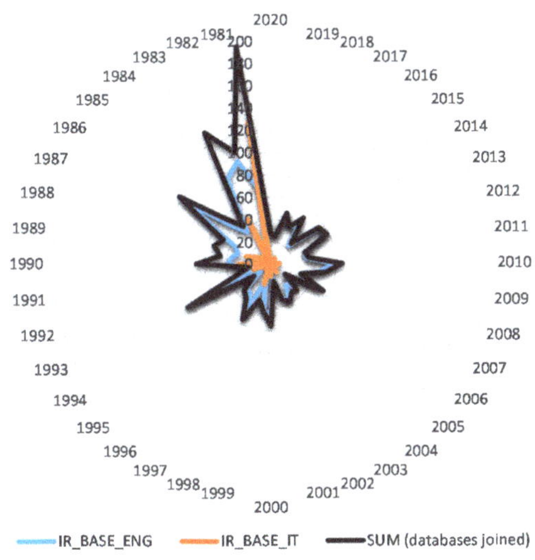

Figure 4. Number of authors involved in the research on 1980 since the earthquake occurrence. The data are represented both for each database taken individually and through the sum of the data contained therein. In the counting, the works published as "Working Group" are considered as published by one author.

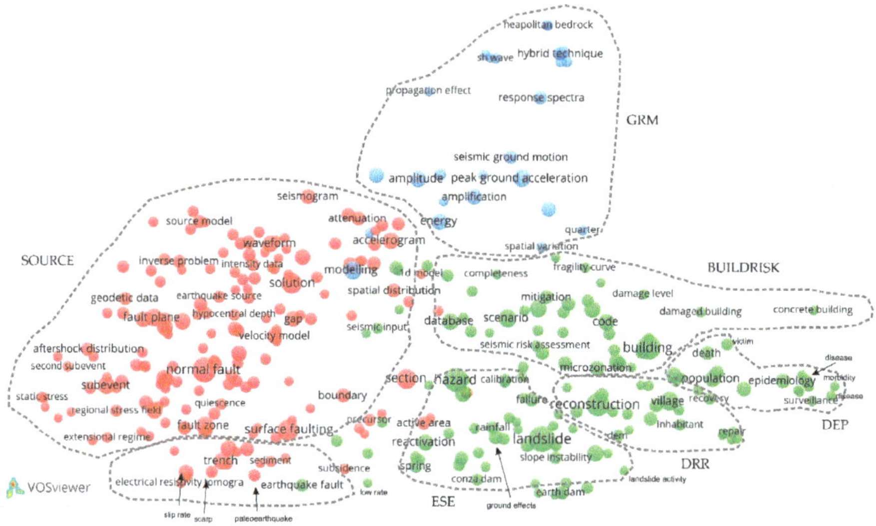

Figure 5. Network visualization by VOSviewer of IR_BASE_ENG database. For each term, the size of the term's label and the size of the term's circle depend on the number of occurrences. The map shows the three main clusters. The first two wider clusters may be divided into some subclusters. For example, the green cluster can, in turn, be divided into four main subclusters whose terms and "boundaries" are roughly drawn manually in the map. Taken overall, the map identifies seven main research fields on the 1980 earthquake. To improve the readability of the map, we added manually some names of terms.

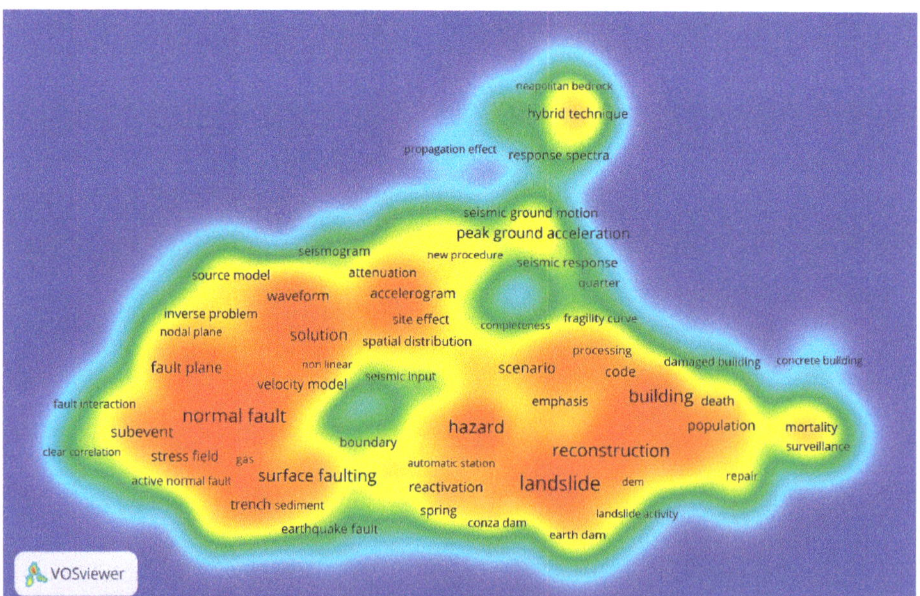

Figure 6. Density visualization by VOSviewer of IR_BASE_ENG database. The map shows the main terms leading the clusters and the connection between them.

4. Discussion

4.1. Cluster 1 (Red) (SOURCE)

The red cluster (Figure 5), led by the word *normal fault* (45 occurrences), may be roughly split into two larger subclusters, which call back on studies concerning the seismic sources through: (1) (mainly) seismological data (the upper and central-lower portion) and (2) (direct) paleoseismological and (indirect) geophysical investigations on the Irpinia fault (lower portion).

4.1.1. Studies Concerning the Seismic Source through Mainly Seismological Data

This subcluster can be identified in the upper and central-lower portion of the red group of terms. The richness of the words, their recurrence, and the mutual relationship of the terms well label both the complexity and high degree of maturity of the studies. These involved also the use of multidisciplinary approaches by integrating seismological, geodetic, and geological data, so showing the great efforts made by the scientific community to fuse multiple data thus building a proper source model.

One of the first works on the seismic source regarded the processing of seismograms recorded at teleseismic distance [21]. The authors gained information on the origin time, location, magnitudes, both mb (6.1) and Ms (6.8), and the geometry of the causative (normal) fault, whose strike direction was found to follow closely the axis of the Apennines in accordance with both the *stress pattern* (4) indicated for this area by focal mechanism analysis of previous earthquakes and neotectonic evidence. The authors also found an anomalous shift in the origin time for the earthquake, so assuming that it might have been caused by a multiple event, which, however, was regarded as not having caused fault breaking at the surface.

In the same year, the *Gruppo di Lavoro Sismometria del Terremoto del 1980* published an important dataset that will then be used in the following years to both constrain the location and extent of the seismogenic source [22]. In addition, the data contained in Berardi et al. [23], which published the accelerometric recordings of the 1980 earthquake, were subsequently the subject of in-depth analyzes to investigate the complexity of the earthquake rupture mechanism.

As it had already emerged immediately after the event, other studies performed in the early 1980s confirmed the normal faulting mechanism of the earthquake, supplying information about the fault features. For example, Del Pezzo et al. [24], analyzed the seismograms from permanent and temporary stations, located the *hypocenter* (13) of the event, confirming the normal faulting fault-plane solution. Furthermore, analyzing more than 600 aftershocks, the authors found an alignment of events over an area having a length of about 70 km. Arca et al. [25], examining the *waveforms* and the *aftershock distribution* (7), confirmed the normal faulting mechanism suggesting an inhomogeneity of the slip. In addition, the same authors proposed the vertical movements as to be modeled by a normal fault segmented in various branches, with an abrupt stop of the rupture in the NW tip, where the *vertical displacement*(s) (6) reached its maximum values. Deschamps and King [26], starting from far field and local data, performed the *waveform modelling* (8), the fault plane and aftershocks analysis also finding that the 1980 earthquake was a normal event with a large component of left-lateral strike slip.

The year later, the same authors [27], performed detailed analysis of over a thousand of aftershocks founding a reverse faulting in the same area that created a large component of normal faulting in the main event, proposing a speculative model to interpret the unusual findings. The same year, Westaway and Jackson (1984) [28], based on a field survey in the epicentral region, recognized as primary effects environmental evidence that had been interpreted in previous years as gravitational phenomena. Indeed, the authors described for the first time, the 1980 *surface faulting* (28), so starting a new leading research line of the following years (see below) aimed at exploring in depth the location as well as the geometry of the surface rupture caused by the 1980 occurrence. The more than 10 km long faulting was considered as consistent with the focal mechanisms of the event, so concluding that deformation associated with earthquakes in the southern Apennines took place in the upper 10–15 km of the Earth's crust on steep planar normal faults.

In order to gain information on the source model, next studies paid more attention to the fusion of different datasets. For example, Westaway (1985) [29] stressed the importance of different data to both study the complex earthquakes and constraint the sources, combined long-period teleseismic and short-period strong motion *waveform modelling* with field survey of *surface faulting* and other data, and pointed out that the faulting had a segmented geometry in which motion occurred on planar normal faults which broke the surface. Crosson et al. (1986) [30] analyzed the aftershocks, the pattern of strong ground motion, the focal mechanism of the mainshock, and two critically placed leveling profiles, arguing that the 1980 faulting resulted in a complex and uncommon pattern of two high-angle subperpendicular direct faults. The work spurred a scientific discussion the year later [31,32]. On the same year, Westaway and Jackson (1984) [33] correlating teleseismic waveforms, local ground acceleration, elevation changes, surface faulting, and aftershocks investigated the three-dimensional *fault geometry* (11) and the timing of the faulting, which caused about 12 km of surface faulting, so updating the data of Westaway and Jackson [28]. The modelling of long-period teleseismic body waves also allowed to question about the hypocenter and focal mechanism, finding a seismogenic normal fault structure to be approximately planar, with a dip of 60°. Moreover, the authors reported a total of six *subevent*(s): i) within 10 s of the origin time of the seismic event, the motion happened on three discrete fault segments extending for 30 km along the strike, ii) a fourth subevent occurred about 13 s after the first motion, and iii) two later fault ruptures also arose about 20 and 40 s after the first motion. Some years later, Bernard and Zollo (1989) [34] using strong motion, leveling data, teleseismic waveform modeling and aftershock studies, analyzed the kinematics of the 1980 normal fault confirming the three main rupture episodes at 0, 20, and 40 s.

The use and fusion of *levelling data* (4) with field observations, the last useful to recognize three main strands forming a 38-km-long northwest trending *fault scarp* (9), were envisaged by Pantosti and Valensise (1990) to delineate the faulting model. The fault scarp, with an average strike of 308°, extended between the north-facing slope of Mt. Cervialto (near Lioni, Avellino) and the Pantano di San Gregorio Magno (Salerno). The scarp was 40–100 cm height [35]. Later on, Blumetti et al. (2002) [6] contributed to identify a set of open fractures and south-southwest-dipping normal fault scarps around Castelgrande, Muro Lucano, and Bella, for a total length of about 8 km. These were interpreted by the authors as possibly primary tectonic effect, probably related to the 40 s event.

In 1992 and 1993, Westaway (1992, 1993) [36,37], relocated the *nucleation* (5) point of the fault, also suggesting an updated sequence of rupture subevents. According to the author, the main shock was characterized by seven rupture subevents, among which were the 20 and 40 s ones. The initial fault rupture nucleated at or near the SE end of the Carpineta fault and propagated NW.

In the same years, Amato et al. (1992) [38], based on the velocity distribution at depth, argued that 40 s rupture was the result of reactivation of an old thrust as a normal fault. Further investigations using the inversion of strong motion *waveform*(s) (13) shed light on the spatiotemporal pattern of rupture process, which propagated northwestward [39,40].

Furthermore, 1993 was the year in which many studies regarding the 1980 earthquake were published as results of the Meeting held in Sorrento in 1990 on the 10th anniversary of the earthquake. The works, already published in a preliminary form in the meeting Proceedings, flowed in the final draft form in the Special Issue of "Annali di Geofisica" [14]. Therefore, several articles were published beyond the two already cited just above [39,40].

Amato and Selvaggi (1993) [41], starting from the relocation of about 600 aftershocks, computed a *velocity model* (12) highlighting that the complexity of the Irpinia mainshock was due to the rupture of the highly heterogeneous medium. Giardini (1993) [42], making use of data recorded at teleseismic distance, determined the main seismological parameters, including a focal mechanism with an almost pure dip-slip mechanism. Pingue at al. (1993) [43] modeled the Irpinia source using geodetic data, but integrating and constraining the results with different data sets useful to overcome some limitations of geodetic data alone. The model described three fault segments each of which related to one of the three main rupture episodes of the main shock. Other studies dealt with the investigation on both

0 s subevent to analyze the propagation rupture [44] and second and third ones to constraint their location and mechanism [45]. In the same year, Vaccari et al. (1993) [46] studied the rupture model by the inversion of accelerometric waveform and the comparison between the model-derived synthetic isoseismal and the observed damage patterns.

In the central-lower portion of the red cluster (Figure 5), we can identify some terms related to studies on the stress change caused by the 1980 earthquake. Nostro et al. (1997) [47] assessed the *static stress change* (4) induced by the normal-faulting 1980 earthquake on nearby seismogenic faults, concluding that the Irpinia earthquake caused an increase on both the fault zone active during the 1990 and 1991 events and faults responsible of 1732 (Irpinia) and 1857 (Basilicata) earthquakes. Belardinelli et al. (1999) [48] computed the dynamic spatiotemporal stress changes caused by the rupture of the first subevent (0 s). The modelling showed that after a transient phase, the stress time history evolved to the final static stress value. In particular, the dynamic stress peaked on the *second subevent* (4) fault plane and it was reached between 7 and 8 s after the rupture initiation on the main fault, with the static stress level on the *second subevent* (20 s) fault plane to be reached nearly after 14s.

Other studies in 2000s, continued to investigate the source and the faulting mechanism using inversion of coseismic *vertical displacement* (6) data [49], *relocation* (4) of the main event, P-wave velocity *inversion (procedure)* (5), and analysis of postseismic *ground deformation* (6) [50].

4.1.1.1. Studies of the Seismic Source by Paleoseismological and Geophysical Investigations

In the lower part of the cluster, we can trace some terms identifying both direct (paleoseismological) and indirect (geophysical) investigations on the Irpinia fault.

The paleoseismological studies started in the late 1980s following the identification of 1980 surface faulting by Westaway and Jackson (1984) [28].

Pantosti et al. (1989) [51] built a thorough mapping of the 35 km-long sector of NW trending NE-facing *scarp(s)* (6) related to the earthquake. In the central part of the surface rupture (Piano di Pecore), the *trench*(es) (14) dug showed a surface displacement of 85 cm, also identifying other three previous paleoearthquakes, beyond the 1980 event.

Two new trenches were excavated in 1990 at Pantano di San Gregorio Magno, an elongated depression located close to the southern end of the Irpinia fault. Four pre-1980 paleoearthquakes were identified, so suggesting the first data on both earthquake recurrence intervals and slip per event, as well as the *slip rate* (12) on the fault (e.g., [52–54]).

In the 2000s, some studies were also oriented to the geophysical investigations of the structure of the Irpinia fault. For example, Improta et al. (2003) [55] performed a high-resolution multifold wide-angle seismic survey carried out across the scarp to investigate the shallow structure of the fault. Bruno et al. (2010) [56] carried out a two-step imaging method to dense wide-aperture data with the goal of imaging the Irpinia fault in its complex geologic setting. Galli et al. (2014a) [57] performed integrated and complex geophysical investigation (electrical resistivity tomography, ground penetrating radar measurements, and horizontal-to-vertical spectral ratio microtremor analysis) along the fault, so supplying a subsurface image of the near-surface fault architecture. Geophysical survey jointly with previous paleoseismological studies allowed the assessment of some fault parameters and the precise locating of the fault trace. Galli et al. (2014b) [58] identified, across the active Mount Marzano Fault System, several clusters of inflection points that, once compared with *historical seismicity* (5) of the area, gathered paleoseismological data along some significant segments of the fault, and previous geophysical investigation (Galli et al., 2014a) [57] allowed to make new considerations about the present *slip rate* and recurrence time for high-magnitude earthquakes.

Vassallo et al. (2016) [59] correlating the ambient noise recorded at broadband stations, found a *low-velocity* anomaly in the area bounded by the two main faults that caused the 1980 earthquake. Furthermore, Lo Re et al. (2016) [60] performed a microgravity survey to precisely locate and better characterize the near-surface geometry of a segment of the Irpinia Fault.

4.2. Cluster 2 (Green)

The green cluster may be subdivided in at least four main subclusters related to research on: (1) Earthquake Secondary Effects (ESE); (2) Disaster Response and Recovery (DRR), (3) Disaster EPidemiology (DEP); and (4) BUILDing and infrastructure damage, vulnerability, seismic RISK assessment and mitigation (including seismic microzonation studies) (BUILDRISK).

4.2.1. Earthquake Secondary Effects (ESE)

The subcluster, in the lower-left portion, includes studies on slope movements, ground cracks, hydrological anomalies, and liquefaction phenomena.

The studies dealing in some way with the *landslide*(s) (52) triggered by the 1980 earthquake cover almost the entire study period analyzed in this article. First works on the subject were those examining the *reactivation* (13) of old landslides in Senerchia, in the Sele River valley [61–63]. The early 1980s also saw studies on liquefaction phenomena [64].

Further research looked at the landslide phenomena in many other localities or areas. We refer to the Sauro torrent, near the village of Stigliano, in the basin of the Agri River [61] Calitri [65–67], an area close to Atella [68], the Upper Valley of Sele River [69], the Valley of Tammaro River [70], and in the Mounts Cervialto and Terminio-Tuoro [71].

While these works, as well as most of the papers published over the following decades, looked at the earthquake-induced landslides in specific areas or towns, Alexander (1981) [72] examined the *hazard*(s) (40), mechanisms and the effects of landsliding provoked by the 1980 earthquake at a wider territorial scale, both in Basilicata and Campania regions. The effects of the induced mass-movement were considered with respect to the slope instability and damage to both settlements and infrastructures. Furthermore, the author also framed the landslide problem within the historical, social, political, and economic factors. As regards the latter, the author underlined as poor conservation of soil in Italy determined that 46% of soil was vulnerable to landslides after flooding, erosion and earthquakes.

About two decades after, Porfido et al. (2002) [73], in line with a wide-ranging analysis of earthquake-induced landslides, performed a systematic overview of the territorial spreading of *ground effect*(s) (9) induced by two strong earthquakes in the southern Italian Apennines, among which just the 1980 event. The authors also shown a likely correlation between maximum distance of effects and length of the reactivated fault zone. Five years later, Porfido et al. (2007) [74] discussed the seismically induced environmental effects extending the study to three strong earthquakes occurred in the Southern Italian Apennines.

In the term map (Figure 5), we can observe a certain closeness between the "landslide" and "spring" terms. This link recalls, e.g., the studies relating to the change in spring flows due to the sliding process triggered by the earthquake [75] or the influence of spring flow increase on the slope failures in the Sele valley [76].

Other terms in the map, such as hazard, indicate the complexity and importance of the landslide studies. Del Prete (1993) [77] discussed two examples of mudslide hazards in the Upper Sele Valley concluding, among other things, that this type of landslide is prone to be reactivated especially due to the seismic shaking. Some years later, Parise and Wasowski (1999) [78] dealt with the assessment of landslide hazard in some areas of the Southern Apennines hit by the 1980 earthquake. The aim was reached by the preliminary building of the landslide activity maps integrating different source of information. The authors, estimating the areal frequency of active landsliding for the last 40 years, argued the significant impact that the 1980 earthquake had on the stability of slopes located close to epicenter. The Upper Sele Valley was also considered by investigations of Capolongo et al. (2002) [79]. The authors evaluated the earthquake-triggered landslide hazard integrating different data into a typical earthquake stability model of slope to assess the landslide potential during the 1980 earthquake in the Valley.

As regards the proximity between *landslide* and *database* (14) this can be justified, e.g., by the building of the list of earthquake-induced ground failures in Italy (Italian acronym CEDIT), which collects data regarding landslides, liquefactions, ground cracks, surface faulting, and ground changes triggered by strong earthquakes (e.g., [80]).

4.2.2. Disaster Response and Recovery (DRR)

The issue of reconstruction was already considered in the early 1980s jointly with the analysis of the emergency phase (e.g., [81–83]).

D'Souza (1982) [84] dealt with the recovery and rebuilding in two communities of Southern Italy, showing that the combination of appropriate aid and effective leadership affect strongly the efficacy for recovery. The same author in 1984 (D'Souza, 1984) [85], analyzing one community living in a village of Salerno province (Campania region), debated that outside aid to the affected area should encourage and sustain indigenous solutions as well as the use of community's own skills and resources. The same year, Alexander (1984) [86] considered the policies implemented by the institutions in two phases, the first involving the relocation of the survivors in temporary prefabricated homes and the second concerning the reconstruction of permanent housings. The author sustained that the financial aids for the rebuilding process was reduced in efficacy due to bureaucratic delays, legal complexities, and inequality in the distribution of economic resources. The author also argued that the occurrence of other natural hazards in the aftermath of the 1980 earthquake had an increasing effect in the formation, in 1982, of a Ministry for Civil Protection with the consequent strengthening of both disaster relief and prevention actions.

Caporale et al. (1985) [87] sustained that there were highly differentiated patterns of recovery in the different villages affected by the earthquake. Therefore, the authors examined the social, economic, and political factors responsible for the differences to draw policy implications for emergency management, reconstruction strategies, and intervention policies.

Some years later Alexander (1989) [88], discussing on how preserving the identity of small settlements during the large-scale phase of reconstruction, analyzed the impact on the urban landscape of temporary shelter and reconstruction. Furthermore, the author argued that inconsistencies in government reconstruction funding were the cause for much of the variability of postdisaster recovery in Italy, concluding that the existing theoretical models were inadequate to predict the development of the rebuilding phase.

Many years later, Chubb (2002) [89] analyzed comparatively the government response to three earthquakes such as the Belice 1968, Friuli 1976, and 1980 one.

Other studies were performed on the reconstruction over the following years. For example, Forino et al. (2015) [90] reflected on the relationship between state-driven developmental policies and postwar territorial transformations in southern Italy. To do this, they also analyzed to what extent the measures for industrial recovery and settlement reconstruction put into the field after the 1980 earthquake in Campania and Basilicata affected soil resource depletion and land degradation. The authors found manifold links between the post-war economic policy and the downward environmental spiral observed in southern Italy.

Recently, Porfido et al. (2017) [91] scrutinized the in situ reconstruction of two villages as well as the rebuilding of another village that was rebuilt far from its original place. Downstream of their analyses, the authors stressed the key role of technical experts both in the built environment and in social as well as ethical context for the proper rebuilding of villages, having particularly the people resilience in mind.

4.2.3. Disaster Epidemiology (DEP)

Looking at the right-terminal portion of the green cluster, close and correlated to the disaster response/recovery subcluster, we can notice a group of terms headed by the word *epidemiology* (10).

These terms and the related ones, reflect empirical studies carried out mainly in the 1980s on the epidemiological consequences of the 1980 earthquake. For example, Greco et al. (1981) [92] performed three typologies of surveys aimed at the: (1) surveillance of hospital admissions of the *survivor*(s),

(2) surveillance of infectious *disease*(s) (4) among the survivors, and (3) statistical analysis of mortality and traumatic injury in the survivor(s). The authors concluded that earthquakes do not necessary cause epidemics nor make the epidemiology of infectious illness worse as respect the predisaster situation. Alexander (1982) [93], analyzing the disease epidemiology of the earthquake and the health service response in the *emergency phase* (4), concluded that there were no serious epidemics and the hospital system was able to cope with the aftermath needs. The author also underlined as the national plan of vaccination of both survivors and rescue volunteers was probably an overreaction of institutions with respect to the dominant epidemiological condition.

Three years later, Bruycker et al. (1985) [94] analyzed the *morbidity* (number of injured) (3) and *mortality* (7) after the 1980 earthquake based on a sample of over 3600 people of near-epicenter *village*(s). They found that *injury* (6) rates were more than five times higher in trapped than in nontrapped victims and chance for escape was vital for *survival* (3) and interconnected with the seismic vulnerability of buildings. The disaster response analysis performed by the authors, also indicated that the emergency phase for medical care was limited to the 3–4 days after the earthquake occurrence.

4.2.4. Building and Infrastructure Damage, Vulnerability, Risk Assessment, and Mitigation (BUILDRISK)

One of the first works regarded the analysis of the damage on different structures, such as *building*(s) (38) and bridge(s), was performed by a Swiss reconnaissance team who surveyed the most affected area about 3 weeks later the earthquake occurrence [95]. In addition, Guepinar et al. (1981) [96] performed a 7-day field trip at the end of January 1981, paying special attention to the damage caused by soil failures or soil–structure interaction.

Postpischl together with other authors published some progress reports on the macroseismic surveys in the affected area starting from 1981 (e.g., [97,98]) until the work of Postpischl et al. (1985) [99] in which the isoseismal maps of the 1980 earthquake were published in an Atlas.

Coburn et al. (1982) [100] described the typical forms of local buildings classifying them. Furthermore, the type and *damage level* (4) analyzed in relation to building type, location, and ground movement were also considered. The researchers found that local damage level correlated well with both instrumental and spectral acceleration. Further analyses on both building damage and behavior of (reinforced) *concrete building* (3) (e.g., [101–103]) were also performed in the following years.

Braga et al. (1982) [104] in order to define the damage probability matrices (DPMs) for the most common types of buildings, performed a statistical study on damaged buildings in 41 municipalities affected by the earthquake. The vulnerability assessment through the DPMs was tested for the first time just after the 1980 earthquake, using the MSK-1976 as reference macroseismic scale. Moreover, the same authors [105], based on their experience on damage survey after the 1980 event discussed a quick method for damage assessment of dwelling buildings. The approach was aimed to setup well-grounded policy for *reconstruction*.

More recently, other studies analyzed statistically large data set of damaged buildings allowing deriving empirical *fragility curve*(s) (4) to define seismic vulnerability of classes of buildings. For example, Rota et al. (2008) [106] derived typological fragility curves starting from a reanalysis of a set of about 150,000 survey building records related to postearthquake data on *damaged building*(s) (4) for the Italian earthquakes of the last three decades. Recently, Del Gaudio et al. (2020) [107], in order to define fragility curves, investigated a set of about 25,000 residential (reinforced) *concrete building*(s) derived from a sample originating from the Web-GIS platform named "Da.D.O." (Observed Damage Database) [108].

Within the subcluster, we find also research related to the microzonation studies of some towns heavily affected by the 1980 earthquake (e.g., [109–112]). After the 1980 event, the new rules imposed that general development urban plans had to be prearranged according to site conditions [109]. However, the first studies on microzonation came back to the early 1980s when the study on seismic microzonation in the emergency phase were performed within the framework of the Italian Geodynamics Project [113,114].

4.3. Cluster 3 (Blue) (GRM)

The third cluster, in the topmost portion of the VOSviewer map, refers mainly to studies concerning the estimations of seismic ground motion of 1980 earthquake and site effect analysis in localities hit by the 1980 earthquake also using in situ geophysical surveys.

For example, Rovelli et al. (1988) [115], analyzing horizontal components of many accelerogram related to moderate and strong normal faulting earthquakes happened in the Italian Central and Southern Apennines, proposed a numerical simulation of *seismic ground motion* (9) peak values as a function of the earthquake size, for seismic risk purpose. Bosco et al. (1992) [116] presented a possible procedure to build artificial acceleration time histories (peak acceleration, response spectra, and envelope curves of the acceleration time history) of bedrock motion consistent with records of past events such as the 1980 Irpinia earthquake. Convertito et al. (2010) [117], using the Irpinia Seismic Network and the 1980 earthquake data, developed a fast procedure to get ground-shaking maps in view of damages and loss assessment for seismic emergency management in high seismic risk areas, such as Campania and Basilicata. They found the approach useful to predict peak ground-motion parameters of high-magnitude events regarding the attenuation relationship. Cultrera et al. (2013) [118] computed hybrid shakemaps for the 1980 earthquake integrating recorded data with deterministic-stochastic method, finding results different from the standard analysis, in the near-fault area, especially. They also found an agreement between peak ground velocity (PGV)-derived Mercalli–Cancani–Sieberg (MCS) intensities and the observed damage.

Nunziata et al. (2000) [119] simulated the *seismic ground motion* generated by the 1980 earthquake in Naples adopting the *hybrid technique* based on the *mode summation* (7) and the *finite difference method*(s) (6). The numerical approach was validated with the 1980 earthquake data recorded at a 10 km-away site. The application to a sample area located in the eastern district of Naples, for which the basic data were well known, allowed modeling the seismic response. Data showed the surficial cover as responsible of an increase in the amplitudes of the signal as respect to the bedrock. Di Giulio et al. (2008) [120] studied the variations of local seismic response for the entire urban area of Benevento (Southern Italy), making a large use of ambient noise recordings, seismic, geological, and geotechnical data. The authors also generated synthetic seismograms of moderate to strong earthquakes among which was the 1980 Irpinia earthquake, finding large amplification at soft soils. Maresca et al. (2012) and Maresca et al. (2018) [121,122] analyzed the effects of surface geology on *seismic ground motion* in Avellino using HVSR (horizontal-to-vertical spectral ratio) free-field peak frequency analysis, also correlating the data with the 1980 urban damage pattern.

5. Statistic of Documents by Topics

A cross-correlated analysis of VOSviewer outputs with the two databases allowed analyzing the total number of documents for each topic (Figure 7) as well as its trend over time (Figure 8). The analysis was also made for the IR_BASE_IT database, consulting the records and manually associating them to each of the cluster(s) identified by the IR_BASE_ENG analysis. Each figure reports data both for the two databases taken separately and fused (SUM).

Figure 7 shows the total number of records for each topic(s), the last including also the records for which an attribution was performed for more than one category. The highest number of documents in IR_BASE_ENG is related to the study of the seismic source (SOURCE) for which 153 (29.9% of 512 documents) documents were published. This follows the groups of documents relating to earthquake secondary effects (ESE) with 88 papers (17.2%), BUILDRISK (84, 16.4%), DRR (52, 10.2%), and GRM (34, 6.6%). Regarding IR_BASE_IT, the largest amount of records in the database is relating to the BUILDRISK topic (32, 25.8% of 124 documents). Next, the documents relating to secondary seismic effects (ESE), with 30 (24.2%) papers are present, followed by those concerning the study of the seismic source (SOURCE, 25, 20.2%). What is more, the database includes only one document related to the disaster epidemiology cluster (DEP).

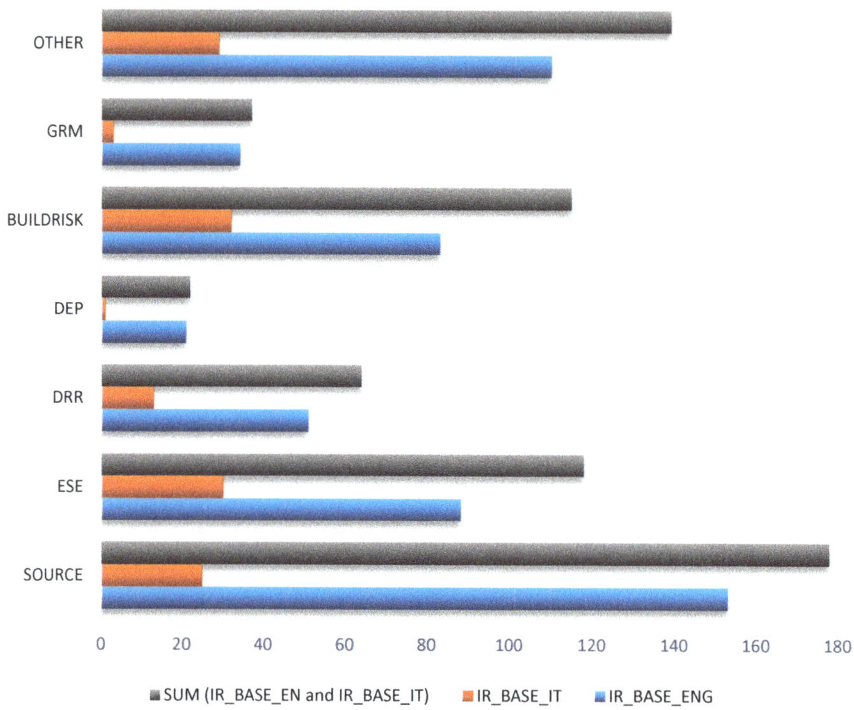

Figure 7. Total number of documents for each database and topic/(sub)cluster.

Looking at the SUM of the two databases, we can see that the relative frequencies of documents in the topics reflect those of IR_BASE_ENG, so confirming that the scientific community addressed efforts mainly in learning about the source mechanism, as discussed in the minireview of the previous sections. Obviously, the two databases also include records that do not mirror the belonging to the (sub)clusters/topics discussed in the previous sections. This is due to the thresholds used to build the term map. These records amount at about 140 (OTHER). They can be referred to different studies such as those on the geological and seismotectonic aspects of the epicentral area of the earthquake, research relating to the study of precursors of the seismic events, papers concerning geotechnical phenomena as well as those dealing with the social and demographic long-term impact of the natural extreme event. However, in order to shed light on details of these research outcomes, further investigations are required.

Figure 8, which should be read in conjunction with Figure 7, represents the temporal trend of documents published yearly by each topic in both the databases (IR_BASE_ENG and IR_BASE_IT) as well as in their joint one (SUM). For clarity, the SUM section also reports the number of documents published yearly, which is directly proportional to the bar heights.

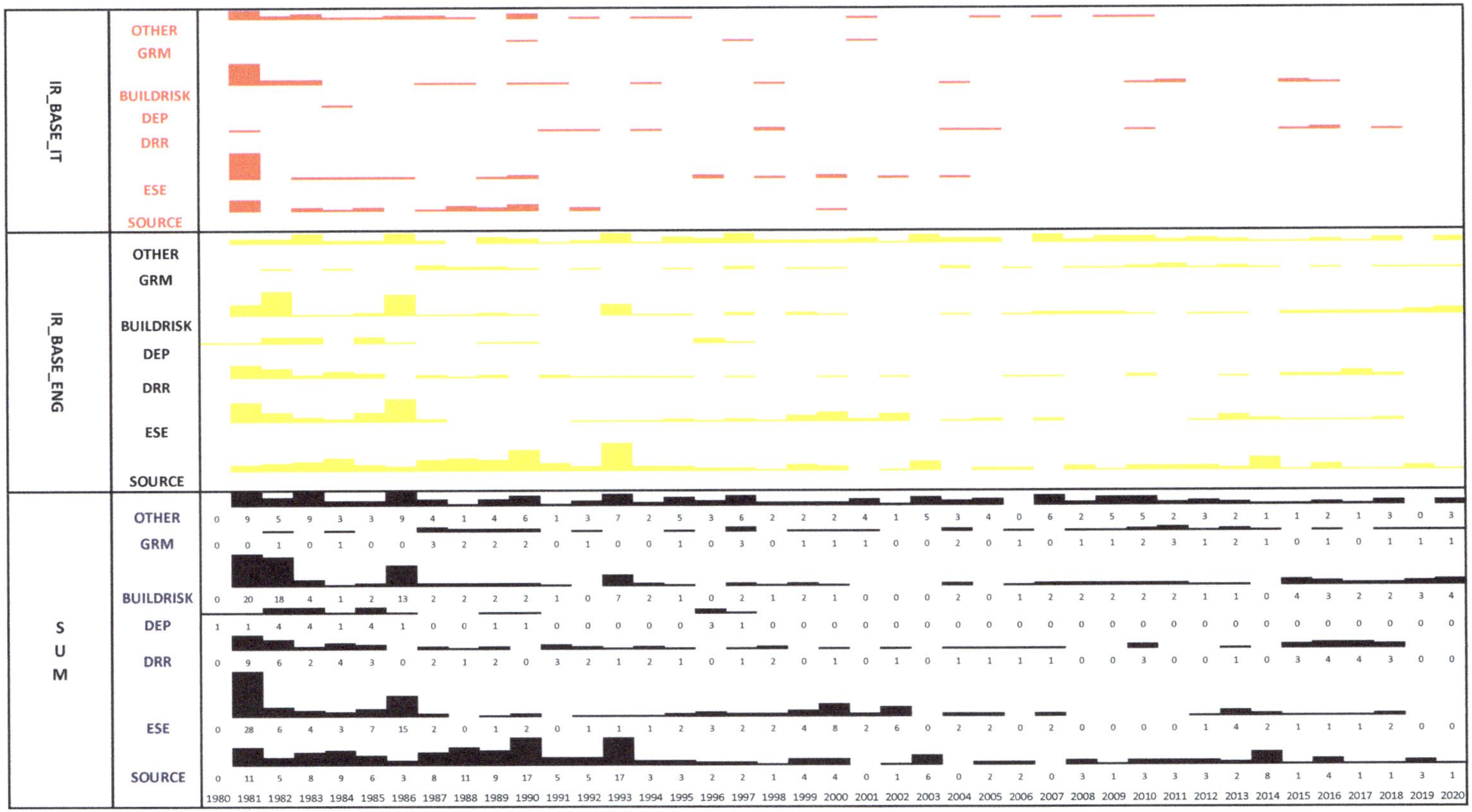

Figure 8. Number of records for each topic over time.

From the three sections of the graph, one can see that for almost all the topics, the documents are mainly distributed by the mid-1990s. In particular, the analysis of the SUM data for the SOURCE field shows a temporal continuity over the entire 40-year long period, even if the number of documents is decreasing over time, with as many as 64.0% of items published by 1993. There are two main peaks, at 1990 and 1993. These can be explained by the works published for the meeting held in Sorrento and the publication of the *Annali di Geofisica* Special Issue (see previous minireview sections, SOURCE). There is less temporal continuity for studies relating to secondary seismic effects (ESE), which show higher concentration in the first half of the 1980s (53.4% of documents published by the year 1986), with special attention to 1981 (28 items, 23.7%). Studies relating to DRR display higher frequency in the first half of the 1980s, but research also continued in the following decades, albeit in a discontinuous way. Epidemiological studies (DEP) almost exclusively characterize the 1980s, because of the predominantly empirical nature of the research. The BUILDRISK topic also shows a higher frequency in the first half of the 1980s, with continuity throughout the following decades. The OTHER studies cover the whole period under examination here, thus confirming that the 1980 earthquake was a full-scale laboratory for investigating several and different aspects.

Due to the high statistical weight of the IR_BASE_ENG records (~80%) compared to the two databases taken cumulatively, the temporal pattern of IR_BASE_ENG can be roughly overlapped on that of SUM. Conversely, the documents of IR_BASE_IT are mainly distributed in the 1980s and concern chiefly papers in the SOURCE (72.0%), ESE (70.0%), BUILDRISK (65.6%), and OTHER (62.1%) topics. Furthermore, the publication of documents is markedly irregular and decreasing for the other three decades. This can be explained by a greater propensity on the part of scientists to publish in international journals for higher visibility of the research.

6. Conclusions

The paper has shed light on the scientific landscape related to the 1980 Irpinia-Basilicata earthquake (Me=6.7 [1]; Mw=6.9 [123]) from its occurrence until now. In order to reach the aim, the authors performed a tailored procedure to build two databases of both international and national (Italian) studies.

Overall, in the almost 40-year period, about 640 documents were published, which cover each year of the time window analyzed. Furthermore, research involved a high number of authors over the years, confirming the high complexity of the natural event that involved many different competencies by researchers.

Publications include studies on the seismic source, environmental effects, earthquake damage, seismic microzonation, disaster response and recovery, disaster epidemiology, ground motion estimations, and other research. Except for the epidemiological studies, which only cover part of the 1980s and mid-1990s, the scientific outputs are concentrated especially in the 1980s and early 1990s. However, investigations characterized also other decades, the studies of seismic source especially, which will probably be a leading research area in the coming years as well.

The study confirms that the earthquake was a significant occasion for the scientific community to grow the knowledge on the seismic phenomena, as well as to learn lessons in view of setting up preventive actions to mitigate the seismic risk.

Limitations

Three main limitations can be referred to this study. First, the unfeasibility to prepare all-inclusive search queries to search for documents might have left out some relevant items. Second, the number of oldest documents may be underestimated, for the 1980s, especially, due to the lower coverage of the citation indexes as well as the frequent lack of abstracts that we found both in Scopus and in Web of Science. This prevents the query from properly interrogating the mother databases from which we extracted the 1980 bibliographic data. Third, the search for the building of IR_BASE_IT database was performed only considering the document titles. However, overall, the documents retrieved and analyzed here can be considered as a reliable cross-section of the main research activities performed on the Irpinia-Basilicata earthquake over time.

Author Contributions: F.T.G. conceived the work and mainly wrote the text. M.R.P. supported the building and analysis of the two databases by bibliographic searches. All authors have read and agreed to the published version of the manuscript.

Funding: This research received no external funding.

Acknowledgments: The authors wish to thank Rocchino Caivano, technician of IMAA-CNR (Research Area of Potenza), for the searches of the bibliographic sources not available on the net by the Document Delivery service, essential for the building of the two databases. We also wish to thank two anonymous reviewers for their helpful suggestions, which allowed us to improve the final version of the manuscript.

Conflicts of Interest: The authors declare no conflict of interest.

Appendix A

Search terms considered to interrogate Scopus and WoS. "November 1980" earthquake*; "1980 november 23"; "23 november 1980"; "november 23 1980"; "23rd November 1980"; "november 23rd, 1980"; "23-11-1980"; Irpinia* 1980; "1980 earthquake*"; 1980 earthquake* Italy; 1980 "South* Ital* earthquake*"; 1980 Basilicata earthquake*; "Irpinia* earthquake*"; "Campania-lucania" earthquake*; "Campania earthquake*"; "campania-basilicata" earthquake*; "irpinia* fault"; "strong earthquake*" Italy; "large earthquake*" Italy; "large earthquake*" South* Apennine; Ital* 1980 "seismic event*"; earthquake* "South* Apennine."

References

1. Guidoboni, E.; Ferrari, G.; Mariotti, D.; Comastri, A.; Tarabusi, G.; Sgattoni, G.; Valensise, G. *CFTI5Med, Catalogo dei Forti Terremoti in Italia (461 a.C.-1997) e Nell'area Mediterranea (760 a.C.-1500)*; Istituto Nazionale di Geofisica e Vulcanologia (INGV): Rome, Italy, 2018. [CrossRef]
2. Commissione Parlamentare D'Inchiesta. 1991—Commissione Parlamentare D'inchiesta Sulla Attuazione Degli Interventi per la Ricostruzione e lo Sviluppo dei Territori Della Basilicata e Della Campania Colpiti dai Terremoti del Novembre 1980 e Febbraio 1981, Relazione Conclusiva e Propositiva, Vol. I, Tomo I. 1991. Available online: http://www.senato.it/leg/10/BGT/Schede/docnonleg/30412.htm (accessed on 20 June 2020). (In Italian)
3. Pantosti, D.; Valensise, G. Faulting mechanism and complexity of the November 23, 1980, Campania- Lucania earthquake, inferred from surface observations. *J. Geophys. Res.* **1990**, *95*, 15319–15341. [CrossRef]
4. Cotecchia, V. Ground deformations and slope instability produced by the earthquake of 23 November 1980 in Campania and Basilicata. *Geol. Appl. Idrogeol.* **1986**, *21*, 31–100.
5. Esposito, E.; Gargiulo, A.; Iaccarino, G.M.; Porfido, S. Distribuzione dei Fenomeni Franosi Riattivati dai Terremoti dell'Appennino Meridionale. In Proceedings of the Censimento Delle Frane del Terremoto del 1980. Proc. Intern. Convention on Prevention of Hydrogeological Hazards, CNR(IRPI), Torino, I, Alba, Torino, Italy, 5–7 November 1996; pp. 409–429. (In Italian)
6. Blumetti, A.M.; Esposito, E.; Ferreli, L.; Michetti, A.M.; Porfido, S.; Serva, L.; Vittori, E. New Data and Reinterpretation of the November 23, 1980, M 6.9, Irpinia-Lucania Earthquake (Southern Apennine) Coseismic Surface Effects. In Proceedings of the International Workshop "Large-Scale Vertical Movements and Related Gravitational Processes" Special Issue, Studi Geologici Camerti, Siena, Italy, 20–21 September 2002; pp. 19–27.
7. Waltman, L.; Van Eck, N.J.; Noyons, E. A unified approach to mapping and clustering of bibliometric networks. *J. Infometr.* **2010**, *4*, 629–635. [CrossRef]
8. Liu, X.; Zhan, F.B.; Hong, S.; Niu, B.; Liu, Y. A bibliometric study of earthquake research: 1900–2010. *Scientometrics* **2012**, *92*, 747–765. [CrossRef]
9. Stahl, B.C.; Heersmink, R.; Goujon, P.; Flick, C.; van den Hoven, J.; Wakunuma, K.; Ikonen, V.; Rader, M. Identifying the ethics of emerging information and communication technologies: An essay on issues, concepts and method. In *Ethical Impact of Technological Advancements and Applications in Society*; Luppicini, R., Ed.; Information Science Reference: Hershey, PA, USA, 2012; pp. 61–79.

10. Gizzi, F.T.; Leucci, G. Global Research Patterns on Ground Penetrating Radar (GPR). *Surv. Geophys.* **2018**, *39*, 1039–1068. [CrossRef]
11. Knapczyk, A.; Francik, S.; Pedryc, N.; Hebda, T. Bibliometric analysis of research trends in engineering for rural development. In Proceedings of the 17th International Scientific Conference Engineering for Rural Development, Jelgava, Latvia, 23–25 May 2018; pp. 700–707. [CrossRef]
12. Krauskopf, E. A bibiliometric analysis of the Journal of Infection and Public Health: 2008–2016. *J. Infect. Public Health* **2018**, *11*, 224–229. [CrossRef]
13. Guz, A.N.; Rushchitsky, J.J. Scopus: A system for the evaluation of scientific journals. *Int. Appl. Mech.* **2009**, *45*, 351–362. [CrossRef]
14. Boschi, E.; Pantosti, D.; Slejko, D.; Stucchi, M.; Valensise, G. Special Issue on the meeting "Irpinia Dieci Anni Dopo". *Ann. Geofis.* **1993**, *36*, 351.
15. Valensise, G. Summary of Contributions on the 23 November 1980, Irpinia earthquake. *Ann. Geofis.* **1993**, *XXXVI*, 345–351.
16. Bartol, T.; Mackiewicz-Talarczyk, M. Bibliometric analysis of publishing trends in fiber crops in Google Scholar, Scopus, and Web of Science. *J. Nat. Fibers* **2015**, *12*, 531–541. [CrossRef]
17. Choudhury, M.; Chatterjee, D.; Mukherjee, A. Global topology of word co-occurrence networks: Beyond the two-regime power-law. In Proceedings of the International Conference on Computational Linguistics, Beijing, China, 23–27 August 2010; pp. 162–170.
18. Van Rijsbergen, C.J. A theoretical basis for the use of co-occurrence data in information retrieval. *J. Doc.* **1977**, *33*, 106–119. [CrossRef]
19. Van Eck, N.J.; Waltman, L. Text mining and visualization using VOSviewer. *ISSI Newsl.* **2011**, *7*, 50–54.
20. Van Eck, N.J.; Waltman, L. Visualizing bibliometric networks. In *Measuring Scholarly Impact: Methods and Practice*; Ding, Y., Rousseau, R., Wolfram, D., Eds.; Springer: Berlin/Heidelberg, Germany, 2014; pp. 285–320.
21. Boschi, E.; Mulargia, F.; Mantovani, E.; Bonafede, M.; Dziewonski, A.M.; Woodhouse, J.H. The Irpinia earthquake of November 23, 1980. *EOS* **1981**, *62*, 330.
22. Scarpa, R.; RAPP, S.R. Sismometria del Terremoto del 23/11/1980. Il terremoto campano-lucano del 23.11. 1980: Elaborazione dati sismometrici. *Rend. Soc. Geol. Ital.* **1981**, *4*, 427–450.
23. Berardi, R.; Berenzi, A.; Capozza, F. *Campania-Lucania Earthquake on 23 November 1980: Accelerometric Recordings of the Main Quake and Relating Processing*; Technical Report, Ente Nazionale per l'Energia Elettrica (ENEL); Contribution to the Annual Convention of the Italian National Research Project on Italian Seismicity: Udine, Italy, 1981; pp. 1–103.
24. Del Pezzo, E.; Iannaccone, G.; Martini, M.; Scarpa, R. The 23 november 1980 Southern Italy Earthquake. *Bull. Seismol. Soc. Am.* **1983**, *73*, 187–200.
25. Arca, S.; Marchioni, A. I movimenti verticali del suolo nelle zone della Campania et della Basilicata interessate dal sisma del novembre 1980. *Boll. Geod. Sci. Affin.* **1983**, *42*, 125–135.
26. Deschamps, A.; King, G.C.P. The Campania-Lucania (southern Italy) earthquake of 23 November 1980. Earth planet. *Sci. Lett.* **1983**, *62*, 296–304.
27. Deschamps, A.; King, G.C.P. Aftershocks of the Campania-Lucania (Italy) earthquake of 23 November 1980. *Bull. Seismol. Soc. Am.* **1984**, *74*, 2483–2517.
28. Westaway, R.; Jackson, J. Surface faulting in the southern Italian Campania-Basilicata earthquake of 23 November 1980. *Nature* **1984**, *312*, 436–438. [CrossRef]
29. Westaway, R. Geometry of faulting in the Campania-Basilicata (Southern Italy) earthquake of 23rd November 1980. *Geophys. J. R. Astron. Soc.* **1985**, *81*, 335.
30. Crosson, R.S.; Martini, M.; Scarpa, R.; Key, S.C. The southern Italy earthquake of 23 November 1980: An unusual pattern of faulting. *Bull. Seismol. Soc. Am.* **1986**, *76*, 381–394.
31. Crosson, R.S.; Martini, M.; Scarpa, R.; Key, S.C. The southern Italy earthquake of 23 November 1980: An unusual pattern of faulting. Reply. *Bull. Seismol. Soc. Am.* **1987**, *77*, 1075–1077.
32. Westaway, R. The Southern Italy earthquake of 23 November 1980—An unusual pattern of faulting—Comment. *Bull. Seismol. Soc. Am.* **1987**, *77*, 1071–1074.
33. Westaway, R.; Jackson, J. The earthquake of 1980 November 23 in Campania–Basilicata (southern Italy). *Geophys. J. R. Astron. Soc.* **1987**, *90*, 375–443. [CrossRef]
34. Bernard, P.; Zollo, A. The Irpinia (Italy) 1980 earthquake: Detailed analysis of a complex normal faulting. *J. Geophys. Res.* **1989**, *94*, 1631–1647. [CrossRef]

35. Pantosti, D.; Valensise, G. Source geometry and long-term behavior of the 1980, Irpinia earthquake fault based on field geologic observations. *Ann. Geofis.* **1993**, *36*, 41–43.
36. Westaway, R. Revised hypocentre and fault rupture geometry for the 1980 November 23 Campania-Basilicata earthquake in southern Italy. *Geophys. J. Int.* **1992**, *109*, 376–390. [CrossRef]
37. Westaway, R. Fault rupture geometry for the 1980 Irpinia earthquake: A working hypothesis. *Ann. Geofis.* **1993**, *36*, 51–69.
38. Amato, A.; Chiarabba, C.; Malagnini, L.; Selvaggi, G. Three-dimensional P-velocity structure in the region of the MS = 6.9 Irpinia, Italy, normal faulting earthquake. *Phys. Earth Planet. Inter.* **1992**, *75*, 111–119. [CrossRef]
39. Cocco, M.; Pacor, F. The rupture process of the 1980 Irpinia, Italy, earthquake from the inversion of strong motion waveforms. *Tectonophysics* **1993**, *218*, 157–177. [CrossRef]
40. Cocco, M.; Pacor, F. Space-time evolution of the rupture process from the inversion of strong-motion waveforms. *Ann. Geofis.* **1993**, *36*, 109–130.
41. Amato, A.; Selvaggi, G. Aftershock location and P-velocity structure in the epicentral region of the 1980 Irpinia earthquake. *Ann. Geofis.* **1993**, *1*, 3–15.
42. Giardini, D. Teleseismic observation of the November 23 1980, Irpinia earthquake. *Ann. Geofis.* **1993**, *36*, 17–25.
43. Pingue, F.; De Natale, G.; Briole, P. Modeling of the 1980 Irpinia earthquake source: Constraints from geodetic data. *Ann. Geofis.* **1993**, *1*, 27–40.
44. Sirovich, L.; Chiaruttini, C. The influence of source complexity on the polarization and azimuthal radiation of S-Waves, and a simplified synthesis of the macroseismic field. *Ann. Geofis.* **1993**, *1*, 81–91.
45. Bernard, P.; Zollo, A.; Trifu, C.I.; Herrero, A. Details of the rupture kinematics and mechanism of the 1980 Irpinia earthquake: New results and remaining questions. *Ann. Geofis.* **1993**, *1*, 71–80.
46. Vaccari, F.; Harabaglia, P.; Suhadolc, P.; Panza, G.F. The Irpinia (Italy) 1980 earthquake: Waveform modelling of accelerometric data and macroseismic considerations. *Ann. Geofis.* **1993**, *1*, 93–108.
47. Nostro, C.; Cocco, M.; Belardinelli, M.E. Static stress changes in extensional regimes: An application to southern Apennines (Italy). *Bull. Seismol. Soc. Am.* **1997**, *87*, 234–248.
48. Belardinelli, M.E.; Cocco, M.; Coutant, O.; Cotton, F. Redistribution of dynamic stress during coseismic ruptures: Evidence for fault interaction and earthquake triggering. *J. Geophys. Res. Solid Earth* **1999**, *104*, 14925–14945. [CrossRef]
49. Amoruso, A.; Crescentini, L.; Scarpa, R. Faulting geometry for the complex 1980 Campania-Lucania earthquake from levelling data. *Geophys. J. Int.* **2005**, *162*, 156–168. [CrossRef]
50. Amoruso, A.; Crescentini, L.; Di Lieto, B.; Scarpa, R. Faulting mechanism of the Campania-Lucania 1980 earthquake, Italy, from high-resolution, 3D velocity structure, aftershock relocation, fault-plane solutions, and post-seismic deformation modeling. *Ann. Geophys.* **2011**, *54*, 806–821.
51. Pantosti, D.; Schwartz, D.P.; Valensise, G. Paleoseismologic and geomorphic observations along the 1980 Irpinia surface fault rupture, southern Apennines (Italy). *EOS Trans. AGU* **1989**, *70*, 1349.
52. D'Addezio, G.; Pantosti, D.; Valensise, G. Paleoearthquakes along the Irpinia fault at Pantano di San Gregorio Magno (Southern Italy). *Alp. Mediterr. Quat.* **1991**, *4*, 121–135.
53. Pantosti, D.; D'Addezio, G.; Cinti, F.R. Paleoseismological Evidence of Repeated Large Earthquakes along the 1980 Irpinia Earthquake Fault. *Ann. Geofis.* **1993**, *36*, 321–330.
54. Pantosti, D.; Schwartz, D.P.; Valensise, G. Paleoseismology along the 1980 surface rupture of the Irpinia fault: Implications for earthquake recurrence in the southern Apennines, Italy. *J. Geophys. Res. Solid Earth* **1993**, *98*, 6561–6577. [CrossRef]
55. Improta, L.; Zollo, A.; Bruno, P.P.; Herrero, A.; Villani, F. High-resolution seismic tomography across the 1980 (Ms 6.0) Southern Italy earthquake fault scarp. *Geophys. Res. Lett.* **2003**, *30*. [CrossRef]
56. Bruno, P.P.; Castiello, A.; Improta, L. Ultrashallow seismic imaging of the causative fault of the 1980, M6.9, southern Italy earthquake by pre-stack depth migration of dense wide-aperture data. *Geophys. Res. Lett.* **2010**, *37*, L19302. [CrossRef]
57. Galli, P.A.C.; Giocoli, A.; Peronace, E.; Piscitelli, S.; Quadrio, B.; Bellanova, J. Integrated near surface geophysics across the active Mount Marzano Fault System (southern Italy): Seismogenic hints. *Int. J. Earth Sci.* **2014**, *103*, 315–325. [CrossRef]
58. Galli, P.A.C.; Peronace, E.; Quadrio, B.; Esposito, G. Earthquake fingerprints along fault scarps: A case study of the Irpinia 1980 earthquake fault (southern Apennines). *Geomorphology* **2014**, *206*, 97–106. [CrossRef]

59. Vassallo, M.; Festa, G.; Bobbio, A.; Serra, M. Low shear velocity in a normal fault system imaged by ambient noise cross correlation: The case of the Irpinia fault zone, Southern Italy. *J. Geophys. Res. Solid Earth* **2016**, *121*, 4290–4305. [CrossRef]
60. Lo Re, D.; Florio, G.; Ferranti, L.; Ialongo, S.; Castiello, G. Self-constrained inversion of microgravity data along a segment of the Irpinia fault. *J. Appl. Geophys.* **2016**, *124*, 148–154. [CrossRef]
61. Guerricchio, A.; Melidoro, G. Pseudo-tectonic mass movements in the southern Apennines. [Movimenti di massa pseudo-tettonici nell'appennino dell'Italia meridionale]. *Geol. Appl. Idrogeol.* **1981**, *16*, 251–294.
62. Maugeri, M.; Motta, E.; Sorriso Valvo, M. Senerchia landslide triggered by the 23 November 1980 earthquake. In Proceedings of the 4th International Congress—International Association of Engineering Geology, New Delhi, India, 10–15 December 1982; Volume 8, pp. 139–149.
63. Bousquet, J.C.; Gars, G.; Lanzafame, G.; Philip, H. Surface breaks of gravitational origin at the time of the Irpinia earthquake-23/11/1980; southern Italy. *Geol. Appl. Idrogeol.* **1983**, *18*, 427–435.
64. Da Roit, R.; Fontanive, A.; Lojelo, L.; Muzzi, F.; Spat, G. Terremoto Campano-Lucano del 23 Novembre 1980: Evidenze di Liquefazione di Terreni non Coesivi Saturi. Comm. ENEA-ENEL per lo Studio dei Problemi Sismici Connessi con la Realizzazione di Impianti Nucleari. In Proceedings of the Conv. Ann. del CNR-PFG su Sismicità d'Italia, stato delle Conoscenze e Qualità della Normativa Sismica, 1981, Udine, Italy, 2–14 May 1982. (In Italian)
65. Del Prete, M.; Liuzzi, G.T. Results of a preliminary study of the Calitri (AV) landslide triggered by the 23 November 1980 earthquake. [Risultati dello studio preliminare della frana di Calitri (AV) mobilitata dal terremoto del 23 Novembre 1980]. *Geol. Appl. Idrogeol.* **1981**, *16*, 153–165.
66. Faccioli, E. Engineering Seismology Aspects of the M-6.5, Southern Italy Earthquake of Nov. 23, 1980: A Preliminary Review. International Conferences on Recent Advances in Geotechnical Earthquake Engineering and Soil Dynamics. 1981, p. 16. Available online: https://scholarsmine.mst.edu/icrageesd/01icrageesd/session08/16 (accessed on 20 May 2020).
67. Hutchinson, J.N.; Del Prete, M. Landslides at Calitri, southern Apennines, reactivated by the earthquake of 23rd November 1980. *Geol. Appl. Idrogeol.* **1985**, *20*, 9–38.
68. Baldassarre, G. Geological effects of the 23 November 1980 earthquake in Atella, outskirts of Basilicata. [Effetti geologici del sisma del 23 Novembre 1980 nella periferia dell'abitato di Atella (Basilicata)]. *Geol. Appl. Idrogeol.* **1981**, *16*, 227–238.
69. Agnesi, V.; Carrara, A.; Macaluso, T.; Monteleone, S.; Pipitone, G.; Sorriso-Valvo, M. Preliminary observations of slope instability phenomena induced by the earthquake of November 1980 on the upper valley of Sele river. [Osservazioni preliminari sui fenomeni di instabilita dei versanti indotti dal sisma del 1980 nell'alta valle del Sele]. *Geol. Appl. Idrogeol.* **1982**, *17*, 79–93.
70. Genevois, R.; Prestininzi, A. Deformations and mass movements induced by the November 23rd, 1980 earthquake in the middle valley of the Tammaro River-BN. [Deformazioni e movimenti di massa indotti dal sisma del 23-11-1980 nella media valle del F. Tammaro (BN)]. *Geol. Appl. Idrogeol.* **1982**, *17*, 305–318.
71. Salvemini, A. Basic aspects of phenomena of instability in the limestone massifs of Monti Cervialto and Terminio-Tuoro (southern Apennines) induced by the earthquake of 23 November 1980. [Aspetti fondamentali dei fenomeni di dissesto nei massicci carbonatici dei Monti Cervialto e Terminio-Tuoro (Appennino Meridionale) indotti dal sisma del 23 novembre 1980]. *Geol. Appl. Idrogeol.* **1982**, *17*, 209–218.
72. Alexander, D. Preliminary assessment of landslides resulting from the earthquake of 23rd November 1980 in Southern Italy. *Disasters* **1981**, *5*, 376–383. [CrossRef]
73. Porfido, S.; Esposito, E.; Vittori, E.; Tranfaglia, G.; Michetti, A.M.; Blumetti, M.; Ferreli, L.; Guerrieri, L.; Serva, L. Areal distribution of ground effects induced by strong earthquakes in the southern Apennines (Italy). *Surv. Geophys.* **2002**, *23*, 529–562. [CrossRef]
74. Porfido, S.; Esposito, E.; Guerrieri, L.; Vittori, E.; Tranfaglia, G.; Pece, R. Seismically induced ground effects of the 1805, 1930 and 1980 earthquakes in the Southern Apennines, Italy. *Boll. Soc. Geol. Ital.* **2007**, *126*, 333–346.
75. Cotecchia, V.; Salvemini, A. Correlation between seismic events and discharge variations at Caposele and Cassano Irpino springs, with particular reference to the 23 November 1980 earthquake. [Correlazione fra eventi sismici e variazione di portata alle sorgenti di Caposele e Cassano Irpino, con particolare riferimento al sisma del 23 Novembre 1980]. *Geol. Appl. Idrogeol.* **1981**, *16*, 167–192.
76. Wasowski, J.; Del Gaudio, V.; Pierri, P.; Capolongo, D. Factors controlling seismic susceptibility of the Sele valley slopes: The case of the 1980 Irpinia earthquake re-examined. *Surv. Geophys.* **2002**, *23*, 563–593. [CrossRef]

77. Del Prete, M. Example of mudslide hazard in the Southern Apennines (Italy). *Ann. Geofis.* **1993**, *36*, 271–276.
78. Parise, M.; Wasowski, J. Landslide activity maps for landslide hazard evaluation: Three case studies from Southern Italy. *Nat. Hazards* **1999**, *20*, 159–183. [CrossRef]
79. Capolongo, D.; Refice, A.; Mankelow, J. Evaluating earthquake-triggered landslide hazard at the basin scale through GIS in the Upper Sele river Valley. *Surv. Geophys.* **2002**, *23*, 595–625. [CrossRef]
80. Martino, S.; Prestininzi, A.; Romeo, R.W. Earthquake-induced ground failures in Italy from a reviewed database. *Nat. Hazards Earth Syst. Sci.* **2014**, *14*, 799–814. [CrossRef]
81. Cavazzani, A. *Social and Institutional Impact of the 1980 Earthquake in Southern Italy: Problems and Prospects of Civil Protection*; The Council of Yugoslav Association of Self-Managed Communities of Interest for Scientific Research: Ljubljana, Slovenia, 1982; pp. 425–436.
82. Ventura, F. The earthquake in Campania and Basilicata (Italy) of 23rd November 1980—The nature of the state's emergency interventions and the future quality of reconstruction. In *Tenth World Conference of Sociology, Mexico*; ISIG Publication: Gorizia, Italy, 1982.
83. Ventura, F. The long-term effects of the 1980 earthquake on the villages of southern Italy. *Disasters* **1984**, *8*, 11. [CrossRef]
84. D'Souza, F. Recovery following the South Italian earthquake, November 1980: Two contrasting examples. *Disasters* **1982**, *6*, 101–109. [CrossRef]
85. D'Souza, F. Recovery following the south Italian earthquake, November 1980: Contrasting examples. *Ekistics* **1984**, *51*, 476–482. [CrossRef]
86. Alexander, D. Housing crisis after natural disaster: The aftermath of the November 1980 southern Italian earthquake. *Geoforum* **1984**, *15*, 489–516. [CrossRef]
87. Caporale, R.; Rossi, I.; Chaeretakis, A. *Reconstruction and Socio-Cultural Systems, A Long-Range Study of Reconstruction Processes Following the November 23, 1980 Earthquake in Southern Italy*; Institute for Italian-American Studies: New York, NY, USA, 1985; Unpublished report.
88. Alexander, D. Preserving the Identity of Small Settlements during Post-Disaster Reconstruction in Italy. *Disasters* **1989**, *13*, 228–236. [CrossRef] [PubMed]
89. Chubb, J. Three earthquakes: Political response, reconstruction, and the institutions: Belice (1968), Friuli (1976), Campania (1980). In *Disastro: Disasters in Italy Since 1860: Culture, Politics, Society*; Dickie, J., Foot, J., Snowden, F.M., Eds.; Palgrave: New York, NY, USA, 2002; pp. 186–210.
90. Forino, G.; Ciccarelli, S.; Bonamici, S.; Perini, L.; Salvati, L. Developmental Policies, Long-Term Land-Use Changes and the Way Towards Soil Degradation: Evidence from Southern Italy. *Scott. Geogr. J.* **2015**, *131*, 123–140. [CrossRef]
91. Porfido, S.; Alessio, G.; Gaudiosi, G.; Nappi, R.; Spiga, E. The resilience of some villages 36 years after the Irpinia-Basilicata (Southern Italy) 1980 earthquake. In Proceedings of the Workshop on World Landslide Forum in Advancing Culture of Living with Landslides, Ljubljana, Slovenija, 29 May–2 June 2017; Mikoš, M., Vilímek, V., Yin, Y., Sassa, K., Eds.; Springer: Cham, Switzerland, 2017; pp. 121–133. [CrossRef]
92. Greco, D.; Faustini, A.; Forastiere, F.; Galanti, M.R.; Magliola, M.E.; Moro, M.L.; Piergentili, P.; Rosmini, F.; Stazi, M.A.; Luzi, S.; et al. Epidemiological surveillance of diseases following the earthquake of 23rd November 1980 in Southern Italy. *Disasters* **1981**, *5*, 398–406. [CrossRef] [PubMed]
93. Alexander, D. Disease epidemiology and earthquake disaster. The example of Southern Italy after the 23 November 1980 earthquake. *Soc. Sci. Med.* **1982**, *16*, 1959–1969. [CrossRef]
94. Bruycker, M.D.; Greco, D.; Lechat, M.F. The 1980 earthquake in Southern Italy—Morbidity and mortality. *Int. J. Epidemiol.* **1985**, *14*, 113–117. [CrossRef] [PubMed]
95. Berger, E.; Studer, J. *Southern Italy Earthquake, November 23, 1980. Reconnaissance Summary*; Earthquake Engineering Research Institute: Oakland, CA, USA, 1981; pp. 88–94.
96. Guerpinar, A.; Vardanega, C.; Ries, E.R. November 23, 1980 Irpinia earthquake (terremoto campano lucano) observations of soil-structure interaction effects. In Proceedings of the 6th International Conference on Structural Mechanics in Reactor Technology. Mechanical and Thermal Problems of Future Fusion Reactor Power Plants, Paris, France, 17–21 August 1981; p. 20. Available online: https://www.scopus.com/inward/record.uri?eid=2-s2.0-0019708742&partnerID=40&md5=01c2a1d6484e65b7b3ecd86fa55993bd (accessed on 13 June 2020).

97. Postpischl, D.; Ferrari, G.; Giovani, L.; Spadea, M.C.; Vecchi, M.; Stucchi, M.; Bottari, A.; Lo Giudice, E.; Baldassarre, R.; Branno, A.; et al. Il terremoto irpino del 23 Novembre 1980; rilievo macrosismico. Stato di avanzamento al 12-5-81. *Rend. Soc. Geol. Ital.* **1981**, *4*, 503–505.
98. Postpischl, D.; Perrone, V. *Il Terremoto del 23/11/1980 in Campania e Lucania: Rilievo Macrosismico*; C.N.R. Progetto Finalizzato Geodinamica: Pisa, Italy, 1981; p. 121.
99. Postpischl, D.; Brano, A.; Esposito, E.G.L.; Ferrari, G.; Marturano, A.; Porfido, S.; Rinaldis, V.; Stucchi, M. The Irpinia earthquake of November 23, 1980. *Atlas Isoseismal Maps Ital. Earthq.* **1985**, *114*, 152–157.
100. Coburn, A.W.; Hughes, R.E.; Nash, D.F.T.; Spence, R.J.S. Damage assessment and ground motion in the Italian earthquake of 23.11.1980. In Proceedings of the 7th European conference on Earthquake Engineering, Athens, Greece, 20–25 September 1982; pp. 101–108.
101. Ammann, W.; Porro, B. Earthquake in southern Italy of 23 November 1980 engineering aspects and interpretation of building damage. In Proceedings of the 7th European Conference on Earthquake Engineering, Athens, Greece, 20–25 September 1982; Volume 1, pp. 151–158.
102. Hojjat, A. Behavior of reinforced concrete buildings during the southern Italy earthquake of November 23, 1980. In Proceedings of the 7th European Conference on Earthquake Engineering, Athens, Greece, 20–25 September 1982; Volume 1, pp. 143–150.
103. Cotecchia, V.; Nuzzo, G.; Salvemini, A.; Tafuni, N.; Pavoncelli, G. Tunnel and structural analysis of large underground structure damaged by the earthquake of November 23, 1980. In Proceedings of the Internazional Sypmosium on Geology Problems in Seismic Areas, Bari, Italy, 13–19 April 1986; Volume 21, pp. 329–352.
104. Braga, F.; Dolce, M.; Liberatore, D. A Statistical Study on Damaged Buildings and Ensuing Review of MSK-76 Scale, Southern Italy November 23 1980 Earthquake. In *Progetto Finalizzato di Geodinamica*; CNR Italia, Pub. No. 503 (7ECEE, Athens): Athens, Greece, 1982; pp. 65–84.
105. Braga, F.; Liberatore, D.; Dolce, M. Fast and reliable damage estimation for optimal relief operations. In *Earthquake Relief in Less Industrialized Areas*; Schuppisser, S., Studer, J.A., Eds.; Balkema: Rotterdam, The Netherlands, 1984; pp. 145–151.
106. Rota, M.; Penna, A.; Strobbia, C. Processing Italian damage data to derive typological fragility curve. *Soil Dyn. Earthq. Eng.* **2008**, *28*, 933–947. [CrossRef]
107. Del Gaudio, C.; Di Ludovico, M.; Polese, M.; Manfredi, G.; Prota, A.; Ricci, P.; Verderame, G.M. Seismic fragility for Italian RC buildings based on damage data of the last 50 years. *Bull. Earthq. Eng.* **2020**, *18*, 2023–2059. [CrossRef]
108. Dolce, M.; Speranza, E.; Giordano, F.; Borzi, B.; Bocchi, F.; Conte, C.; Di Meo, A.; Faravelli, M.; Pascale, V. Observed damage database of past Italian earthquakes: The Da. DO WebGIS. *Boll. Geofis. Teor. Appl.* **2019**, *60*, 141–164.
109. Scire, E.; Siro, L.; Stucchi, M.; Gaiazzi, M. Geo- seismic investigations and urban planning after the Irpinia-Basilicata 1980 earthquake: Part 2, the case of S. Angelo dei Lombardi (Italy). *Geol. Appl. Idrogeol.* **1986**, *21*, 387–397.
110. Scire, E.; Siro, L.; Stucchi, M.; Gaiazzi, M. Geo-seismic investigations and urban planning after the Irpinia-Basilicata 1980 earthquake: Part 1, the case of Caposele and Conza della Campania. *Geol. Appl. Idrogeol.* **1986**, *21*, 441–450.
111. Chiocchini, U.; Cherubini, C. Seismic microzoning of the Lioni village destroyed by the November 23rd 1980 earthquake (Irpinia, Campano-Lucano Apennine). *Geol. Appl. Idrogeol.* **1986**, *21*, 341–362.
112. Maugeri, M.; Carrubba, P. Microzonation for ground motion during the 1980 Irpinia earthquake at Calabritto, Italy. In Proceedings of the 14nd International Conference on Soil Mechanics and Foundation Engineering, Hamburg, Germany, 6–12 September 1997; pp. 81–96.
113. Siro, L. Microzonation in emergency: A short technical report. The Italian Geodynamics Project, Publication Number 503, National Research Council of Italy. Presented at the VII ECEE, Athens, Greece, 6–7 September 1982; pp. 53–63.
114. Siro, L. Emergency microzonations by Italian Geodynamics Project after November 23, 1980 earthquake. In Proceedings of the 3rd International Conference on Microzonation, Seattle, WA, USA, 28 June–1 July 1982; Volume 3, pp. 1417–1427.
115. Rovelli, A.; Bonamassa, O.; Cocco, M.; Di Bona, M.; Mazza, S. Scaling laws and spectral parameters of the ground motion in active extensional areas in Italy. *Bull. Seismol. Soc. Am.* **1988**, *78*, 530–560.

116. Bosco, G.; Dolce, M.; Marino, M. Artificial accelerograms consistent with the 1980 Irpinia earthquake. In Proceedings of the Tenth World Conference on Earthquake Engineering, Madrid, Spain, 24 July 1992; pp. 895–900.
117. Convertito, V.; De Matteis, R.; Cantore, L.; Zollo, A.; Iannaccone, G.; Caccavale, M. Rapid estimation of ground-shaking maps for seismic emergency management in the Campania Region of southern Italy. *Nat. Hazards* **2010**, *52*, 97–115. [CrossRef]
118. Cultrera, G.; Ameri, G.; Saraò, A.; Cirella, A.; Emolo, A. Ground-motion simulations within ShakeMap methodology: Application to the 2008 Iwate-Miyagi Nairiku (Japan) and 1980 Irpinia (Italy) earthquakes. *Geophys. J. Int.* **2013**, *193*, 220–237. [CrossRef]
119. Nunziata, C.; Costa, G.; Marrara, F.; Panza, G.F. Validated Estimation of Response Spectra for the 1980 Irpinia Earthquake in the Eastern Area of Naples. *Earthq. Spectra* **2000**, *16*, 643–660. [CrossRef]
120. Di Giulio, G.; Improta, L.; Calderoni, G.; Rovelli, A. A study of the seismic response of the city of Benevento (Southern Italy) through a combined analysis of seismological and geological data. *Eng. Geol.* **2008**, *97*, 146–170. [CrossRef]
121. Maresca, R.; Nardone, L.; Pasquale, G.; Pinto, F.; Bianco, F. Effects of surface geology on seismic ground motion deduced from ambient-noise measurements in the town of Avellino, Irpinia region (Italy). *Pure Appl. Geophys.* **2012**, *169*, 1173–1188. [CrossRef]
122. Maresca, R.; Nardone, L.; Gizzi, F.T.; Potenza, M.R. Ambient noise HVSR measurements in the Avellino historical centre and surrounding area (southern Italy). Correlation with surface geology and damage caused by the 1980 Irpinia-Basilicata earthquake. *Measurement* **2018**, *130*, 211–222. [CrossRef]
123. Rovida, A.; Camassi, R.; Gasperini, P.; Stucchi, M. (Eds.) *Catalogo Parametrico dei Terremoti Italiani (CPTI11)*; Istituto Nazionale di Geofisica e Vulcanologia (INGV): Milano, Bologna, 2011. [CrossRef]

geosciences

MDPI

Article

The M_S 6.9, 1980 Irpinia Earthquake from the Basement to the Surface: A Review of Tectonic Geomorphology and Geophysical Constraints, and New Data on Postseismic Deformation

Alessandra Ascione [1], Sergio Nardò [1] and Stefano Mazzoli [2,*]

[1] Department of Earth Sciences, Environment and Georesources (DiSTAR), University of Naples Federico II, 80126 Naples, Italy; alessandra.ascione@unina.it (A.A.); sergio.nardo@unina.it (S.N.)
[2] School of Sciences and Technology, Geology Division, University of Camerino, 62032 Camerino, Italy
* Correspondence: stefano.mazzoli@unicam.it

Received: 30 October 2020; Accepted: 7 December 2020; Published: 9 December 2020

Abstract: The M_S 6.9, 1980 Irpinia earthquake occurred in the southern Apennines, a fold and thrust belt that has been undergoing post-orogenic extension since ca. 400 kyr. The strongly anisotropic structure of fold and thrust belts like the Apennines, including late-orogenic low-angle normal faults and inherited Mesozoic extensional features besides gently dipping thrusts, result in a complex, overall layered architecture of the orogenic edifice. Effective decoupling between deep and shallow structural levels of this mountain belt is related to the strong rheological contrast produced by a fluid-saturated, shale-dominated mélange zone interposed between buried autochthonous carbonates—continuous with those exposed in the foreland to the east—and the allochthonous units. The presence of fluid reservoirs below the mélange zone is shown by a high V_P/V_S ratio—which is a proxy for densely fractured fluid-saturated crustal volumes—recorded by seismic tomography within the buried autochthonous carbonates and the top part of the underlying basement. These crustal volumes, in which background seismicity is remarkably concentrated, are fed by fluids migrating along the major active faults. High pore fluid pressures, decreasing the yield stress, are recorded by low stress-drop values associated with the earthquakes. On the other hand, the mountain belt is characterized by substantial gas flow to the surface, recorded as both distributed soil gas emissions and vigorous gas vents. The accumulation of CO_2-brine within a reservoir located at hypocentral depths beneath the Irpinia region is not only interpreted to control a multiyear cyclic behavior of microseismicity, but could also play a role in ground motions detected by space-based geodetic measurements in the postseismic period. The analysis carried out in this study of persistent scatterer interferometry synthetic aperture radar (PS-InSAR) data, covering a timespan ranging from 12 to 30 years after the 1980 mainshock, points out that ground deformation has affected the Irpinia earthquake epicentral area in the last decades. These ground motions could be a result of postseismic afterslip, which is well known to occur over years or even decades after a large mainshock such as the 23 November 1980, M_S 6.9 earthquake due to cycles of CO_2-brine accumulation at depth and its subsequent release by $M_w \geq 3.5$ earthquakes, or most likely by a combination of both. Postseismic afterslip controls geomorphology, topography, and surface deformation in seismically active areas such as that of the present study, characterized by ~M 7 earthquakes. Yet, this process has been largely overlooked in the case of the 1980 Irpinia earthquake, and one of the main aims of this study is to fill such the substantial gap of knowledge for the epicentral area of some of the most destructive earthquakes that have ever occurred in Italy.

Keywords: 1980 Irpinia earthquake; active faults; postseismic afterslip; southern Apennines

1. Introduction

The Irpinia M_S 6.9 earthquake occurred on 23 November 1980 [1], producing vast damage and causing about 3000 fatalities. It nucleated on an approximately 60 km long, NW-SE-striking normal fault system with at least three main rupture episodes at 0 s, 20 s, and 40 s (Figure 1).

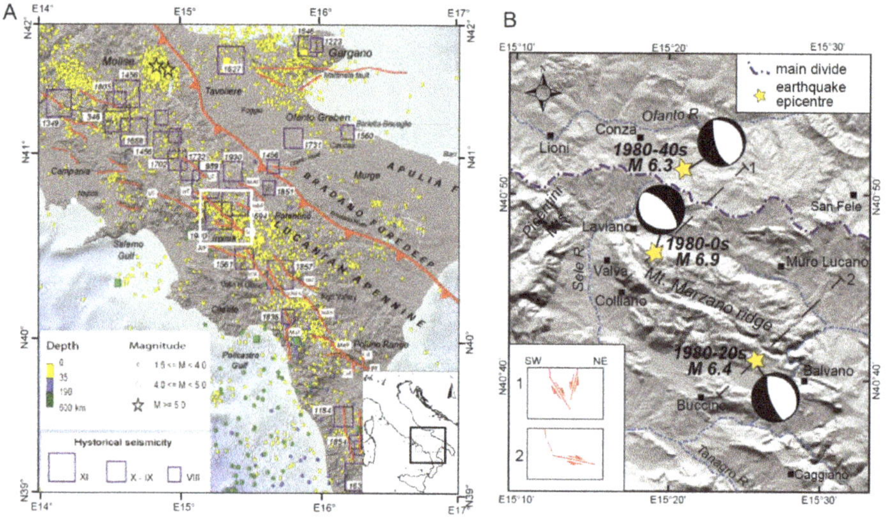

Figure 1. (**A**): Historical and instrumental seismicity of southern Italy (modified after Frepoli et al. [2]). The white frame indicates the location of the map in diagram B. (**B**): Epicentral locations for the three main rupture episodes at 0 s, 20 s, and 40 s, with Mw 6.9, 6.4, and 6.3, of the 1980 Irpinia earthquake; focal mechanisms from Westaway and Jackson [3]. Inset in the lower left corner are schematic sections (location in the map), indicating the fault system activated with the 1980 earthquake (redrawn from Barnard and Zollo [1]).

A crustal extensional stress regime controls earthquake generation processes in the Apennine mountain chain. Background seismicity is characterized, during the postseismic to (possibly) inter-seismic period, by micro-earthquakes ($M_L < 3.5$) approximately confined within the same volume wherein the faults that caused the 1980 earthquake and its aftershocks were located [4–6]. The surface faults exposed in this area are decoupled from the seismically active deep-seated structures, which, in turn, reactivate inherited basement faults [7]. Effective decoupling between deep and shallow structural levels is related to the strong rheological contrast produced by the fluid-saturated, clay-rich mélange zone interposed between the foreland Apulia Platform carbonates and the allochthonous units, which include carbonate platform (i.e., Apennine Platform) and basin (i.e., Internal and Lagonegro) successions (Figure 2).

The identification of active faults and, more in general, the reconstruction of the active tectonic setting on a regional scale are crucial to the assessment and mitigation of seismic hazard and related phenomena (e.g., ground shaking, surface ruptures, landsliding, etc.), and of hazard resulting from surface deformation (e.g., flooding, subsidence, etc.). The definition of the active tectonic setting of an area requires the identification and geometric characterization of potentially active structures and the estimation of rates of activity of the identified faults. An important contribution to such definition is provided by the combination of data deriving from instrumental seismicity, from the historical record of seismicity, and from paleoseismological techniques. Despite providing information crucial to the seismic hazard prevention, such a combined approach is, however, partial due to the relatively narrow time window (in general, $\leq 10^3$ y) it explores. In fact, taking into account that rates of deformation may be either slow (e.g., a long quiescence may separate large earthquakes) and/or uneven both spatially (i.e., along strike of single structures over short timespans) and temporally (on single structures over larger timespans), short-term records may fail sampling active deformation. This may result in both an underestimation of active fault segments and an incomplete outline of the surface deformation scenario.

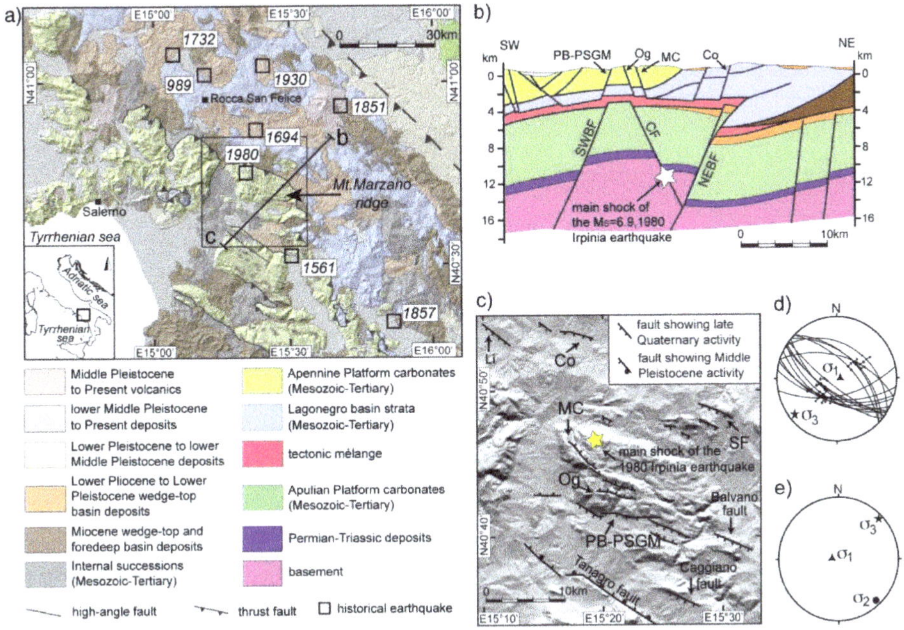

Figure 2. Tectonic setting of the 1980 Irpinia earthquake. (**a**) Geological sketch map of the southern Apennines, showing location of seismic stations and main historical and instrumental earthquakes. (**b**) Cross-section (after Ascione et al. [7]). Star shows hypocenter of the M_s 6.9, 1980 Irpinia earthquake; SWBF: SW Boundary Fault; CF: Central Fault; NEBF: NE Boundary Fault. (**c**) Outcropping Quaternary fault arrays (modified after Ascione et al. [7]). PB-PSGM: Piani di Buccino-Pantano di San Gregorio Magno fault zone; Og: Mount Ogna fault zone; MC: Mount Marzano–Mount Carpineta fault zone; Li: Lioni fault; Co: Conza fault; SF: San Fele fault. (**d**) Results of cumulative stress inversion from outcropping active fault segments (from Ascione et al. [7]). (**e**) Results of cumulative stress inversion from earthquake focal mechanism, data are from De Matteis et al. [4] (modified from Amoroso et al. [8]).

A more comprehensive definition of the deformation scenario is obtained if information on geometry and kinematics of deep-seated active fault segments, on spatial distribution and intensity of seismic shaking, and on the behavior (in terms of recurrence and magnitude of surface offset) of single fault segments is paralleled with observations encompassing larger timescales (e.g., the last 300–400 ky, i.e., 'intermediate timescales' according to Burbank and Anderson [9]). The latter, although less detailed on the history of single structures, may provide significant insights on the spatial distribution, pattern, and average rates of surface deformation. In fact, deformation accumulates through time, leaving an increasing imprint in the landscape which may be detected by an analysis of topography and may be temporally constrained by the stratigraphic record. Although it has become common to relate the motions during earthquakes to the Quaternary geological structures and topography in the epicentral regions, the deformation preserved in structural and topographic features represents the combined effects of entire earthquake cycles rather than only the coseismic period during which earthquake slip occurs. Therefore, in order to understand the formation and evolution of geology and topography over successive earthquake cycles, it is necessary to make observations that cover as wide a range of the cycle as possible.

The 23 November 1980, M_S = 6.9 Irpinia earthquake, the strongest and most destructive (I_0 = X MCS) seismic event of the last decennia in southern Italy, affected the Mt. Marzano area with widespread coseismic ruptures (e.g., [1,10–17]). The 1980 Irpinia earthquake strongly impacted the state of knowledge on seismicity of Italy, primarily because it was the first Italian earthquake for which coseismic surface faulting was assessed at two sites (namely, the Piano di Pecore and Pantano di San Gregorio Magno sites) by means of just posteventum surveys [18,19]. In addition, with further studies, a 38 km long alignment of surface faulting was recognized, in agreement with the inferred length of the rupture at depth [12]. The history of surface faulting in the last millennia at Piano di Pecore was assessed with the first paleoseismological study in the Italian territory [13].

In the last 40 years, many studies have provided constraints to seismicity and tectonics active in the 1980 event epicentral areas. According to such studies, it appears that both the earthquake complexity and the historical and present-day seismicity represent the response to a complex active tectonics scenario and suggest the existence of an 'active' graben-like structure defined by at least two antithetic major faults, which were originally identified by Barnard and Zollo [1] (Figure 1). However, the pattern, distribution, and localization of active structures, is still debated and, in some instances, controversial, with major implications on both cumulative deformation rates and related hazard. In fact, the active tectonic framework of the Irpinia earthquake epicenter area has been related to one or more structures substantially following the surface ruptures originally identified by Pantosti and Valensise [12,20–22], while a more articulated pattern composed by several normal faults arrays showing substantially coeval late Quaternary activity and spanning over the seismologically identified graben structure has been identified by detailed geomorphological analyses integrated by age constraints on the Quaternary deposits [7]. Further constraints on fault activity were obtained by Ciotoli et al. [23] using soil gas trace and carrier species as potential tracers of fault systems in the Piano di Pecore, a small-sized tectono-karstic basin located in the northern Mt. Marzano massif, which was affected by coseismic surface faulting of the 1980 Irpinia earthquake. The coseismic rupture zone in Piano di Pecore was investigated by Ciotoli et al. [23] in order to define the relationship between shallow distribution of gases and the main strike of the fault and verify whether the degassing process is still active along this fault trace. In fact, crustal discontinuities, such as fractures and faults of various dimensions, facilitate degassing flux from the Earth to the hydrosphere and the atmosphere. For this reason, the chemical composition and transport of soil gases within fault zones have been the subject of extensive investigations, including fault tracing and seismic surveillance as a precursor for geochemical anomalies to seismotectonic activity, due to their potential [24–35]. Faults can be described as weakened zones composed of highly fractured materials, gouge, and fluids. Active faults favor gas leaks because they usually increase the permeability of rocks and even their overlying soils. Gas anomalies at active faults can be either 'direct leak anomalies' where the gas measured corresponds to the deep gas phase or 'secondary anomalies' linked to the different mineralogical and hydrological behavior of the fault [29,36–38]. Anomalous gas emissions primarily occur in two different environments: Volcanic areas, where gas seepage is located both at central vents and often in large distal areas, and seismically active zones, where evidence of preferential degassing occurs near active faults. In the seismic zones, degassing has been shown to occur mainly as advective fluxes through soils of fractured areas and/or as free nonmixed gas phase from thermo-mineral springs due to pressure drop during ascent of the fluid to the surface [39]. The gas emission over seismically active faults corresponds to a long-term permanent phenomenon (with respect to earthquake recurrence times), which indicates that active faults are characterized by a high permeability and act as preferential conduits in the crust [39]. Several gases with different origins and contrasting behaviors in soil have been documented for detecting a fracture network and characterizing its extension and shape [37–40]. Because fluid transfer in the crust is strongly promoted by fractures, high geochemical contrasts are expected in faulted zones (e.g., [27]). The composition and distribution of gases in the soil pores are affected by surface features, such as pedological, biogenic, and meteorological factors. Several phenomena, like variation of the groundwater table, meteorological changes, soil porosity/permeability, and the degree of fracturing, may alter original gas concentrations for a single gas. However, these are thought to have a subordinated effect on gas leakage from deep fault-related features [36,38,40], and can be further reduced by studying a number of soil gases with different origins and contrasting behaviors. In fact, even in a restricted fractured area, gas distribution in soil can display contrasting patterns. It has been shown that the contrasting permeability in fault gouges and intensely sheared zones generate complex geochemical patterns in soil atmospheres [40]. This characteristic has already been used to search for active faults mainly using Rn emanation, but sometimes also using H_2, He and CO_2 distribution in soils.

The idea that CO_2 overpressure controls earthquake nucleation in the Apennines [41] is consistent with large CO_2 and CH_4 emissions at surface [23,42], while the coexistence of CO_2 and brine is well documented in several wells [43]. The presence of liquid and gas fluid phases in a fault volume has important consequences on seismicity production. In fact, the presence of fluid within the fault gauge may enhance seismicity due to lubrication mechanisms. Seismicity may also be enhanced by an increase of pore pressure in the country rocks embedding the faults. In the Irpinia region, modelling of microearthquake spectra has provided a rather low average seismic radiation efficiency [44], thus implying that rupture lubrication mechanisms are not favored. Therefore, it may be envisaged that the dominant mechanism triggering and controlling microseismicity in the Irpinia region is the pore pressure increase induced by fluid diffusion in the host rock medium [45,46]. In particular, at hypocentral depths, gases may significantly increase the pore pressure compared to liquids, up to a level for which it equals the lithostatic pressure [47]. According to Amoroso et al. [8], the high-resolution 3-D P- and S-wave velocity models of the Irpinia fault zone highlight a significant fluid accumulation within a 15 km wide volume of highly fractured rock located between SW and NE boundary faults as indicated in Figure 2b. A high V_P/V_S ratio recorded by seismic tomography points to the presence of fluid reservoirs below the mélange zone [8], consisting of brine-CO_2/CH_4 or CO_2-CH_4 mixtures [48]. The background micro-seismicity was therefore attributed to pore pressure changes in fluid-filled cracks surrounding major faults, which can trigger the episodic nucleation of moderate to large earthquakes. Analyses of micro-earthquake sequences revealed that they are primarily concentrated in densely fractured and limited regions characterized by horizontal, NE-SW-directed extension. These fault zones could be the source of repeated earthquakes due to the internal mechanical readjustments from local stress release and/or fluid migration along the fault damage zone [49].

2. Tectonic Framework

The southern Apennines form part of the Alpine-Apennine orogenic system, which derived from the convergence of the African and Eurasian plates in Late Cretaceous to Quaternary times (e.g., [50,51] and references therein). The Apennine accretionary wedge is composed of both ocean-derived [52,53] and continental margin-derived tectonic units. The latter include Mesozoic-Tertiary carbonate platform/slope successions (Apennine Platform) and pelagic basin successions (Lagonegro), stratigraphically covered by Neogene foredeep and wedge-top basin sediments (e.g., [54]) (Figure 2a). At the surface, the structure is characterized by low-angle tectonic contacts separating the Apennine Platform carbonates in the hanging wall and the Lagonegro Basin successions in the footwall [55]. These tectonic contacts consist of both thrusts—in part reactivated during extensional stages—and newly formed low-angle normal faults [56]. The Apennine accretionary wedge is tectonically superposed onto the buried Apulian Platform, which has a thickness of 6 km to 8 km and consists of a Mesozoic-Tertiary shallow-water carbonate succession, continuous with the outcropping in the foreland to the NE ([57] and references therein). The detachment between the allochthonous units and the buried Apulian Platform unit is marked by a mélange zone of variable thickness, locally reaching ca. 1500 m [58] (Figure 2b). The buried Apulian Platform is characterized by reverse-fault-related, open, long-wavelength folds that form the hydrocarbon traps for the significant oil discoveries in southern Italy [57]. Deep exploration wells also document the local occurrence of CO_2 gas caps in the top part of structural culminations made of fractured Apulian Platform carbonates, while waters of variable salinity occur below the gas caps (where these are present) and along the sides of the culminations [43].

Geophysical evidence has shown that the crystalline basement is involved in deep-seated reverse faulting [59–61]. The associated deformation is represented by significant vertical offsets along steep reverse faults, with relatively limited horizontal displacements.

Crustal shortening ceased in the Middle Pleistocene (e.g., [62]), when NE-SW-oriented horizontal extension became dominant over the whole orogen. Extensional faults postdating and dissecting the thrust belt (e.g., [63]) are also responsible for the active tectonics and seismogenesis in the southern Apennines (e.g., [64–66]).

3. Geological Framework of the Study Area

The study area is dominated by the Mt. Marzano morphostructural high, with a maximum elevation of 1579 m. The Mt. Marzano massif is composed of a pile of slope facies limestones and dolostones more than 2000 m thick, spanning in age from the Late Triassic to the Early Miocene [67]. The carbonates of Mt. Marzano massif, as well as those of platform facies forming the backbone of the Picentini and Alburni Mts. massifs (located to the west and south of Mt. Marzano, respectively) and of minor ridges interposed between the different massifs, are related to the Apennine Platform. Surface and subsurface (San Gregorio Magno 1 and Contursi 1 deep wells [68,69]) data from the Mt. Marzano area show that the Apennine Platform carbonates are tectonically sandwiched between the underlying Lagonegro basin strata (outcropping to the north of the Mt. Marzano massif) and the overlying Upper Cretaceous-Burdigalian, basinal Parasicilide unit, representing the deformed distal portion of the Apulian foreland palaeomargin [70,71] and forming part of the Internal successions in Figure 2a. The latter are covered by Burdigalian-Langhian foredeep and wedge-top basin deposits (Figure 2) [70,72]. In the study area, wedge-top basin deposits of Pliocene age (i.e., Zanclean and Piacentian) [54,72,73] consist of marine sediments (clays, sands, and conglomerates), locally passing upward to lacustrine and fluvial deposits, outcropping on top of the Mt. Marzano massif and in the adjacent southern (Tanagro river valley) and northern topographic lows (Figure 2a). Coeval, clayey-to-shallow marine deposits (related to the Ofanto wedge-top basin) occupy the northernmost part of the study area, i.e., the large Ofanto River valley.

In the Mt. Marzano area, the occurrence of a structural high of the Apulian Platform below the outcropping Apennine Platform carbonates and the underlying Lagonegro basin strata has been evidenced by seismic reflection profiles and gravity data (Figure 2b) (e.g., [7,74–76]). Similar to the structural traps of the oil fields of the Basilicata region south of our study area [57], the Apulian Platform positive structure in the Mt. Marzano area appears to consist of a large inversion feature controlled by multiple reactivation of a major SW-dipping fault of probable Triassic original age. According to this interpretation, substantial structural relief of the Apulian Platform in this area resulted from Late Pliocene to Early Pleistocene shortening and inversion, having been only weakly modified by subsequent late Quaternary extension. The top of the Apulian Platform carbonates, which is overlain by thick deposits over than 600 m thick including Messinian evaporites and mélange units (probably involving Lower Pliocene clastics), lies at a depth of 3977 m in the San Gregorio Magno 1 well [68], whereas in the Contursi 1 well [69], the top of the mélange, is found at a depth of c. 3100 m.

The fold and thrust structure is dissected by extensional faults, which control continental depocentres such as those occupying the Tanagro River valley [77] and the Piani di Buccino (hereinafter, PB) and Pantano di San Gregorio Magno (hereinafter, PSGM) basins [78,79]. The Tanagro basin, which hosts several terraced alluvial units framed in the late Early Pleistocene to late Middle Pleistocene timespan [77], formed along a major WNW-ESE-striking, NE-dipping normal fault (Tanagro fault) [77,78]. This fault has recorded repeat activity over the Middle Pleistocene. The formation of the Tanagro basin predates those of the PB and PSGM (Figure 2), which occurred in the late part of the Middle Pleistocene, and of several minor basins within the Mt. Marzano ridge [7]. Within the Mt. Marzano massif, high-angle WNW-ESE-trending strike-slip faults and east-trending, NE-dipping normal faults offset the Triassic-Miocene succession and the overlying Pliocene deposits [72]. Such faults have been sealed by several remnants of Pliocene–Early Pleistocene erosional surfaces molded in the top surface of the massif and graded down to the SSE, with an overall stair-like arrangement, from around 1500 m to 600–700 m [78]. The erosional surfaces, which evidence progressive surface lowering that has affected the Mt. Marzano topographic high in Pliocene–Early Pleistocene times, predate the extensional faulting, which has been active in the Mt. Marzano area since the Middle Pleistocene. Extensional tectonics activated major WNW-ESE trending, both NE and SW-dipping faults, which caused the formation of fault-bounded mountain fronts and of the previously mentioned intramontane basins filled with lacustrine and/or alluvial successions. Extensional faulting in the area spanning from Mt. Marzano to the Alburni Mts. has been active in late Quaternary times, as it is shown by widespread geomorphological-geological and paleoseismological evidence (discussed in the following sections).

4. Morphotectonic Features of the Mt. Marzano Area

The Mt. Marzano and surrounding area are characterized by widespread evidence of Quaternary extensional tectonics. Stratigraphic and morphotectonic evidence allows unravelling faulting chronology. This points to a progressive younging, from the SW to the NE, of vertical motion initiation over Middle to Late Pleistocene times in the area spanning from the Alburni Mts. mountain front to the Mt. Marzano massif. In the whole area, in addition, several faults show geomorphological/stratigraphical evidence for Late Pleistocene-Holocene activity.

The oldest Quaternary extensional tectonics was recorded in the Tanagro river valley, to the SW of the Mt. Marzano massif. The Tanagro river valley follows a Quaternary continental basin formed along a major WNW-ES-striking, NE-dipping normal fault, namely the Tanagro fault [77] shown in Figure 2c. Seismic reflection profiles have [80] highlighted the occurrence of several NE-dipping splays synthetic to the major NE-dipping Tanagro fault and of a horst-and-graben structure in the southern sector of the valley, i.e., across the Alburni Mts.–Mt. S. Giacomo–Mt. Marzano transect. The sedimentary record in the Tanagro valley basin provides evidence for repeated activity of the Tanagro fault over the Middle Pleistocene. This is shown by (i) the backtilting of early Middle Pleistocene alluvial fans in the SE area; (ii) SW-tilting of late Middle Pleistocene travertine-alluvial body and related terraces in the northern Tanagro valley—NTV (see Figure 3 for location of terraces in the NTV); and (iii) fluvial terraces located in the hanging wall block at the northern termination of the Tanagro Fault, pointing to renewed aggradation (following dissection of the travertine-alluvial body) in response to damming of the Sele-Tanagro rivers valley along the Tanagro fault [77].

Post-late Middle Pleistocene activity of the Tanagro fault has not yet been proven by either geomorphic or stratigraphic evidence. On the other hand, evidence for younger, Late Pleistocene-Holocene faulting characterizes the Caggiano fault that bounds the southern Tanagro valley toward the NE (location in Figure 2c). This is shown by the occurrence of a post-glacial midslope bedrock fault scarp running along the southern slope of Mt. S. Giacomo. Holocene historical activity of this fault is also evidenced by paleoseismological investigations [81]. Based on morphostratigraphical evidence, alluvial deposition in the NTV depocenter predates the formation of the PB and PSGM basins. In fact, the formation of the PB and PSGM interrupted S-flowing fluvial paths and related sedimentary (alluvial fan) inputs from the Mt. Marzano massif to the NTV, as it is inferred from the paleo-alluvial fan related to the Palomonte wind gap (Figure 3), which is cut into the PB southern border.

Surface and well data show that the PB and PSGM are filled with fine-grained deposits (mainly sandy-silts and clays), passing laterally into alluvial fans and slope debris. In those basins, lacustrine/marshy environments were present in historical times and persisted until they were reclaimed in the XX century [79]. The PB and PSGM basins are bounded toward the NE by a major WNW–ESE-trending fault zone consisting of N100° to N120° en-echelon normal faults, N70°-striking transfer faults (probably reactivating pre-existing inherited structures) [7], and minor antithetic faults toward the SW. The occurrence of a major northern fault zone is consistent with the N-ward thickening of the PB fill up to about 100 m [7]. The thickness of the PSGM basin fill, which exceeds the 62 m depth of a core drilled at the basin axis, is not well constrained. The carbonate bedrock becomes shallower (about 35 m deep) toward the eastern basin termination, as shown by geophysical investigations [82]. Geophysical investigations have also imaged the WNW-ESE-trending northern and southern (i.e., the NE-dipping Mt. Difesa Ripa Rossa; location in Figure 3) normal faults down to a depth of c. 300 m. A lateral variation in the amount of offset along the PSGM northern fault zone is made evident by the occurrence of a bedrock high between the PSGM and PB basins.

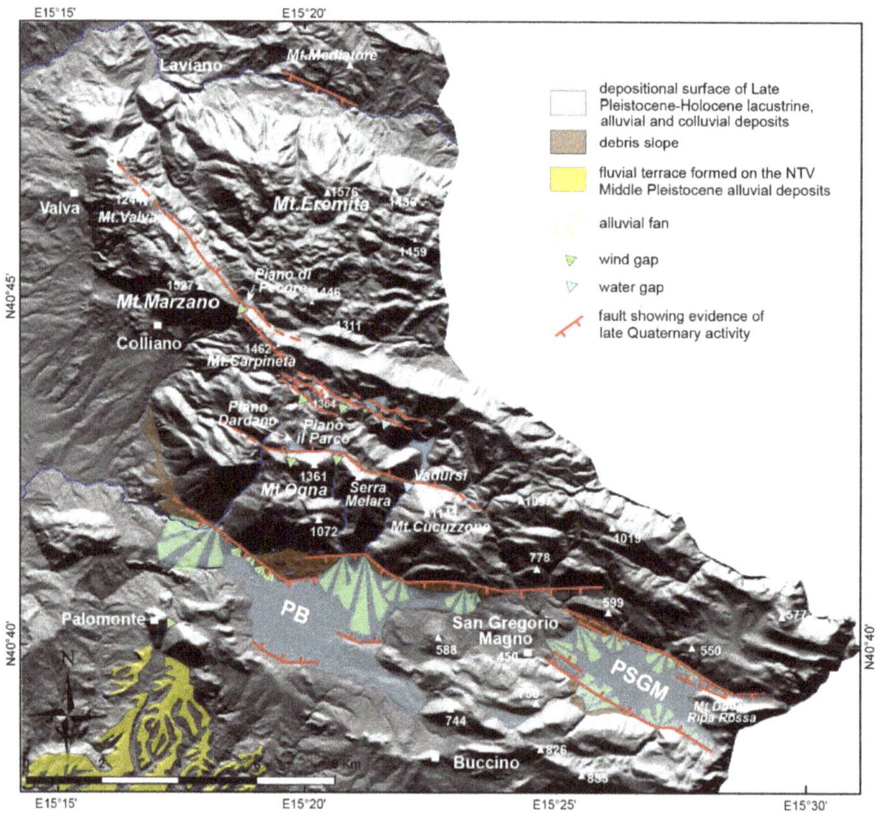

Figure 3. Faults showing evidence of late Quaternary activity in the Mt. Marzano ridge area (redrawn after Ascione et al. [7]).

Scraps around a few meters high affect Holocene lacustrine terraces and alluvial fans within the PSGM and PB basins. Furthermore, bedrock fault scarps and wineglass valley cross-profiles, as well as faulted Upper Pleistocene-Holocene slope breccia and alluvial fan deposits, occur along the entire northern fault-bounded mountain front. To the east, a debris slope is offset along the Balvano fault. The heights and estimated ages of both bedrock and alluvial fault scarp along the northern fault bounded mountain front of both PB and PSGM basins suggest that fault slip rate only slightly exceeds the accumulation rate within the two basins (ranging from 0.1 to 0.37 mm/a, with a mean value of 0.2 mm/a in the last 240 ka in the PSGM [7]). The slip rate estimated so far is consistent with the value (<0.5 mm/a) independently estimated by slip profiles analysis of fault scarps in the northern mountain front [20].

Evidence for Late Pleistocene-Holocene fault activity in the Mt. Marzano massif is widespread further to the north and east of the PB-PSGM major fault (Figure 2c). Among the faults active since the final part of the Middle or the Late Pleistocene are the NE-dipping Mt. Valva–Mt. Marzano–Mt. Carpineta and the Mt. Ogna–Mt. Cucuzzone fault zones, consisting of a succession of fault strands with average NW-SE to WNW-ESE strike (Figure 3). Offsets accumulated by these younger faults have not significantly affected topography of the Mt. Marzano massif, which culminates on the Mt. Eremita peak (1579 m) located in the northernmost part of the massif. Recent activity of NE-dipping faults in the Mt. Marzano ridge have interacted with surface processes, causing widespread valley damming (evidenced by several wind gaps) and the formation of alluvial basins (Figures 3 and 4). Recent activity has also affected the longitudinal profiles of the streams that dissect the ridge (Ascione et al. 2013).

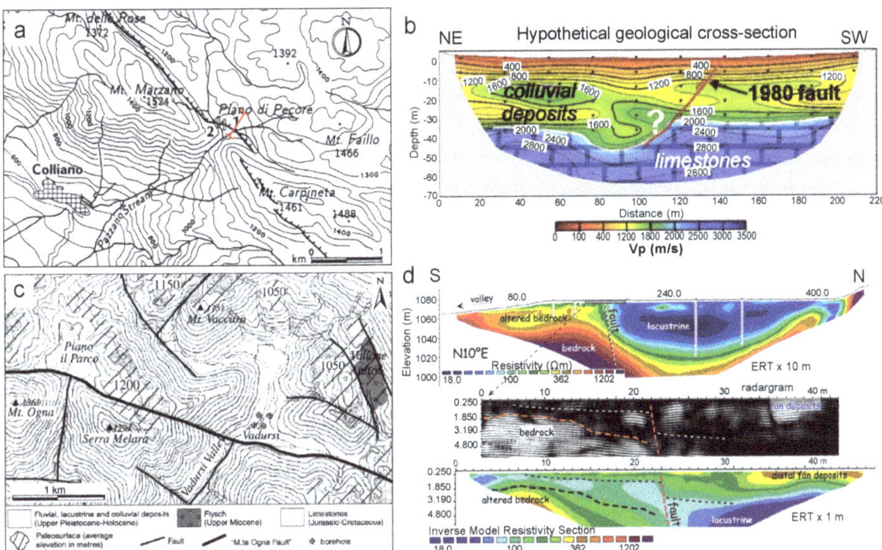

Figure 4. (**a**) Location of the Piano di Pecore basin, created by fault damming of the Pazzano stream with an indication of the coseismic N-facing fault scarp (from Pantosti et al. [13]). (**b**) Seismic tomography and inferred geological cross-section across the Piano di Pecore basin, along the red trace in diagram a, showing thickening of the basin fill toward the southern basin border (from Ascione et al. [78], modified). (**c**) Geological sketch map of the Mt. Ogna area, with the Piano il Parco and Vadursi basins, created by damming of S-flowing stream valleys (from Ascione et al. [78], modified). (**d**) Geophysical investigations of the Piano il Parco basin, showing thickening of the basin fill toward the southern basin border (from Galli et al. [83], modified).

Due to the carbonate nature of the rocks that form the backbone of the Mt. Marzano ridge, the dammed valleys evolved as karst basins in several instances. One of such basins is occupied by the Piano di Pecore plain. The Piano di Pecore basin is an undissected karst basin filled with continental deposits with an inferred thickness around 30–40 m (Figure 4). The basin fill, based on surface evidence and on shallow excavations (paleoseismological trenches, see Section 6), is composed of slope debris, limited to the basin margins, as well as fine-grained colluvial, lacustrine-marshy, and volcanoclastic sediments. The formation of the Piano di Pecore basin occurred in response to the damming of a S-flowing stream (Pazzano stream) by the NE-dipping Mt. Valva–Mt. Marzano–Mt. Carpineta fault zone. This is inferred from the occurrence, at the SW border of the basin, of a 200 m deep wind gap cut across the Mt. Marzano-Mt. Carpineta ridge (Figure 3) [78]. Evidence for recent (Holocene) activity of the Mt. Marzano–Mt. Carpineta fault is provided by the 5 m vertical separation between the undissected basin floor and the wind gap bottom, whereas the youngest tectonic activity has been evidenced by the 1980 coseismic rupture (see Section 6). Tomographic investigations allowed the identification of the fault activated with the 1980 earthquake in the Piano di Pecore subsurface by a bedrock throw of approximately 20 m, which affects the top of the carbonate bedrock of the basin (Figure 4). The bedrock throw has most probably overestimated the total fault offset which, based on the vertical separation between the basin bottom and the adjacent wind gap bottom, has been evaluated at about 40 m, consistent with the 40–50 m vertical separation between displaced erosional surface cut on both the hanging wall and the footwall blocks of the entire Mt. Valva–Mt. Marzano–Mt. Carpineta fault zone [78].

5. Seismicity

Active tectonics evidence the intense seismicity affecting the Mt. Marzano area, which falls within the epicentral area of some of the strongest historical earthquakes of southern Italy, i.e., those with intensity I ≥ X MCS occurred in 989, 1694, in 1930, 1962 and 1980 (Figure 2a) (e.g., [11,14–16,84]). Among such events, the M_S = 6.9 1980 Irpinia earthquake dramatically struck the Mt. Marzano and the neighboring area, which recorded widespread coseismic ruptures and a large number of secondary geological effects, including landslides, ground cracks, liquefaction, and variations in the discharge rate of major carbonate springs [14]. The area, which was also struck by a seismic sequence with a M_L = 4.9 mainshock [85] in 1996, is presently affected by subdued background seismicity with M_L ranging between 1 and 3.3 (Figure 5) [4,86–88]. In July 2020, the NW termination of the extensional fault system (in the Rocca San Felice area; location in Figure 2a) was affected by a minor seismic sequence (with maximum M_L = 3.0 recorded for two shocks), including 43 events distributed over a NW-SE-trending, narrow, ca. 5 km long zone [87]. Using the 3D velocity model described by Amoroso et al. [8], 36 selected events were relocated, depicting a fault zone dipping 50–60° toward the NE [87]. The related focal mechanisms show fault plane solutions characterized by dominate dip-slip, normal fault kinematics with a subordinate strike-slip component. The seismological dataset associated with this seismic sequence, being generally consistent with the seismogenic sources of the Irpinia region, has recorded fault activity in the proximity of the NW tip of the regional seismogenic structure. One of the largest nonvolcanic natural emissions of low-temperature CO_2 rich gases ever measured on Earth, i.e., the Mefite d'Ansanto (e.g., [89]), is located in this area. This is consistent with previous studies suggesting that the densely fractured 'process zone' surrounding fault tips represent major conduits for fluid flow, giving rise to strong gas emissions at the surface [23]. Boreholes in this area also indicate the occurrence of a CO_2 gas cap at the top of an antiformal trap involving fractured Apulian carbonate reservoir rocks, while saline water variably occurs along the sides of the structural trap and below the gas cap [43].

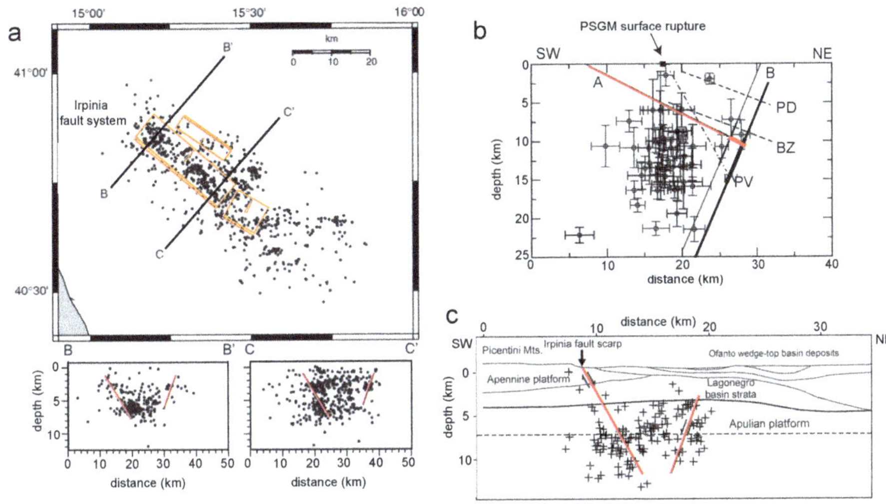

Figure 5. (**a**) Seismicity of the Irpinia region from August 2005 to April 2011 in map view (in orange, seismogenic sources from DISS working group [88]) and in cross-sections, along the profiles reported in the map (the red lines represent the projection of the fault segments activated with the 1980 earthquake) (from De Matteis et al. [4], modified). (**b**) Comparison of models for the 20 s fault, and projection of aftershocks (crosses indicate uncertainty) localized on a vertical N31E cross-section cutting across the PSGM; A and B (effectively slipped area: thick line): Amoruso et al. [90]; BZ: Bernard and Zollo [1]; PD: Pingue and De Natale [91]; PV: Pantosti and Valensise [12] (from Amoruso et al. [90], modified). (**c**) Gravity profile, and the fault segments and aftershocks [92] (crosses indicate uncertainty) of the 1980 earthquake (partly redrawn from Improta et al. [74]).

6. The 1980 Irpinia Earthquake

As already mentioned above, the 1980 Irpinia earthquake was characterized by a complex source mechanism, consisting of three major subevents at 0 s, 20 s and 40 s, with M_W 6.9, 6.4, and 6.3 respectively (Figure 1) (e.g., [1,3]). The mainshock originated in the 8–13 km depth range [93]. Bernard and Zollo [1] defined the three normal faulting ruptures as follows: (i) The M_S = 6.9 mainshock, which nucleated on the NE-dipping Mt. Marzano and Picentini Mts. segments (about 20 km long); (ii) the 20 s subevent, which nucleated about 15 km southwest of the first event on a circa 20 km long normal fault; (iii) the 40 s subevent, located in the Ofanto basin area, which was associated with a SW-dipping normal fault antithetic to the first activated fault (Figures 1B and 5). According to Nostro et al. [94], the 40 s subevent might have reactivated a fault segment which ruptured during the Me = 6.9 [95], 1694 earthquake, which struck an area overlapping with that hit by the 1980 event (e.g., [14,15]). Pingue and De Natale [91] estimated an 80° SW dip for the Ofanto fault. On the other hand, the NE-dipping fault segment responsible for the Mt. Marzano mainshock is constrained in a 53°–63° dip range by seismological data [93], whereas dip values of 60° (at depth) to 70° (at surface) have been proposed based on breakout and log analysis of the S. Gregorio Magno 1 well [96]. Probably due to the scarce source coverage [97], the localization and mechanism of the 20 s subevent are debated. In fact, the 20 s nucleation has been associated with: (i) A deep-seated, NE-dipping, low angle (20°) fault [1]; (ii) a fault plane dipping 60° to the NE [12]; and (iii) a SW-dipping fault antithetic to that of the mainshock (and roughly aligned with the 40 s fault; Figure 5b), which is suggested to represent a reactivation of the southernmost part of the fault activated with the 1694 earthquake [90]. On the other hand, based on absence of record of the 1694 event in trenches dug along the 1980 ruptures, it has been proposed that the fault responsible for the 1694 earthquake was similar to and longer than the 40 s fault [98].

However, in the paleoseismological trenches (see below), no record of other large earthquakes that struck the area, e.g., those that occurred in 1930 and 1962, was found.

The Mt. Marzano–Picentini Mts. fault segments and the antithetic Ofanto fault define the boundaries of both a NW-trending zone of aftershocks (which coincide with a high P-wave velocity zone) [92,99] and a volume affected by background microseismicity on subparallel predominantly normal faults (Figure 5a) [4]. The 1980 earthquake affected the epicentral area with widespread coseismic surface ruptures and ground deformation, which have been detected by geodetic levelling surveys [1,22,91].

Post-eventum surveys in the region recording the largest intensity ($I_0 \geq$ IX) have reported more than 3000 ruptures, with those with a vertical offset showing a marked N120° main trend [100]. Among such ruptures, the circa 2 km long, up to 50 cm high, WNW-ESE trending, NE-facing scarp formed in the PSGM basin (Mt. Difesa Ripa Rossa scarp [19]; location in Figure 3) and the mainly NW-SE trending, NE-facing scarps in the Mt. Marzano–Piano di Pecore area [18] were interpreted as surface faulting (Figure 6).

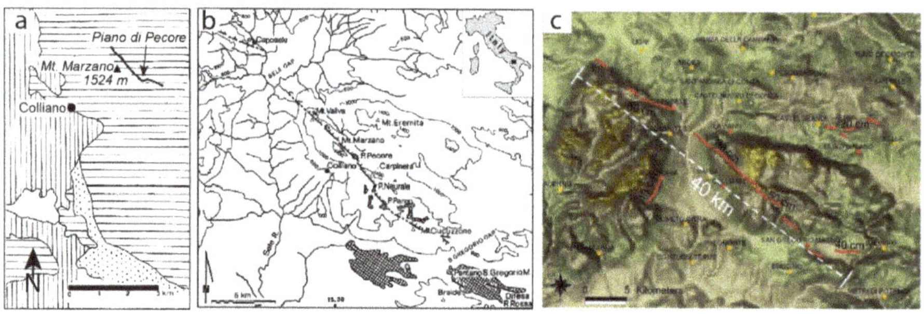

Figure 6. (**a**) Post-eventum mapping of the 1980 Irpinia earthquake surface faulting in the Mt. Marzano peak—Piano di Pecore plain area (modified from Cinque et al. [18]). (**b**) Mapping of the 1980 Irpinia earthquake coseismic surface ruptures from [12]. (**c**) Distribution of the 1980 Irpinia earthquake coseismic surface faulting, after [10,12,14,18,19,101,102] (from Serva et al. [15], modified).

In particular, a high scarp ranging from several tens of centimeters to around 1 m, running across the NE slope of the Mt. Valva–Mt. Marzano ridge and the floor of the Piano di Pecore plain and continuing toward the SE for some hundreds of meters, was recognized [18]. The scarp was characterized, for most of its length, by a NW-SE trend and a NE dip. A roughly E-W trending, N-facing scarp was also recognized (Figure 6). The latter scarps, and those affecting the SW slope of Mt. Carpineta ridge (the downthrown block being the uphill side), were interpreted as part of a 10 km long succession of right-stepping strands of the 1980 surface faulting, and the PSGM scarp was related to a further fault segment [8]. Coseismic faulting was related to three main strands separated by gaps (i.e., the Sele valley and San Gregorio Magno gaps), with changes in strike of about 10° [12]. Such fault strands, interpreted as part of a 38 km long, NE-dipping scarp with an average N128° strike ('Irpinia fault'), are identified with: (i) The Picentini Mts. footslope, with a N125° trend; (ii) the N135° trending Mt. Marzano–Mt. Valva scarp coupled with the N135°–N110° trending Mt. Carpineta scarp; and (iii) the N130°–N120°-striking PSGM scarp (Figure 5). The Mt. Marzano–Mt. Valva–Mt. Carpineta and Picentini Mts. segments have been associated with the mainshock and the PSGM strand with the 20 s subevent. It has also been suggested that the Mt. Valva–Mt. Marzano–Mt. Carpineta and the Picentini Mts. strands are part of single, continuous fault having a subdued topographic expression as it crosses the Sele River Valley, which is occupied by rocks with a low resistance to erosion [20]. The 40 s subevent has been associated [101] with circa 8 km long, up to 30 cm high, SW-facing scarps exposed between Santomenna and Muro Lucano (Figure 6c). Paleoseismological investigations carried out in the Piano di Pecore site revealed a mean slip rate around 0.3 mm/a [13], and a 0.17–0.4 mm/a slip rate in the last 20 ky was estimated for the antithetic, NE-dipping, Mt. Difesa Rossa fault [103].

Recently, Ascione et al. [7] identified several fault arrays within the epicentral area of the 1980 earthquake (Figure 2c). Within the structures belonging to the identified fault arrays, those developed in the area spanning from the northern Mt. Marzano massif (i.e., the Marzano fault array) to the Ofanto river valley—being characterized by a NE dip in the Mt. Marzano area and a SW dip to the NE of such a massif—are considered as the surface expression of the deep-seated graben-like structure which was activated with the 1980 earthquake (e.g., [1,90,91]). Their spatial distribution overlaps the belt affected by the 1980 earthquake aftershocks (e.g., [92]) and by the present-day low magnitude seismicity, which occurs on subparallel, predominantly normal faults (Figure 2e) [4]. A major SW-dipping active fault system was identified to the SW of the earthquake epicenter, at the northern boundary of the S. Gregorio Magno—Buccino basins (Figure 2c) [7]. Stress inversion from surface faults and from instrumental earthquake focal mechanisms show a consistent pattern of NE-SW roughly horizontal maximum extension (Figure 2d), which is compatible with the T axis obtained from the 1980 main shock and with results of cumulative stress inversion from earthquake focal mechanism data (Figure 2e). Surface structures are decoupled from the seismically active deep-seated structures, which, in turn, reactivate inherited basement faults (Figure 2b) [7]. Decoupling between deep and shallow structural levels is related to the strong rheological contrast produced by the fluid-saturated, clay-rich mélange zone interposed between the foreland Apulian Platform carbonates and the allochthonous units, which include carbonate platform (i.e., Apennine Platform) and basin (i.e., Internal and Lagonegro) successions (Figure 2b). Based on such an interpretation, the surface expression of the deep SW Boundary Fault (SWBF) is represented by the PB-PSGM fault zone, that of the Central Fault (CF) by the Mount Marzano–Mount Carpineta (MC) and Mount Ogna (Og) fault zones and that of the NE Boundary Fault (NEBF) by the Conza (Co) and S. Fele fault strands (Figure 2b,c).

Amoroso et al. [8], thanks to the installation of dense, high dynamic range, seismic network operated by INGV (Istituto Nazionale di Geofisica e Vulcanologia) and AMRA (Analisi e Monitoraggio dei Rischi Ambientali) in the area struck by the 1980 Irpinia earthquake, recorded a massive waveform dataset of micro-earthquakes with magnitude larger than about 1 from August 2005 through April 2011. The researchers analyzed V_P and V_S wave velocities in the upper crust of the Irpinia fault system and the related microseismicity distribution, which appears to be confined within an uplifted block including the main normal fault rupture of the 1980 earthquake (Figure 7). They observed high V_P/V_S and low $V_P \times V_S$ values in the region where intense microseismicity is located, which suggests fluid accumulation within a ~15 km wide rock volume. Further studies [48] have provided evidence for a composition of the fluids permeating the subsurface of the 1980 Irpinia earthquake region dominated by CO_2 and brine.

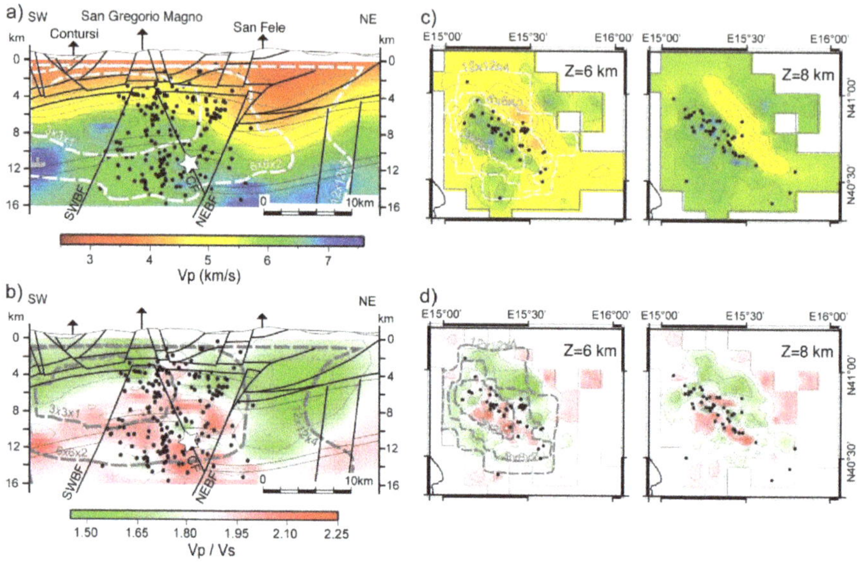

Figure 7. Three-dimensional tomographic model of the 1980 Irpinia earthquake region. (**a**) V_P velocity model and microearthquake locations projected onto the cross-section located in Figure 2a (refer to Figure 2b for the tectonic contacts and geological units). Dashed curves delimit the well-resolved regions of the model. Each curve corresponds to a different resolution scale obtained from estimating the resolvability function for each model parametrization used in the multiscale approach, as explained in detail by Amoroso et al. [8]. Star shows the hypocenter of the M_S 6.9, 1980 Irpinia earthquake. (**b**) V_P/V_S ratio for the same depth section as in (a). (**c**) Horizontal slice through the P-wave tomographic model at depths (Z) of 6 km and 8 km. (**d**) Horizontal slice showing V_P/V_S ratio at depths (Z) of 6 km and 8 km (modified from Amoroso et al. [8]).

7. Postseismic Deformation

Recent studies suggest that a significant proportion of the topography and geomorphology in tectonically active areas is controlled by deformation in the postseismic time period following earthquakes [104–107]. The deformation during the parts of the earthquake cycle that do not involve slip during earthquakes may be analyzed by means of satellite data to map the motion of the Earth's surface (known as space-based geodetic measurements, e.g., GPS and InSAR). At present, the length of the earthquake cycle (decades to millennia) rules out examining an entire cycle on a given fault using these space-based geodetic methods. However, even imaging parts of the coseismic, postseismic, and interseismic periods—using the decades-long archive that is now available for exploitation—may provide useful insights into the geometry and mechanisms of the deformation. Within this framework, it is crucial to compare the strain in the earthquake cycle—including that produced by postseismic afterslip—with the geology and topography of the study area by means of tectonic geomorphology analyses, which clearly provide insights into the deformation integrated over much longer timescales with respect to a single earthquake cycle.

Construction of Ground Deformation Maps Based on PSs Mean Vertical Velocity

In this section, we present the procedure followed to obtain two radar vertical mean velocity ground deformation maps that cover the region around the epicentral area of the 1980, 6.9 Mw, Irpinia earthquake. The maps were constructed using Permanent Scatterers—PS datasets processed by images recorded along both the ascending and descending orbits by ERS 1/2 (1992–2000) and Envisat (2003–2010) satellites. The PS datasets analyzed in this study are listed in Tables 1 and 2.

Table 1. ERS Permanent Scatterers datasets used in this study.

PS ERS Original Dataset	Number of Pss
ERS_T358_F819_CL003_CAPOSELE_A	38,864
ERS_T358_F819_CL002_BENEVENTO_A	45,672
ERS_T494_F2781_CL001_POTENZA_D	75,178

Table 2. ENVISAT Permanent Scatterers datasets used in this study.

PS ENVISAT Original Datasets	Number of PSs
ENVISAT_T86_F816_CL001_FOGGIA_A	391,518
ENVISAT_T358_F801_CL001_SALERNO_A	248,986
PST2009_ENVISAT_T358_F819_CL001_BENEVENTO_A	532,482
ENVISAT_T265_F2781_CL001_AVELLINO_D	463,801

The datasets listed in Tables 1 and 2 were obtained from the Geoportale Nazionale of the Italian Ministry of Environment (MATTM) database (www.pcn.minambiente.it/mattm) and preprocessed with PSInSAR by the TRE Company and PSP-DIFSAR by the e-GEOS Company. Details on the signal processing have been described by the authors of [108,109]. The PS position values for C-band SAR satellites, such as the ERS and ENVISAT satellites, may be affected by a measurement error of ±3 mm. However, the ERS and ENVISAT PS data from the MATTM database are of high quality. Any single pixel in a radar image is selected only if its temporal and geometrical decorrelation values are extremely low. In fact, the coherence values (ranging from 0 to 1) of the PS of the datasets that cover the investigated area are ≥0.6, which means that the PS coherence values are high. In other words, the analyzed PS are stable with respect to the radar and, thus, their motions are measured with high precision [109,110]. Consistently, the standard deviations of mean velocity values of single PS, calculated over the timespans of the entire records (i.e., the 1991–2000 and 2003–2010 timespans, respectively), are ≤0.5 mm/y for about the 90% of both datasets.

A GIS geospatial data analysis was used to construct, by means Arcgis 10.7® software (ESRI, Redlands, CA, USA), a series of raster maps synthesizing the spatial distribution of various parameters at different timespans. For the geospatial analysis, the Inverse Distance Weighting interpolator (IDW) statistics were used. In the first steps, the entire original ERS and ENVISAT PS datasets (listed in Tables 1 and 2) were processed. Afterward, the PS datasets were analyzed in order to select, from the original datasets, subsets of data (hereinafter labelled PS normal subsets) by excluding outlier data (see below).

Due to the near-polar orientation of SAR orbits, components of ground-motion-oriented N-S are substantially undetectable, while E-W-oriented components contribute to the motion detected by SAR satellites along its Line of Sight (LoS). Starting from PS motion recorded along SAR both ascending and descending orbits, the displacement/velocity in the vertical plane (z) oriented E-W may be reconstructed through the evaluation of the vertical (Dz or Vz) and horizontal (E-oriented, Deast or Veast) components of PS displacement, or velocity using the following equations by the authors of [111–113]:

$$Dz = (D_{LoSd} + D_{LoSa})/2\cos\phi \quad (1)$$

$$Deast = (D_{LoSd} - D_{Losa})/2\mathrm{sen}\phi \quad (2)$$

$$Vz = (V_{LoSd} + V_{LoSa})/2\cos\phi \quad (3)$$

$$Veast = (V_{LoSd} - V_{Losa})/2\mathrm{sen}\phi \quad (4)$$

where D_{LoSd} and D_{LoSa} are PS displacement values oriented along the descending and ascending LoS, respectively; V_{LoSd} and V_{LoSa} are PS velocity values oriented along the descending and ascending LoS, respectively; and ϕ is the LoS angle of incidence, which is around 23° for ERS and ENVISAT satellites.

The SAR data analysis was focused on a key area including the Mt. Marzano and Conza sites, where surface evidence of the main NE-dipping fault responsible for the 1980 Irpinia earthquake mainshock and the antithetic SW-dipping fault responsible for the 40-s subevent occurred [7,22]. The perimeters of the areas of interest (shown in Supplementary Figure S1) were traced along the regions where the ascending and the descending original datasets and subsets are consistently overlapped. This is a necessary condition to operate, subsequently, on the interpolated raster data in order to derive the horizontal and vertical components of the ground motions by applying the relationship [3]. As a consequence, the parameters of the analyzed ERS and ENVISAT datasets are slightly different (Supplementary Figure S2).

The IDW analysis, with cell size 50 × 50 m, was applied to the ERS PS 'native' ascending and descending datasets (i.e., the datasets that include both normal and outlier PSs) in order to construct the LoS-oriented mean velocity maps, which are reported in Supplementary Figure S3. Afterward, starting from the maps of the ascending (V_{LoSa}) and descending (V_{LoSd}) mean velocities, by applying Equation [3], the vertical component of mean velocity (Vz), or ground deformation, was constructed for the 1992–2000 timespan (Figure 8).

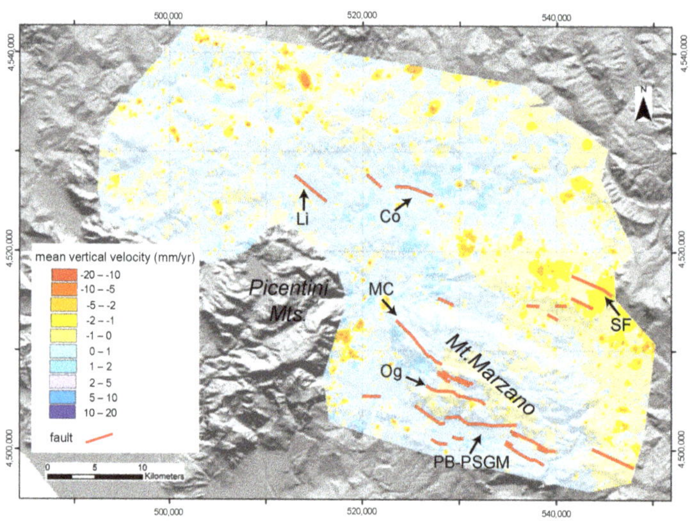

Figure 8. IDW interpolation (cell size 50 × 50 m) of ERS PS 'native' datasets (i.e., the datasets that include both normal and outlier PSs), showing the vertical mean velocity/ground deformation in the 1992–2000 timespan. Fault/fault array labelling as in Figure 2b.

The maps of the LoS-oriented (ascending and descending) mean velocities in the 2003–2010 timespan, constructed using the 'native' ENVISAT datasets, are reported in Supplementary Figure S4. Through the relationship [3], the vertical component of mean velocity for the ENVISAT PS original dataset was obtained (Figure 9).

A geospatial analysis was then carried out by applying both the *Hot Spot Analysis* (Getis-Ord Gi*) and the *Cluster and Outlier Analysis* (Anselin Local Moran's I) mapping tools to all the PS subsets without applying any filter based on coherence values. The data analysis started with the extraction of 'PS normal subsets' from the 'native' datasets through the outliers' boundaries (fence) evaluation (Supplementary Figure S5 and Table S1). Afterward, the PSs (of both the ERS and ENVISAT datasets) falling inside the perimeters shown in Supplementary Figure S2 were selected. The data included in the investigated perimeters form three new subsets for the ERS data, and four new subsets for the ENVISAT data, all identified with the label '*marzano*' (Supplementary Figures S6 and S7).

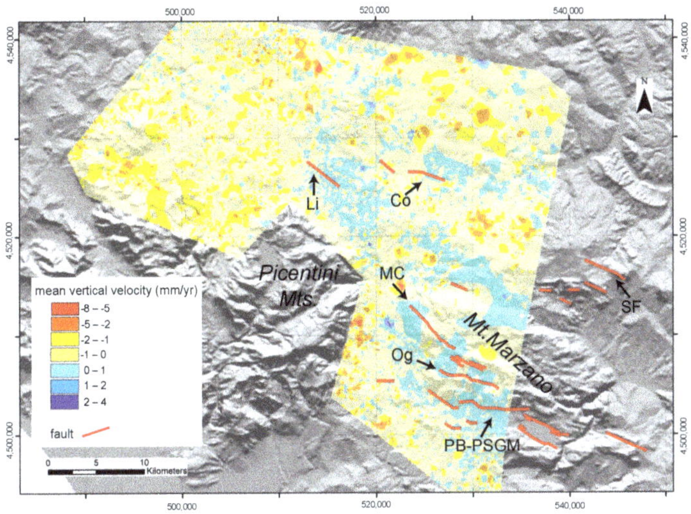

Figure 9. IDW interpolation (cell 50 × 50 m) of ENVISAT PS 'native' datasets (i.e., the datasets that include both normal and outlier PSs), showing the vertical mean velocity/ground deformation in the 2003–2010 timespan. Fault/fault array labelling as in Figure 2b.

Following, in part, the procedure by [114], both the *Hot Spot Analysis* (Getis-Ord Gi*) and the *Cluster and Outlier Analysis* (Anselin Local Moran's I) mapping tools for all 'marzano' were applied to the PS subsets (the subsets arising from the *Hot Spot Analysis* are labelled with 'HS,' and those from the *Cluster and Outlier Analysis* with 'CO'). The label '2k' of the subsets' names indicates the 2000 m *Threshold Distance* that is '*a cutoff distance for Inverse Distance option. Features outside the specified cutoff for a target feature are ignored in analyses for that feature.*' The label 'IDW' is referred to the Conceptualization of Spatial Relationships that is '*nearby neighbouring features have a larger influence on the computations for a target feature than features that are far away.*' The mapping results are shown in Supplementary Figures S8 and S9 for both of the used tools. Inspecting the diagrams in Supplementary Figures S8 and S9, it is evident that both mapping tools give a similar distribution of representative PSs (the red and blue points), and there is a better data visualization in the maps created with the *Cluster and Outlier Analysis* (CO maps). For these reasons, the CO maps were used for any subsequent elaboration.

To operate with a less but still significant number of PSs, we further removed, from the 'CO' subsets, the PSs defined as 'not significant,' 'high outlier' (HL), and 'low outlier' (LH). Examples of maps constructed using ERS and ENVISAT descending subsets composed of PSs classified as 'not significant' (nsig) are shown in Supplementary Figures S10 and S11. The remaining PSs, classified as (HH) and (LL), were selected for the new subsets (identified by the 'nout' label; Supplementary Tables S2–S5), which were used to construct new ground deformation maps for the investigated area (see below). Applying the IDW interpolation (cell size 50 × 50 m) to data of the PS subsets labelled 'CO_2k_IDW_nout', maps of the LoS-oriented ascending and descending mean velocities (V_{LoSa} and V_{LoSd}) were constructed (Supplementary Figures S12 and S13). Afterward, Equation (3) was applied to both the ERS and ENVISAT LoS-oriented ascending and descending velocities to construct maps of the vertical component of mean velocity/ground deformation in the 1992–2000 and 2003–2010 timespans, which are shown in Figures 10 and 11, respectively.

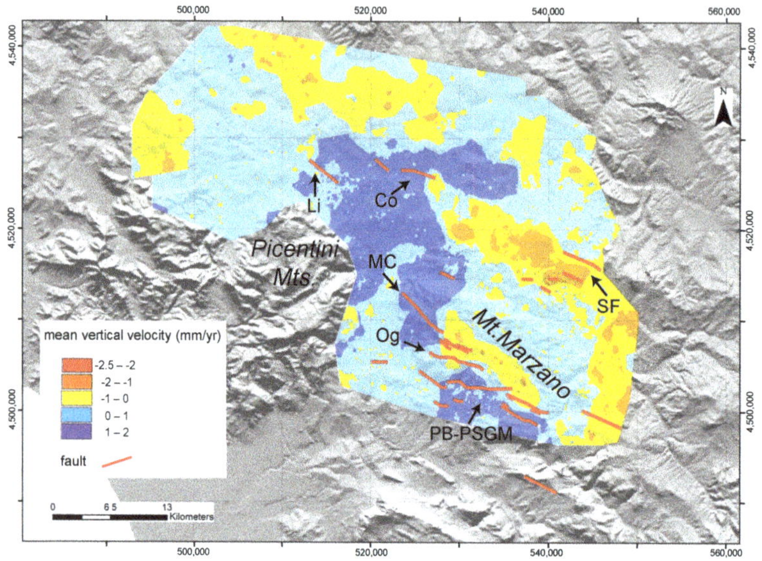

Figure 10. ERS data: Vertical mean velocity/ground deformation map 1992–2000, IDW interpolation (cell size 50 × 50 m). Data statistical selection with the *Cluster and Outlier Analysis*, i.e., map constructed using the ERS subsets labelled 'CO_2k_IDW_nout' (which includes only PS classified as HH and LL). Fault/fault array labelling as in Figure 2b.

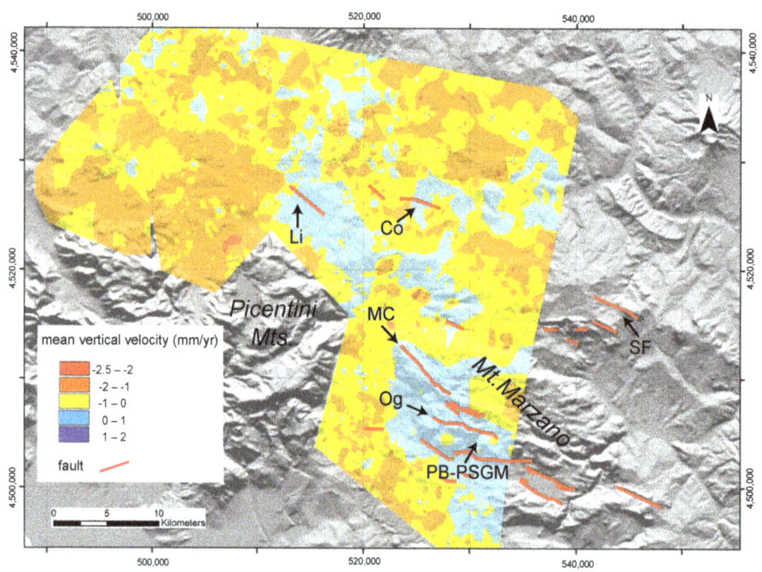

Figure 11. ENVISAT data: Vertical mean velocity deformation map 2003–2010, IDW interpolation (cell 50 × 50 m). Data statistical selection with *Cluster and Outlier Analysis*, i.e., map constructed using the ENVISAT subsets labelled 'CO_2k_IDW_nout' (which includes only PS classified as HH and LL). Fault/fault array labelling as in Figure 2b.

8. Discussion of Postseismic Deformation

Two types of processing were used in this study. One technique (i.e., IDW analysis of 'native' PS datasets) requires few steps of analysis, whereas the other one (based on the *Cluster and Outlier Analysis*) involves a more complex sequence of steps. The maps obtained by means of these two types of processing are quite similar in terms of spatial distribution of vertical (up/down) motion orientations (Figures 8–11).

Both ERS maps show that the area of interest, is for the most, part slightly uplifted (0–1 mm/y mean velocity class). However, the ERS vertical mean velocity deformation map for the timespan of 1992–2000, obtained with the *Cluster and Outlier Analysis* (CO map of Figure 10), shows the presence of areas characterized by slightly higher positive velocity (1–2 mm/y mean velocity class) in the 1992–2000 timespan more impressively than the map of Figure 8. These areas are located in the areas corresponding to the surface projections of the footwall blocks to the major blind seismogenic faults identified as (i) Central Fault (CF) (particularly, the northern and southern parts of such block) and (ii) NE Boundary Fault (NEBF) [8]. The CF and NEBF correspond to the seismogenic structures activated with the main shock (0 s) and 40 s shock of the 1980 Irpinia earthquake [1], respectively. Furthermore, the central part of the sector bounded by the two structures described above, which has been identified as a major graben structure based on both seismological and levelling data [1,96], is also characterized by similar, slightly higher positive velocity.

In addition, two elongated bands, NW-SE-oriented and characterized by slight subsidence (0 to −1 mm/y or −1 to −2 mm/y mean velocity classes), are identified in the surface projection of hanging wall blocks of the CF and NEBF. These bands form narrow (circa 4–5 km wide) belts following the surface projection of the CF and NEBF major structures.

In the maps derived by ENVISAT PS datasets, the area of interest is, for the most part, subject to subsidence (the most represented is the 0 to −1 mm/y mean velocity class), but areas that in the ERS-based maps show uplift are still visible as slightly uplifting.

In the maps derived from the 'native' datasets (Figures 8 and 9), localized orange or blue spots that are widespread in almost the entire analyzed area can be noticed. Comparison of spatial distribution of orange/blue spots with surface geology features shown in the schematic map of Figure 2a suggests that the spots are most probably related to gravitational phenomena, which affect the shale/clay-dominated formations (e.g., internal units and Pliocene wedge-top basin deposits) that have been cropped out in most of the investigated region.

The tectonic setting of the 1980 Irpinia earthquake region is shown in Figure 2. As already mentioned previously, the Irpinia earthquake was characterized by a complex source mechanism, associated with at least three normal faulting ruptures on distinct fault segments [1] involving the activation of both the Central Fault and the NE Boundary Fault shown in Figure 2b.

Like Peltzer et al. [115], we analyzed intermediate- and near-field postseismic surface displacements following the Irpinia 1980 earthquake using processed ERS 1/2 and ENVISAT SAR data with PS-InSAR technique, covering time intervals between 1992 and 2010. The PSs interpolated maps revealed transient displacement patterns that were either not observed or only partially captured by other geodetic techniques. In particular, the PSs interpolated maps depict vertical displacements of the ground surface. Analysis of the range change maps, which cover an 18-year timespan, started only 12 years after the 1980 earthquake.

Both the ERS and ENVISAT vertical mean velocity deformation maps (Figures 8–12) show that the surface projection of the footwall block to the northeast-dipping CF has uplifted over the analyzed timespan. Such an uplift is more marked during the timespan covered by the ERS satellites (mean velocity of about 1–2 mm/y) than in the 2003–2010 timespan (Figures 12 and 13) On the other hand, the surface projection of the hanging wall block of the CF, near field of the fault itself, is characterized by slow subsidence (in the −1 to 0 mm/y mean velocity range; Figure 12). Likewise, the surface projection of the footwall block of southwest-dipping NEBF has uplifted, more markedly (mean velocity of about 1–2 mm/y) during the timespan covered by the ERS surveys, while the hanging wall block is characterized by slow subsidence (Figures 12 and 13). Uplift, more evident in the timespan covered by the ERS surveys, characterizes the sector located at the northwestern termination of the surface projection of the CF (Figure 12).

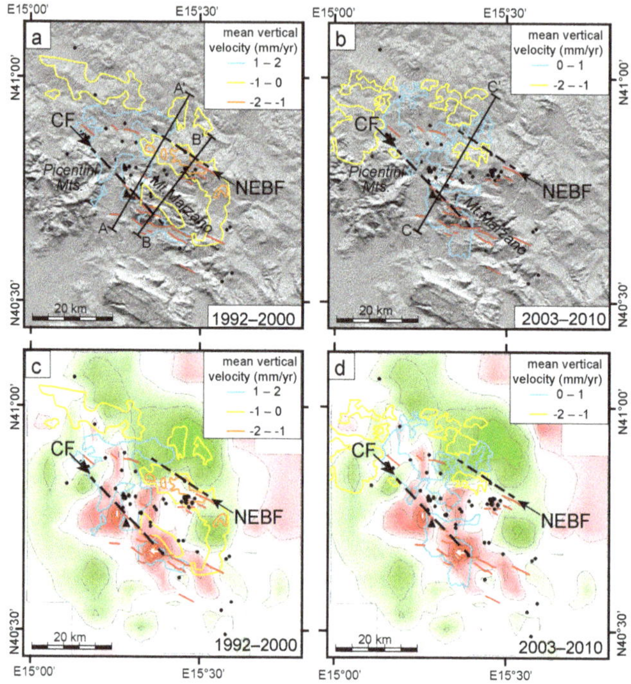

Figure 12. Multitemporal ground deformation patterns plotted against the pattern of active faults at the surface (red segments, from Ascione et al. [7]) and the projections at the surface of the deep-seated CF and NEBF (dashed black lines, from Amoroso et al. [8]; see Figure 2), which correspond to the structures activated with the 1980 Irpinia earthquake at 0 s and 40 s, respectively. The ground deformation patterns result from analyses of the ERS and ENVISAT datasets selected by the *Cluster and Outlier Analysis*. (**a**,**c**) Ground deformation in the 1992–2000 timespan. (**b**,**d**) Ground deformation in the 2003–2010 timespan. The map in the background of diagrams (**c**,**d**) corresponds to the spatial distribution of the V_P/V_S ratio in a horizontal slice at 6 km depth through the P-wave tomographic model, with micro-earthquakes (black dots) recorded from 2005 to 2011, as in Figure 7 (from Amoroso et al. [8], modified). Traces of ground deformation profiles of Figure 13 are also shown in diagrams (**a**,**b**).

Subsidence of the surface projection of the hanging wall to the NEBF, more pronounced in the 1992–2000 timespan along the southeastern profile (Figure 13), suggests that postseismic afterslip occurred along this major seismogenic fault segment at depth. Surface deformation around the CF shows a more articulated pattern. The southeastern portion of the surface projection of this fault is characterized by a pattern consistent with postseismic afterslip (footwall uplift and hanging wall subsidence; Figure 12a,b; Figure 13). However, to the NW, uplift also characterizes the surface projection of the hanging wall block of the CF. Such a behavior has no straightforward interpretation. As high VP/VS values occur at depth in the region of 'anomalous' uplift in the CF hanging wall (Figure 12c,d), a correlation of subdued, positive ground deformation with fluid accumulation at depth may be hypothesized.

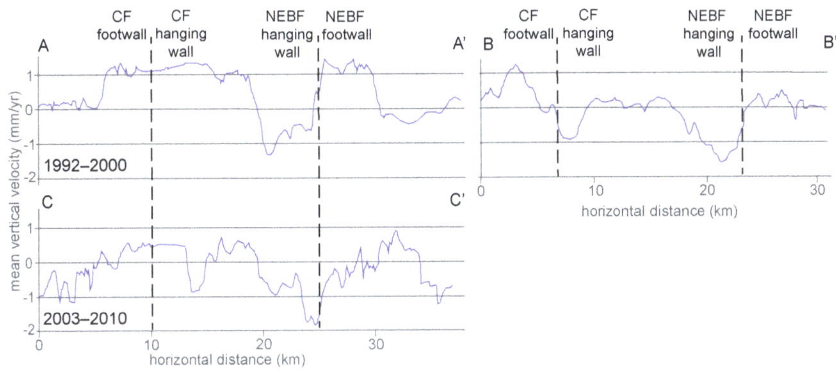

Figure 13. Postseismic ground deformation profiles across the 1980 Irpinia earthquake epicentral area (see Figure 12a for locations of profiles A–A' and B–B' and Figure 12b for location of profile C–C'). The dashed black lines indicate the locations of the projections at the surface of the deep-seated CF and NEBF, which correspond to the structures activated with the 1980 Irpinia earthquake at 0 s and 40 s, respectively.

9. Concluding Remarks

The analysis of PS-InSAR data from the region struck by the strongest Italian earthquake of the last century, the MS 6.9 1980 Irpinia earthquake, has shown that ground deformation has affected this area in the last decades. The PS-InSAR data analysis covers a timespan ranging from 12 to 30 years after the earthquake. The analysis showed that cumulative deformation is consistent with coseismic deformation inferred from both seismological data (rupture mechanisms of the three main shocks which occurred in a 40 s timespan) [1], levelling data [1,91] and coseismic surface faulting [10,12,18,19]. In addition, it is consistent with evidence of Late Quaternary active faults at the surface (e.g., [7,22]). In particular, evidence for continuing uplift of the footwall—and subsidence of the hanging wall blocks—of the two major faults activated with the 23th November earthquakes has been identified.

The results of PS-InSAR data show that postseismic deformation was still occurring 30 years after the earthquake. The slight decrease in the uplift rate from the timespan surveyed by the ERS satellite to that covered by the ENVISAT could represent the effect of decay in the time fault of creep/low energy earthquakes over decades after the earthquake and of the consequent decrease in postseismic deformation. Within this framework, we considered the ERS and ENVISAT datasets, both of which are high quality and perfectly consistent in terms of long-term trends. Therefore, the observed decay of postseismic deformation was interpreted as real and not induced by the different satellites. In addition, the PS-InSAR data analysis also showed that the region in the mid part between the two main structures activated with the 1980 earthquakes is currently affected by slow uplift. Based on spatial superposition of this region with the rock volume (extending from ~4 km depth downward) that has been identified as saturated of fluids, such uplift may be interpreted as the response, at the surface, to the accumulation of fluid (particularly CO_2, based on the widespread evidence discussed in this review paper) at depth. Comparison between the ERS- and ENVISAT-based maps showed that uplift has decreased in the analyzed 20-year time window. However, it is not possible to assess whether such a decrease represents effective diminishing uplift—and possibly decreasing fluid input at depth—or only a stage in a fluctuating long-term process. Previous studies on postseismic deformation elsewhere in the world have suggested that a progressive decay of surface deformation over a 20- to 30-year timespan is compatible with the postseismic afterslip process (e.g., [104]). On the other hand, recent results on high-resolution analysis of microseismicity suggest a multiyear cycle behavior related to CO_2-brine accumulation and sealing within a reservoir at hypocentral depths beneath the Irpinia region [116]. According to these studies, when the pressure in the deep CO_2-brine reservoir increases, the seismicity tends to progressively be distributed over fault zones. The process culminates with Mw ≥ 3.5 earthquakes, which appears to lead to the creation efficient fracture porosity and related permeability, allowing the fluids to escape and migrate toward the surface. The sequences are then followed by a slow decrease in the stress level until a new stress loading cycle starts. These cycles of stress loading related to fluid accumulation in the subsurface [116] clearly point out that there may be further mechanisms besides postseismic afterslip which is able to account for at least some of the ground motions detected by our InSAR analysis.

The 1980 Irpinia earthquake provides a fundamental lesson on earthquake faulting in the Apennines. Fault segmentation occurs at various scales starting from the seismogenic source. The occurrence of at least three major subevents at 0 s, 20 s, and 40 s implies the activation of multiple deep portions/segments of the crustal-scale seismogenic structure. The strongly anisotropic structure of the allochthonous cover, including various thrust sheets overlying a shale-dominated and fluid-saturated mélange zone resting on top of a more rigid substratum represented by the Apulian Platform, produced an even more complex network of shallow, relatively small (<10 km long) fault strands. This, in turn, resulted in a rather discontinuous and apparently chaotic pattern of surface ruptures, as discussed in this review paper. Within this framework, results from paleoseismological studies obtained from a single fault strand should be used, taking into account that earthquake faulting in the Apennines is clearly a three-dimensional process involving crustal volumes rather than just two-dimensional fault planes. As earthquake-related slip is variably and complexly distributed among various surface fault strands, recurrence intervals and slip rates obtained from a particular fault segment may not be representative of the activity of the seismogenic structure controlling seismicity and the characteristic earthquake for that region. This makes the 'hunting' for the faults that produced historical earthquakes particularly challenging.

Supplementary Materials: The following are available online at http://www.mdpi.com/2076-3263/10/12/493/s1, Figure S1: Maps showing the original PS datasets, Figure S2: Spatial selection of PS data, Figure S3: LoS-oriented ascending and descending mean velocity maps in the 1992–2000 time span, Figure S4: LoS-oriented ascending and descending mean velocity maps in the 2003–2010 time span, Figure S5: Comparison between 'PS normal subsets' and 'native' datasets, Figure S6: The three ERS '*marzano*' subsets, Figure S7: The four ENVISAT "Marzano" subsets, Figure S8: Comparison of the results obtained applying the Cluster and Outlier Analysis vs. the Hot Spot Analysis to the ERS PS data subsets, Figure S9: Comparison of the results obtained applying the Cluster and Outlier Analysis vs. the Hot Spot Analysis to the ENVISAT PS data subsets, Figure S10: ERS_T494_F2781_CL001_POTENZA_NORM_D_marzano_CO_2k_IDW and mean velocity map of PSs classified as "not significant", Figure S11: PST2009_ENVISAT_T265_F2781_CL001_AVELLINO_DESCE_NORM_marzano_CO_2k_IDW_nsig and mean velocity map of the PSs classified as "not significant". Table S1: Comparison between the number of PS of the original datasets and those of the subsets obtained through the outliers exclusion process to perform statistical normal distributions of PSs, Table S2: Number of ERS PSs in the subsets deriving from the selection with Cluster and Outlier Analysis mapping tool, Table S3: Number of ENVISAT PSs in the subsets deriving from the selection with Cluster and Outlier Analysis, Table S4, ERS PSs selection with Cluster and Outlier Analysis, Table S5: ENVISAT PSs selection with Cluster and Outlier Analysis.

Author Contributions: Conceptualization, A.A. and S.M.; methodology, A.A. and S.N.; software, S.N.; validation, A.A., S.N. and S.M.; formal analysis, S.N.; investigation, A.A., S.N. and S.M.; resources, A.A., S.N. and S.M.; data curation, A.A. and S.N.; writing—original draft preparation, A.A., S.N. and S.M.; writing—review and editing, A.A. and S.M.; visualization, A.A. and S.N.; supervision, A.A. and S.M.; project administration, A.A. and S.M.; funding acquisition, A.A. and S.M. All authors have read and agreed to the published version of the manuscript.

Funding: This research received no external funding.

Acknowledgments: We wish to thank Aldo Zollo, Gaetano Festa, Matteo Picozzi and Antonio Emolo for fruitful discussions. Thoughtful and constructive reviews by two Anonymous Reviewers are gratefully acknowledged.

Conflicts of Interest: The authors declare no conflict of interest.

References

1. Bernard, P.; Zollo, A. The Irpinia (Italy) 1980 earthquake: Detailed analysis of a complex normal faulting. *J. Geophys. Res.* **1989**, *94*, 1631–1647. [CrossRef]
2. Frepoli, A.; Maggi, C.; Cimini, G.B.; Marchetti, A.; Chiappini, M. Seismotectonic of Southern Apennines from recent passive seismic experiments. *J. Geodyn.* **2011**, *51*, 110–124. [CrossRef]

3. Westway, R.; Jackson, J. The earthquake of the 1980 November 23 in Campania-Basilicata (southern Italy). *Geophys. J. Int.* **1987**, *90*, 375–443. [CrossRef]
4. De Matteis, R.; Matrullo, E.; Rivera, L.; Stabile, T.A.; Pasquale, G.; Zollo, A. Fault Delineation and Regional Stress Direction from the Analysis of Background Microseismicity in the southern Apennines, Italy. *Bull. Seismol. Soc. Am.* **2012**, *102*, 1899–1907. [CrossRef]
5. Matrullo, E.; De Matteis, R.; Satriano, C.; Amoroso, O.; Zollo, A. An improved 1-D seismic velocity model for seismological studies in the Campania–Lucania region (Southern Italy). *Geophys. J. Int.* **2013**, *195*, 460–473. [CrossRef]
6. De Landro, G.; Amoroso, O.; Stabile, T.A.; Matrullo, E.; Lomax, A.; Zollo, A. High-precision differential earthquake location in 3-D models: Evidence for a rheological barrier controlling the microseismicity at the Irpinia fault zone in southern Apennines. *Geophys. J. Int.* **2015**, *203*, 1821–1831. [CrossRef]
7. Ascione, A.; Mazzoli, S.; Petrosino, P.; Valente, E. A decoupled kinematic model for active normal faults: Insights from the 1980, MS = 6.9 Irpinia earthquake, southern Italy. *Geol. Soc. Am. Bull.* **2013**, *125*, 1239–1259. [CrossRef]
8. Amoroso, O.; Ascione, A.; Mazzoli, S.; Virieux, J.; Zollo, A. Seismic imaging of a fluid storage in the actively extending Apennine mountain belt, southern Italy. *Geophys. Res. Lett.* **2014**, *41*, 3802–3809. [CrossRef]
9. Burbank, D.W.; Anderson, R.S. *Tectonic Geomorphology*, 2nd ed.; Wiley-Blackwell: Hoboken, NJ, USA, 2001.
10. Westaway, R.; Jackson, J. Surface faulting in the southern Italian Campania-Basilicata earthquake of 23 November 1980. *Nature* **1984**, *312*, 436–438. [CrossRef]
11. Porfido, S.; Esposito, E.; Luongo, G.; Maturano, A. I terremoti del XIX secolo dell'Appennino Campano-Lucano. *Memorie della Società Geologica Italiana* **1988**, *41*, 1105–1116.
12. Pantosti, D.; Valensise, G. Faulting mechanism and complexity of the November 23, 1980, Campania-Lucania Earthquake, inferred from surface observations. *J. Geophys. Res.* **1990**, *95*, 15319–15341. [CrossRef]
13. Pantosti, D.; Schwartz, D.P.; Valensise, G. Paleoseismology along the 1980 surface rupture of the Irpinia fault: Implications for earthquake recurrence in the Southern Apennines, Italy. *J. Geophys. Res.* **1993**, *98*, 6561–6577. [CrossRef]
14. Pofido, S.; Esposito, E.; Michetti, A.M.; Blumetti, AM.; Vittori, E.; Tranfaglia, G.; Guerrieri, L.; Ferreli,: L.; Serva, L. Areal distribution of ground effects induced by strong earthquakes in the Southern Apennines (Italy). *Surv. Geophys.* **2002**, *23*, 529–562. [CrossRef]
15. Serva, L.; Esposito, E.; Guerrieri, L.; Porfido, S.; Vittori, E.; Comerci, V. Environmental effects from five historical earthquakes in southern Apennines (Italy) and macroseismic intensity assessment: Contribution to INQUA EEE Scale Project. *Quat. Int.* **2007**, *173–174*, 30–44. [CrossRef]
16. INGV-Database Macrosismico Italiano. 2011. Available online: https://emidius.mi.ingv.it/DBMI11/ (accessed on 29 October 2020).
17. Istituto Nazionale di Geofisica e Vulcanologia Catalogo Parametrico dei Terremoti Italiani. Available online: https://emidius.mi.ingv.it/CPTI/ (accessed on 29 October 2020).
18. Cinque, A.; Lambiase, S.; Sgrosso, I. Su due faglie nell'alta valle del Sele legate al terremoto del 23.11.1980. *Rend. Della Soc. Geol. Ital.* **1981**, *4*, 127–129.
19. Bollettinari, G.; Panizza, M. Una 'faglia di superficie' presso San Gregorio Magno in occasione del sisma del 23-XI-1980. *Rend. Della Soc. Geol. Ital.* **1981**, *4*, 135–136.
20. Papanikolaou, I.D.; Roberts, G.P. Geometry, kinematics and deformation rates along the active normal fault system in the southern Apennines: Implications for fault growth. *J. Struct. Geol.* **2007**, *29*, 166–188. [CrossRef]
21. Cowie, P.A.; Roberts, G.P.; Bull, J.M.; Visini, F. Relationships between fault geometry, slip rate variability and earthquake recurrence in extensional settings. *Geophys. J. Int.* **2012**, *189*, 143–160. [CrossRef]
22. Galli, P.; Peronace, E. New paleoseismic data from the Irpinia Fault. A different seismogenic perspective for southern Apennines (Italy). *Earth-Sci. Rev.* **2014**, *136*, 175–201. [CrossRef]
23. Ciotoli, G.; Bigi, S.; Tartarello, C.; Sacco, P.; Lombardi, S.; Ascione, A.; Mazzoli, S. Soil gas distribution in the main coseismic surface rupture zone of the 1980, Ms = 6.9, Irpinia Earthquake (southern Italy). *J. Geophys. Res. Solid Earth* **2014**, *119*, 2440–2461. [CrossRef]
24. Eremeev, A.E. Use of radial compression of thick-walled tube for measuring high pressures. *Meas. Tech.* **1973**, *16*, 860–862. [CrossRef]
25. Durrance, E.M.; Gregory, R.G. *Fracture Mapping in Clays: Soil Gas Surveys at Down Ampney, Gloucestershire*; DOE Report: DOE/RW/88081; Department of Energy: Washington, DC, USA, 1988.

26. Duddridge, G.A.; Grainger, P.; Durrance, E.M. Fault detection using soil gas geochemistry. *Q. J. Eng. Geol. Hydrogeol.* **1991**, *24*, 427–435. [CrossRef]
27. Hickman, S.R.; Sibson, R. and Bruhn, Introduction to special section: Mechanical involvement of fluids in faulting. *J. Geophys. Res.* **1995**, *100*, 12831–12840. [CrossRef]
28. Lombardi, S.; Etiope, G. *The Refinement of Soil Gas Analysis as a Geological Investigative Technique*; Nuclear Science and Technology Series Report EUR 16929: 200; European Commission: Luxembourg, 1996.
29. Ciotoli, G.; Guerra, M.; Lombardi, S.; Vittori, E. Soil gas survey for tracing seismogenic faults: A case study in the Fucino basin, Central Italy. *J. Geophys. Res.* **1998**, *103*, 23781–23794. [CrossRef]
30. Ciotoli, G.; Etiope, G.; Guerra, M.; Lombardi, S. The detection of concealed faults in the Ofanto Basin using the correlation between soil-gas fracture surveys. *Tectonophysics* **1999**, *301*, 321–332. [CrossRef]
31. De Gregorio, S.; Diliberto, I.S.; Giammanco, S.; Gurrieri, S.; Valenza, M. Tectonic control over large-scale diffuse degassing in eastern Sicily (Italy). *Geofluids* **2002**, *2*, 273–284. [CrossRef]
32. Braubon, J.C.; Rigo, A.; Toutain, J.P. Soil gas profiles as a tool to characterise active tectonic areas: The Jaut Pass example (Pyrenees, France). *Earth Planet. Sci. Lett.* **2002**, *196*, 69–81. [CrossRef]
33. Annunziatellis, A.; Ciotoli, G.; Lombardi, S.; Nolasco, F. Short- and long-term gas hazard: The release of toxic gases in the Alban Hills volcanic area (central Italy). *J. Geochem. Explor.* **2003**, *77*, 93–108. [CrossRef]
34. Lewicki, J.L.; Connor, C.; St-Amand, K.; Stix, J.; Spinner, W. Self-potential, soil CO2 flux, and temperature on Masaya volcano, Nicaragua. *Geophys. Res. Lett.* **2003**, *30*, 15. [CrossRef]
35. Chiodini, G.; Cardellini, C.; Caliro, S.; Avino, R.; Donnini, M.; Granieri, D.; Morgantini, N.; Sorrenti, D.; Frondini, F. The hydrothermal system of Bagni San Filippo (Italy): Fluids circulation and CO2 degassing. *Ital. J. Geosci.* **2020**, *139*, 383–397. [CrossRef]
36. Fu, C.C.; Yang, T.F.; Walia, V.; Cheng, C.-H. Reconnaissance of soil gas composition over the buried fault and fracture zone in southern Taiwan. *Geochem. J.* **2005**, *39*, 427–439. [CrossRef]
37. Walia, V.; Yang, T.F.; Lin, S.J.; Kumar, A.; Fu, C.C.; Chiu, J.M.; Chang, H.H.; Wen, K.L.; Chen, C.H. Temporal variation of soil gas compositions for earthquake surveillance in Taiwan. *Radiat. Meas.* **2013**, *50*, 154–159. [CrossRef]
38. Ciotoli, G.; Lombardi, S.; Annunziatellis, A. Geostatistical analysis of soil gas data in a high seismic intermontane basin: Fucino Plain, central Italy. *J. Geophys. Res. Solid Earth* **2007**, *112*, B5. [CrossRef]
39. Toutain, J.-P.; Baubron, J.C. Gas geochemistry and seismotectonics: A review. *Tectonophysics* **1999**, *304*, 1–27. [CrossRef]
40. Annunziatellis, A.; Beaubien, S.E.; Bigi, S.; Ciotoli, G.; Coltella, M.; Lombardi, S. Gas migration along fault systems and through the vadose zone in the Latera caldera (central Italy): Implications for CO_2 geological storage. *Int. J. Greenh. Gas Control* **2008**, *2*, 353–372. [CrossRef]
41. Miller, R.L.; Tegen, I.; Perlwitz, J.P. Surface radiative forcing by soil dust aerosols and the hydrologic cycle. *J. Geophys. Res.* **2004**, *109*, D04203. [CrossRef]
42. Chiodini, G.; Cardellini, C.; Amato, A.; Boschi, E.; Caliro, S.; Frondini, F.; Ventura, G. Carbon dioxide Earth degassing and seismogenesis in central and southern Italy. *Geophys. Res. Lett.* **2004**, *31*. [CrossRef]
43. Inversi, B.; Scrocca, D.; Montegrossi, G.; Livani, M.; Petracchini, L.; Brandano, M.; Brilli, M.; Giustini, F.; Recanati, R.; Gola, G.; et al. 3D geological modelling of a fractured carbonate reservoir for the study of medium enthalpy geothermal resource in the Southern Apennines (Campania Region, Italy). In Proceedings of the European Geothermal Congress 2013, Pisa, Italy, 3–7 June 2013.
44. Zollo, A.; Orefice, A.; Convertito, V. Source parameter scaling and radiation efficiency of microearthquakes along the Irpinia fault zone in southern Apennines, Italy. *J. Geophys. Res. Solid Earth* **2014**, *119*, 3256–3275. [CrossRef]
45. Dvorkin, J.; Mavko, G.; Nur, A. Overpressure detection from compressional and shear wave data. *Geophys. Res. Lett.* **1999**, *26*, 3417–3420. [CrossRef]
46. Vanorio, T.; Virieux, J.; Capuano, P.; Russo, G. Three-dimensional seismic tomography from P wave and S wave microearthquake travel times and rock physics characterization of the Campi Flegrei Caldera. *J. Geophys. Res. Solid Earth* **2005**, *110*. [CrossRef]
47. Hantschel, T.; Kauerauf, A.I. *Foundamentals of Basin and Petroleum Systems Modeling*; Springer: Berlin, Germany, 2009.

48. Amoroso, A.; Russo, G.; De Landro, G.; Zollo, A.; Garambois, S.; Mazzoli, S.; Parente, M.; Virieux, J. From Velocity and Attenuation Tomographies to Rock Physical Modeling: Inferences on fluid-driven earthquake processes at the Irpinia fault system in Southern Italy. *Geophys. Res. Lett.* **2017**, *44*, 6752–6760. [CrossRef]
49. Stabile, T.A.; Amoroso, O.; De Matteis, R.; Maercklin, N.; Matrullo, E.; Orefice, A.; Pasquale, G.; Satriano, C.; Zollo, A. Anatomy of a microearthquake sequence on an active normal fault. *Sci. Rep.* **2012**, *2*, 1–7. [CrossRef] [PubMed]
50. Mazzoli, S.; Helman, M. Neogene Patterns of Relative Plate Motion for Africa-Europe: Some Implications for Recent Central Mediterranean Tectonics. In *Active Continental Margins—Present and Past*; Springer: Berlin/Heidelberg, Germany, 1994.
51. Turco, E.; Macchiavelli, C.; Mazzoli, S.; Schettino, A.; Pierantoni, P.P. Kinematic evolution of Alpine Corsica in the framework of Mediterranean mountain belts. *Tectonophysics* **2012**, *579*, 193–206. [CrossRef]
52. Ciarcia, S.; Mazzoli, S.; Vitale, S.; Zattin, M. On the tectonic evolution of the Ligurian accretionary complex in southern Italy. *Geol. Soc. Am. Bull.* **2012**, *124*, 463–483. [CrossRef]
53. Vitale, S.; Ciarcia, S. Tectono-stratigraphic and kinematic evolution of the southern Apennines/Calabria–Peloritani Terrane system (Italy). *Tectonophysics* **2013**, *583*, 164–182. [CrossRef]
54. Mazzoli, S.; Szaniawski, R.; Mittiga, F.; Ascione, A.; Capalbo, A. Tectonic evolution of Pliocene–Pleistocene wedge-top basins of the southern Apennines: New constraints from magnetic fabric analysis. *Can. J. Earth Sci.* **2012**, *49*, 492–509. [CrossRef]
55. Mazzoli, S.; D'Errico, M.; Aldega, L.; Corrado, S.; Invernizzi, C.; Shiner, P.; Zattin, M. Tectonic burial and 'young' (<10 Ma) exhumation in the southern Apennines fold-and-thrust belt (Italy). *Geology* **2008**, *36*, 243–246. [CrossRef]
56. Mazzoli, S.; Ascione, A.; Buscher, J.T.; Pignolasa, A.; Valente, E.; Zattin, M. Low-angle normal faulting and focused exhumation associated with late Pliocene change in tectonic style in the southern Apennines (Italy). *Tectonics* **2014**, *33*, 1802–1818. [CrossRef]
57. Shiner, P.; Beccacini, A.; Mazzoli, S. Thin-skinned versus thick- skinned structural models for Apulian carbonate reservoirs: Constraints from the Val d'Agri Fields, S Apennines, Italy. *Mar. Pet. Geol.* **2004**, *21*, 805–827. [CrossRef]
58. Mazzoli, S.; Barkham, S.; Cello, G.; Gambini, R.; Mattioni, L.; Shiner, P.; Tondi, E. Reconstruction of continental margin architecture deformed by the contraction of the Lagonegro Basin, southern Apennines, Italy. *J. Geol. Soc. Lond.* **2001**, *158*, 309–319. [CrossRef]
59. Improta, L.; Corciulo, M. Controlled source non-linear tomography: A powerful tool to constrain tectonic models of the southern Apennines orogenic wedge, Italy. *Geology* **2006**, *34*, 941–944. [CrossRef]
60. Speranza, F.; Chiappini, M. Thick-skinned tectonics in the external Apennines, Italy: New evidence from magnetic anomaly analysis. *J. Geophys. Res. Solid Earth* **2002**, *107*, B11. [CrossRef]
61. Steckler, M.S.; Akhter, H.; Seeber, L. Collision of the Ganges–Brahmaputra Delta with the Burma Arc: Implications for earthquake hazard. *Earth Planet. Sci. Lett.* **2008**, *273*, 367–378. [CrossRef]
62. Patacca, E.; Scandone, P. Late thrust propagation and sedimentary response in the thrust-belt—foredeep system of the Southern Apennines (Pliocene-Pleistocene). In *Anatomy of an Orogen: The Apennines and Adjacent Mediterranean Basins*; Vai, G.B., Martini, I.P., Eds.; Springer: Dordrecht, The Netherlands, 2001. [CrossRef]
63. Butler, R.W.H.; Mazzoli, S.; Corrado, S.; De Donatis, M.; Di Bucci, D.; Gambini, R.; Naso, G.; Nicolai, C.; Scrocca, D.; Shiner, P.; et al. Applying thick-skinned tectonic models to the Apennine thrust belt of Italy—Limitations and implications. *AAPG Mem.* **2004**, *82*, 647–667.
64. Cello, G.; Guerra, I.; Tortorici, L.; Turco, E.; Scarpa, R. Geometry of the neotectonic stress field in southern Italy: Geological and seismological evidence. *J. Struct. Geol.* **1982**, *4*, 385–393. [CrossRef]
65. Caiazzo, C.; Ascione, A.; Cinque, A. Late Tertiary–Quaternary tectonics of the Southern Apennines (Italy): New evidences from the Tyrrhenian slope. *Tectonophysics* **2006**, *421*, 23–51. [CrossRef]
66. Ascione, A.; Caiazzo, C.; Cinque, A. Recent faulting in Southern Apennines (Italy): Geomorphic evidence, spatial distribution and implications for rates of activity. *Boll. Della Soc. Geol. Ital.* **2007**, *126*, 293–305.
67. ISPRA. *Foglio 468—Eboli della Carta Geologica d'Italia alla scala 1:50.000*; ISPRA: Roma, Italy, 2014. Available online: https//www.ispraambiente.gov.it/Media/carg/448_ERCOLANO/Foglio.html (accessed on 30 October 2020).
68. Patacca, E.; Scandone, P. Geology of the southern Apennines. *Boll. Della Soc. Geol. Ital.* **2007**, *7*, 75–119.
69. ViDEPI Project. Ministero dello Sviluppo Economico 2010. Available online: https://unmig.mise.gov.it/index.php/it/ (accessed on 29 October 2020).

70. Ciarcia, S.; Vitale, S.; Di Stasio, A.; Iannace, A.; Mazzoli, S.; Torre, M. Stratigraphy and tectonics of an Internal Unit of the southern Apennines: Implications for the geodynamic evolution of the peri-Tyrrhenian mountain belt. *Terra Nova* **2009**, *21*, 88–96. [CrossRef]
71. Matano, F.; Nocera, S.D.; Criniti, S.; Critelli, S. Geology of the Epicentral Area of the November 23, 1980 Earthquake (Irpinia, Italy): New Stratigraphical, Structural and Petrological Constrains. *Geosciences* **2020**, *10*, 247. [CrossRef]
72. Amato, A.; Cinque, A.; Santangelo, N.; Santo, A. Il bordo meridionale del massiccio del Monte Marzano e la valle del Fiume Bianco: Geologia e geomorfologia. *Studi Geol. Camerti* **1992**, *1992*, 191–200.
73. Ascione, A.; Ciarcia, S.; Di Donato, V.; Mazzoli, S.; Vitale, S. The Pliocene-Quaternary wedge-top basins of southern Italy: An expression of propagating lateral slab tear beneath the Apennines. *Basin Res.* **2012**, *24*, 456–474. [CrossRef]
74. Improta, L.; Bonagura, M.; Capuano, P.; Iannaccone, G. An integrated geophysical investigation of the upper crust in the epicentral area of the 1980, Ms= 6.9, Irpinia earthquake (Southern Italy). *Tectonophysics* **2003**, *361*, 139–169. [CrossRef]
75. Cippitelli, G. The CROP-04 seismic profile: Interpretation and structural setting of the Agropoli-Barletta Geotraverse, in Results of the CROP Project, Sub-Project CROP-04 Southern Apennines (Italy). *Ital. J. Geosci.* **2007**, *7*, 267–281.
76. Mazzotti, A.; Stucchi, E.; Fradelizio, G.L.; Zanzi, L.; Scandone, P. Re-processing of the CROP-04 seismic data. *Ital. J. Geosci.* **2007**, *7*, 141–153.
77. Ascione, A.; Cinque, A.; Tozzi, M. La valle del Tanagro (Campania): Una depressione strutturale ad evoluzione complessa. *Studi Geol. Camerti* **1992**, *1992*, 209–219.
78. Ascione, A.; Cinque, A.; Improta, L.; Villani, F. Late Quaternary faulting within the Southern Apennines seismic belt: New data from Mt. Marzano area (Southern Italy). *Quat. Int.* **2003**, *101–102*, 27–41. [CrossRef]
79. Aiello, G.; Ascione, A.; Barra, D.; Munno, R.; Petrosino, P.; Russo Ermolli, E.; Villani, F. Evolution of the late Quaternary San Gregorio Magno tectono-karstic basin (southern Italy) inferred from geomorphological, tephrostratigraphical and palaeoecological analyses: Tectonic implications. *J. Quat. Sci.* **2007**, *22*, 233–245. [CrossRef]
80. Amicucci, L.; Barchi, M.R.; Montone, P.; Rubiliani, N. The Vallo di Diano and Auletta extensional basins in the southern Apennines (Italy): A simple model for a complex setting. *Terra Nova* **2008**, *20*, 475–482. [CrossRef]
81. Galli, P.; Bosi, V.; Piscitelli, S.; Giocoli, A.; Scionti, V. Late Holocene earthquakes in southern Apennine: Paleoseismology of the Caggiano fault. *Int. J. Earth Sci.* **2006**, *95*, 855–870. [CrossRef]
82. Troiano, A.; Di Giuseppe, M.G.; Petrillo, Z.; Patella, D. Imaging 2D structures by the CSAMT method: Application to the Pantano di S. Gregorio Magno faulted basin (Southern Italy). *J. Geophys. Eng.* **2009**, *6*, 120–130. [CrossRef]
83. Galli, P.A.C.; Giocoli, A.; Peronace, E.; Piscitelli, S.; Quadrio, B.; Bellanova, J. Integrated near surface geophysics across the active Mount Marzano Fault System (Southern Italy): Seismogenic hints. *Int. J. Earth Sci.* **2014**. [CrossRef]
84. Pondrelli, S.; Salimbeni, S.; Ekström, G.; Morelli, A.; Gasperini, P.; Vannucci, G. The Italian CMT dataset from 1977 to the present. *Phys. Earth Planet. Inter.* **2006**, *159*, 286–303. [CrossRef]
85. Cocco, M.; Chiarabba, C.; Di Bona, M.; Selvaggi, G.; Margheriti, L.; Frepoli, A.; Lucente, F.P.; Basili, A.; Jongmans, D.; Campillo, M. The April 1996 Irpinia seismic sequence: Evidence for fault interaction. *J. Seismol.* **1999**, *3*, 105–117. [CrossRef]
86. De Matteis, R.; Romeo, A.; Pasquale, G.; Iannaccone, G.; Zollo, A. 3D Tomographic imaging of the southern Apennines (Italy): A statistical approach to estimate the model uncertainty and resolution. *Stud. Geophys. Geod.* **2010**, *54*, 367–387. [CrossRef]
87. Irpinia Seismic Network. Seismic Sequence Rocca San Felice, South Apennines, Italy 04-06/07/2020. Available online: http://isnet.unina.it/wp-content/uploads/2020/07/Sequence_July20_fullReport.pdf (accessed on 29 October 2020).
88. DISS Working Group. *Database of Individual Seismogenic Sources (DISS), Version 3.2.1: A Compilation of 450 Potential Sources for Earthquakes Larger than M 5.5 in Italy and Surrounding Areas*; Istituto Nazionale di 451 Geofisica e Vulcanologia: Roma, Italy, 2018. [CrossRef]
89. Chiodini, G.; Granieri, D.; Avino, R.; Caliro, S.; Costa, A.; Minopoli, C.; Vilardo, G. Non-volcanic CO_2 Earth degassing: Case of Mefite d'Ansanto (southern Apennines), Italy. *Geophys. Res. Lett.* **2010**, *37*. [CrossRef]
90. Amoruso, A.; Crescentini, L.; Scarpa, R. Faulting geometry for the complex 1980 Campania-Lucania earthquake from levelling data. *Geophys. J. Int.* **2005**, *162*, 156–168. [CrossRef]

91. Pingue, F.; De Natale, G. Fault mechanism of the 40 s subevent of the 1980 Irpinia (Southern Italy) earthquake from leveling data. *Geophys. Res. Lett.* **1993**, *20*, 911–914. [CrossRef]
92. Amato, A.; Selvaggi, G. Aftershock location and P-wave velocity structure in the epicentral region of the 1980 Irpinia earthquake. *Ann. Geofis.* **1993**, *36*, 3–15.
93. Giardini, D. Teleseismic observation of the November 23 1980, Irpinia earthquake. *Ann. Geofis.* **1993**, *XXXVI*, 1.
94. Nostro, C.; Cocco, M.; Belardinelli, M.E. Static stress change in extensional regimes: An application to southern Apennines (Italy). *J. Geophys. Res.* **1997**, *87*, 234–248.
95. Serva, L. The earthquake of September 8, 1694 in Campania–Lucania. *Atlas Isoseismal Maps Ital. Earthq.* **1985**, *2A*, CNR-PFG.
96. Pierdominici, S.; Mariucci, M.T.; Montone, P. A study to constrain the geometry of an active fault in southern Italy through borehole breakouts and downhole logs. *J. Geodyn.* **2011**, *52*, 279–289. [CrossRef]
97. Cocco, M.; Pacor, F. The rupture process of the 1980, Irpinia, Italy earthquake from the inversion of strong motion waveforms. *Tectonophysics* **1993**, *218*, 157–177. [CrossRef]
98. Pantosti, D.; D'Addezio, G.; Cinti, F.R. Paleoseismological evidence of repeated large earthquakes along the 1980 Irpinia earthquake fault. *Ann. Geofis.* **1993**, *36*, 321–330.
99. Chiarabba, C.; Amato, A. From tomographic images to fault heterogeneities. *Ann. Geofis.* **1994**, *37*, 1481–1494.
100. Carmignani, L.; Cello, G.; Cerrina Feroni, A.; Funiciello, R.; Kälin, O.; Meccheri, M.; Patacca, E.; Pertusati, P.; Plesi, G.; Salvini, F.; et al. Analisi del campo di fratturazione superficiale indotta dal terremoto Campano-Lucano del 23/11/1980. *Rend. Della Soc. Geol. Ital.* **1981**, *4*, 451–465.
101. Blumetti, A.M.; Esposito, E.; Ferreli, L.; Michetti, A.M.; Porfido, S.; Serva, L.; Vittori, E. New data and reinterpretation of the November 23, 1980, M 6.9, Irpinia-Lucania earthquake (Southern Apennine) coseismic surface effects. *Studi Geol. Camerti* **2002**, 19–27.
102. Esposito, E.; Luongo, G.; Maturano, A.; Porfido, S. I terremoti recenti dal 1980 al 1986 nell'Appennino meridionale. *Mem. Della Soc. Geol. Ital.* **1988**, *41*, 1117–1128.
103. D'Addezio, G.; Pantosti, D.; Valensise, G. Paleoearthquakes along the Irpinia fault at Pantano di San Gregorio Magno (Southern Italy). *IL Quat.* **1991**, *4*, 121–136.
104. Copley, A.; Hollingsworth, J.; Bergman, E. Constraints on fault and lithosphere rheology from the coseismic slip and postseismic afterslip of the 2006 Mw 7.0 Mozambique earthquake. *J. Geophys. Res.* **2012**, *117*, B03404. [CrossRef]
105. Gourmelen, N.; Amelung, F. Postseismic Mantle relaxation in the Central Nevada Seismic Belt. *Science* **2005**, *310*, 1473–1476. [CrossRef] [PubMed]
106. Nishimura, T.; Thatcher, W. Rheology of the lithosphere inferred from postseismic uplift following the 1959 Hebgen Lake earthquake. *J. Geophys. Res. Solid Earth* **2003**, *108*, B8. [CrossRef]
107. Rousset, B.; Barbot, S.; Avouac, J.P.; Hsu, Y.J. Postseismic deformation following the 1999 Chi-Chi earthquake, Taiwan: Implication for lower-crust rheology. *J. Geophys. Res.* **2012**, *117*, B12405. [CrossRef]
108. Costantini, M.; Minati, F.; Ciminelli, M.G.; Ferretti, A.; Costabile, S. Nationwide ground deformation monitoring by Persistent Scatterer Interferometry. In Proceedings of the 2015 IEEE International Geoscience and Remote Sensing Symposium (IGARSS), Milan, Italy, 26–31 July 2015; pp. 1472–1475.
109. Costantini, M.; Ferretti, A.; Minati, F.; Falco, S.; Trillo, F.; Colombo, D.; Novali, F.; Malvarosa, F.; Mammone, F.; Vecchioli, F.; et al. Analysis of surface deformations over the whole Italian territory by interferometric processing of ERS, Envisat and COSMO-SkyMed radar data. *Remote Sens. Environ.* **2017**, *202*, 250–275. [CrossRef]
110. Ferretti, A.; Prati, C.; Rocca, F. Permanent Scatters in SAR interferometry. *IEEE Trans. Geosci. Remote Sens.* **2001**, *39*, 8–20. [CrossRef]
111. Lündgren, P.; Casu, F.; Manzo, M.; Pepe, A.; Berardino, P.; Sansosti, E.; Lanari, R. Gravity and magma induced spreading of Mount Etna volcano revealed by satellite radar interferometry. *Geophys. Res. Lett.* **2004**, *31*, L04602. [CrossRef]
112. Manzo, M.; Ricciardi, G.P.; Casua, F.; Ventura, G.; Zeni, G.; Borgström, S.; Berardino, P.; Del Gaudio, C.; Lanari, R. Surface deformation analysis in the Ischia Island (Italy) based on spaceborne radar interferometry. *J. Volcanol. Geotherm. Res.* **2006**, *151*, 399–416. [CrossRef]
113. Lanari, R.; Casu, F.; Manzo, M.; Zeni, G.; Berardino, P.; Manunta, M.; Pepe, A. An Overview of the Small BAseline Subset Algorithm: A DInSAR Technique for Surface Deformation Analysis. In *Deformation and Gravity Change: Indicators of Isostasy, Tectonics, Volcanism, and Climate Change*; Wolf, D., Fernández, J., Eds.; Birkhäuser Verlag: Basel, Switzerland, 2007; pp. 637–661.

114. Perrone, G.; Morelli, M.; Piana, F.; Fioraso, G.; Nicolò, G.; Mallen, L.; Cadoppi, P.; Baestro, G.; Tallone, S. Current tectonic activity and differential uplift along the Cottian Alps/Po Plain boundary (NW Italy) as derived by PS-InSAR data. *J. Geodyn.* **2013**, *66*, 65–78. [CrossRef]
115. Peltzer, G.; Rosen, P.; Rogez, F.; Hudnut, K. Poroelastic rebound along the Landers 1992 earthquake surface rupture. *J. Geophys. Res. Solid Earth* **1998**, *103*, 30131–30145. [CrossRef]
116. Picozzi, M.; Bindi, D.; Zollo, A.; Festa, G.; Spallarossa, D. Detecting long-lasting transients of earthquake activity on a fault system by monitoring apparent stress, ground motion and clustering. *Sci. Rep.* **2019**, *9*, 16268. [CrossRef] [PubMed]

geosciences

MDPI

Article

Geology of the Epicentral Area of the November 23, 1980 Earthquake (Irpinia, Italy): New Stratigraphical, Structural and Petrological Constrains

Fabio Matano [1,*], Silvio Di Nocera [2], Sara Criniti [3] and Salvatore Critelli [3]

1 Consiglio Nazionale delle Ricerche, ISMAR, 80133 Naples, Italy
2 Università Federico II, 80138 Naples, Italy; silviodinocera0@gmail.com
3 Dipartimento di Biologia, Ecologia e Scienze della Terra, Università della Calabria, 87036 Rende (CS), Italy; sara.criniti88@gmail.com (S.C.); salvatore.critelli@unical.it (S.C.)
* Correspondence: fabio.matano@cnr.it

Received: 15 May 2020; Accepted: 23 June 2020; Published: 25 June 2020

Abstract: The geology of the epicentral area of the 1980 earthquake (Irpinia-Lucania, Italy) is described with new stratigraphic, petrographic and structural data. Subsurface geological data have been collected during the studies for the excavation works of the Pavoncelli bis hydraulic tunnel, developing between Caposele and Conza della Campania in an area that was highly damaged during 1980 earthquake. Our approach includes geological, stratigraphic, structural studies, and petrological analyses of rock samples collected along the tunnel profile and in outcropping sections. Stratigraphic studies and detailed geological and structural mapping were carried out in about 200 km^2 wide area. The main units cropping out have been studied and correlated in order to document the effects of tectonic changes during the orogenic evolution on the foreland basin systems and the sandstone detrital modes in this sector of the southern Apennines. The multi-disciplinary and updated datasets have allowed getting new insights on the tectono-stratigraphic evolution and stratigraphic architecture of the southern Apennines foreland basin system and on the structural and stratigraphic relations of Apennines tectonic units and timing of their kinematic evolution. They also allowed to better understand the relationships between internal and external basin units within the Apennine thrust belt and its tectonic evolution.

Keywords: geological map; tunnel geology; sandstone petrology; foreland basin evolution; 1980 Irpinia-Basilicata earthquake; southern Apennines; Italy

1. Introduction

The southern Apennines divide area between Sele river and Ofanto river valleys (southern Italy) represents a key sector for analyzing the paleogeographic conditions and the geodynamic evolution of the Apennine thrust belt. This area is located among the Irpinia, Cilento and Lucania sectors (Figure 1) and is characterized by very high seismic hazard due to the recurrence of strong earthquakes in the last centuries [1–3]. Indeed, the 23rd November 1980 Irpinia-Lucania, earthquake (MS 6.9—I_{max} X MCS) [4–8] has been the most devastating seismic event occurred in southern Apennines in the 20th century in terms of loss of human life and destruction of cultural heritage (Figure 2). The earthquake caused also severe structural damages to major civil engineering works throughout Irpinia sector, as for example to the Pavoncelli hydraulic tunnel located between Conza dam and Caposele springs, and relevant hydrogeological changes to several springs (i.e., Sanità spring at Caposele showed an abnormal, temporary increase in discharge from 4.35 to 7.32 m^3/s in January 1981) [9–11]. Moreover, a strong ground deformation was caused by the seismic shakings between 30 and 74 cm in height, as resulting by topographic leveling carried out in 1981 near Sella di Conza area [5,7,9,10,12–15]. The Pavoncelli water supply tunnel was built in 1911 [16] and runs just through the epicentral area, sustaining relevant damage consisting of crushing of sidewalls and roof and in vertical to transversal ruptures of the design section [10].

The high level of occurred damage was due to both the ineffectiveness of the technical rules in force at that time for engineering works and buildings and the poor geological knowledge of the stricken territory regarding to several aspects (seismology, stratigraphy, tectonics, and geomorphology). For example, during the 1980s–early 1990s, the regional geological framework of this area was based on the geological surveys made mainly during 1960s–1970s for the old geological sheets at 1:100,000 scale [17,18] that were summarized in the Bonardi et al. [19] geological map. A great improvement for the regional geology framework was done during late 1910s–2010s within the national CARG (nuova CARtografia Geologica) project for the new geological map of Italy at 1:50,000 scale [20,21], whose results were summarized in the Bonardi et al. [22] geological map.

In recent years, new relevant data have been collected in the study area, thanks to the geological study and investigations for the project for the works of the Pavoncelli-bis hydraulic tunnel that was designed for replacing the old Pavoncelli tunnel, which was strongly damaged during the 1980 earthquake.

The main aim of the paper is further analyzing the relationships between internal and external basin units of southern Apennine orogenic belt and constrain its paleogeographic and tectonic evolution. Indeed, the multi-disciplinary updated datasets have allowed getting new insights on the tectono-stratigraphic evolution and stratigraphic architecture of the southern Apennines foreland basin system and on the structural and stratigraphic relations of Apennines tectonic units and timing of their kinematic evolution.

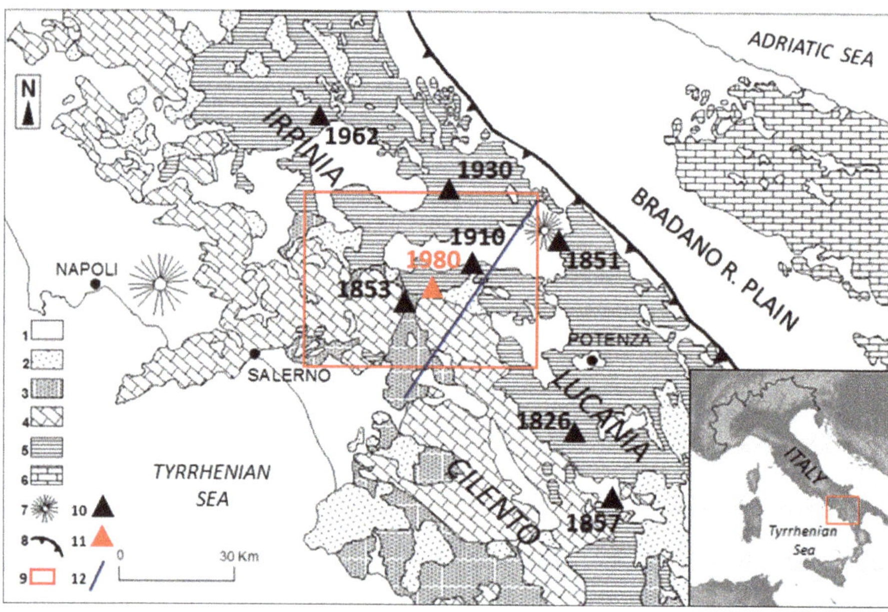

Figure 1. Geological sketch-map of the Southern Apennines (modified after [23]) and study area location (inlet). Legend: **1.** Quaternary and Pliocene deposits; **2.** Miocene deposits; 3. Liguride Complex; **4.** Apennine carbonate platform units; **5.** Molise-Lagonegro basin units; **6.** Apulian carbonate platform units; **7.** Volcanoes; **8.** orogenic front; **9.** study area; **10.** Historical earthquake epicenters [1,2]; **11.** 1980 earthquake epicenter [1,2]; **12.** Trace of geological sketch section.

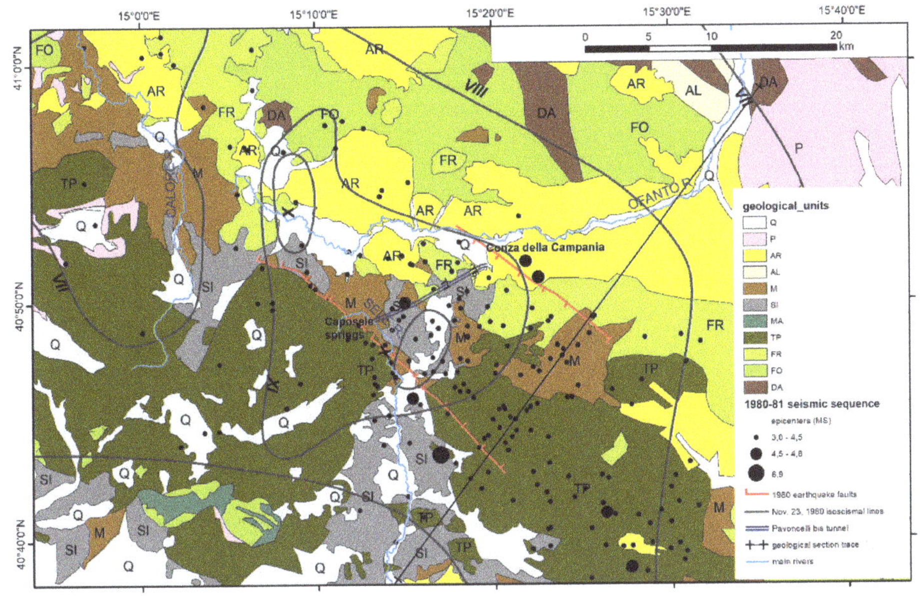

Figure 2. Geological map of the study area [17,18,20,21] and seismic effects of 1980 earthquake. Legend: **Q**, Quaternary deposits; **P**, pyroclastic and volcanic deposits; **AR**, Ariano unit (Pliocene); **AL**, Altavilla unit (Late Miocene); **M**, flysch units (Middle-Late Miocene); **SI**, Sicilide and Liguride units; **MA**, Monte Croce unit; **TP**, Apennine carbonate platform units; **FR**, Frigento unit (Lagonegro basin); **FO**, Fortore unit (Lagonegro basin); **DA**, Daunia unit. Isoseismal lines by Postpischl et al. [24]; earthquake faults by [6,7,12,14].

2. Geological Setting

The southern Italy thrust belt preserves key geodynamic signatures from the Pangaea breakup to the Tethys opening and its closure during the growth of the mountain chains. These paleogeographic domains include the inner domain (i.e., Paleozoic continental-crust sections), the oceanic realm (the Flysch Basin domain), and the external domains including foreland successions (carbonate platform to basin) and under plate stratigraphy (e.g., [25] and bibliography therein). The stratigraphic, tectonic, and paleogeographic relationships between internal and external basin units are an unravelled issue in the scientific debate about southern Apennines geology [22,26–29].

The "Liguride Complex Auct." and the "Sicilide Complex Auct." successions [30] represent remnants of deposition in a Meso-Cenozoic remnant ocean basin related to the western subduction of the Adria oceanic lithosphere beneath the Iberia or Mesomediterranean microplate [25,31,32]. The Liguride Complex records the consumption of the oceanic crust and the accretionary processes along the Adria margin, and the continental accretionary processes of the Calabrian Terranes [33]. The subduction has been active for all the Paleogene and earliest Miocene, producing an accretionary prism [31], and a diffuse calcalkaline volcanism in western Sardinia.

In response of E-NE accretionary processes along the Adria plate, immense volume of turbiditic sedimentation took place in the Apenninic domain during early to middle Miocene. Here, the foreland basin system developed over deformed remnant-ocean basin terranes during the early-to-middle Miocene, over Lagonegro basin and inner Apennine platform units during the upper Miocene, and finally over the previous deformed units and the western margin of the Apulia platform during the Pliocene to Quaternary [33–36]. The foreland basin system migrated in time, and siliciclastic and carbonate deposits, filling the wedge-top and the foredeep, were derived from progressive unroofing of the Calabrian crustal block or from erosion of the forebulge [25,32,37,38].

The Tyrrhenian back-arc extension of the last 10 Ma is responsible for the fragmentation and dispersion of pieces of the Iberian and European plates (Calabria, Sardinia, Corsica), and increased the displacement of the accretionary prism over the Adria plate, the eastward migration of the magmatic arcs, and the roll-back of the Adriatic lithosphere [39–41].

More in detail, the Southern Apennine consists of a north-east verging thrust and fold belt, interposed between the back-arc Tyrrhenian basin, to the west, and the undeformed Apulian-Adriatic foreland, to the east. On the whole, its structural setting is the result of mainly compressive tectonic events, related to the subduction followed by the roll-back of the Adria plate, followed by new tectonic phases related to the opening, since the late Miocene, of the Tyrrhenian sea [40–42]. Structural complexity characterizes the whole orogenic evolution of the EastNorthEast-verging fold-and-thrust belt system, but the major imbrications result from the post-Messinian thrusting and refolding of more ancient structures; this breaching [43] developed after the duplexing of the deep Apulian units [27]. Finally, the belt is widely affected by Plio-Quaternary strike-slip and extensional faults.

Several model and interpretations of stratigraphic and paleogeographic settings have been proposed referring to the different carbonate platform and pelagic basin successions cropping out in mountain belts. In detail, along the axial belt of the southern Apennines several basinal successions crop out that are mainly made of calcareous-mudstone sequences; they have been alternatively assigned to different tectonic units (i.e., Liguride, Parasicilide, Sicilide, Sannio, Lagonegro I and II), belonging to both Molise, Lagonegro, and Lucanian basins, following different interpretations and paleogeographic models [22,23,25–30,33,41,44–54].

Stratigraphic successions formed by the Argille Variegate Group (Argille Varicolori inferiori, Sant'Arcangelo Fm. and Argille Varicolori Superiori) followed by Tufiti di Tusa Fm. or Corleto Sandstone Fm. were deposited in external Lucanian Basin and can be referred to Sicilide Unit. The very similar stratigraphic successions, formed by Argille Varicolori Fm. and Corleto Perticara Fm., followed by the Tufiti di Tusa Fm. or Paola Doce Fm. conformably passing to Numidian Flysch Fm., are referred to the central axial sector of the Lagonegro basin, namely to the Fortore unit. The key difference is that the Sicilide Unit was deformed (in the southern Apennine sector) by orogenic phases in the early Miocene, before the deposition of the Numidian Flysch that unconformably cover them. The Fortore Unit was deformed later during the middle-late Miocene and so the Numidian Flysch conformably lays within the sequences [22,23,27]. Due to the uncertainty in the reconstruction of the stratigraphic succession, the Sannio unit has been considered both of internal or external origin, and their interpretation has been finally proposed by Di Nocera et al. [26] as northern inner part of the Lagonegro-Molise basin.

In this paper, we refer to the pre-orogenic paleogeographic model proposed by Pescatore et al. [54,55], Di Nocera et al. [27] and Critelli [25] for the southern Apennines sector. The model shows four main paleogeographic realms:

Lucanian Basin; it is the westernmost, oceanic basin where the stratigraphic successions of the Liguride and Parasicilide Complex [31,56] and the Sicilide Complex [30] are deposited.

- Apenninic carbonate Platform: this is an isolated platform bank between the oceanic realm of the Lucanian Ocean, to the west, and the deep-marine Lagonegro-Molise basin, to the east; this platform consists of thick, shallow-marine carbonate strata from the Late Triassic to early Miocene. The mostly continuous Mesozoic succession is formed by carbonate deposits related to a carbonate platform and its margins [57–59] with several evidences of platform emersion or drowning during Cretaceous. Main tectonic units are Alburno-Cervati, Mt. Picentini, Mt. Marzano, Mt. Maddalena, Mt. Croce units.
- Lagonegro-Molise Basin: this is located eastward of the carbonate platform and characterized by three different sectors, i.e., western, central, and eastern ones [29,60]. The Lagonegro II [61,62], Monte Arioso [29], Frigento [26], and Sannio [27,55,60] units refer to the internal, western portions of the basin; they are formed by calciclastic deposits produced by the erosion of the western margin of the Apenninic platform, interbedded with hemipelagic clays and marls. The Lagonegro I [61,62],

Groppa d'Anzi [29], and Fortore [54,55,60] units refer to the axial sector of the basin and are formed by cherty calcareous-mudstone sequences. The Campomaggiore, Daunia, and Vallone del Toro units [27,63–65] refer to the eastern sectors and are characterized by clays and marls with calciclastic layers produced by erosion of the western margin of the Apulian platform.

- Apulian carbonate Platform: this includes a more than 7000 m thick successions formed by Permian siliciclastics, Late Triassic evaporitic sequences (Anidridi di Burano Fm.), followed by Mesozoic and Cenozoic to earliest Pleistocene carbonate deposits of neritic environment [66,67].

In the Southern Apennines synorogenic sequences, carbonate sand and siliciclastic turbidite strata are widespread on diverse paleogeographic domains including the frontal portions of the Mesomediterranean microplate (Paludi and Stilo-Capo d'Orlando formations, late Oligocene to early Miocene), the Lucanian ocean basin (siliciclastic and carbonate turbidites of Monte Sant'Arcangelo, Corleto, Albanella, Colle Cappella, and Tufiti di Tusa Formations, early Miocene), and the Africa-Adriatic plate (Numidian Sandstone and Serra Palazzo formations, early-middle Miocene). Moreover, there are several sequences of foreland basin systems (Cilento Group, Tortonian-to early Pliocene clastics) that were first filled by deep-marine quartzarenite and carbonatoclastic turbidite sand derived from craton and foreland areas and were later filled by deep-marine quartzolithic and quartzofeldspathic turbidite sands. Minor volcaniclastic turbidite sandstone strata are locally deposited; they are derived from active volcanic arc source (e.g., Oligo-Miocene Sardinian Volcanic Arc; e.g., [25,32,68].

The carbonate turbidites are dominant in the Monte Sant'Arcangelo Formation and they are composed of abundant intrabasinal and minor extrabasinal carbonate grains. The Cilento Group (Pollica fm., S. Mauro fm.) is the largest exposed siliciclastic basin sequence in southern Italy [38,69]. It is a typical mixed siliciclastic/calciclastic turbiditic suite, over 2000 meters in thickness, including thick layers of resedimented marls (megabeds) and olistostromes made up of extrabasinal materials emplaced by catastrophic marine gravity flows [32]. Three main contemporaneous submarine fans developed in the early-middle Miocene, in a syntectonic wedge-shaped basin [32,70]. The middle Miocene to Pliocene southern Apennine clastics can be grouped in five key intervals [23,27,71,72]:

- Numidian Sandstone, mostly formed by Langhian quartzarenites and conformable Serravallian post-Numidian successions, formed by mixed quartzofeldspathic sandstones and calciclastic arenaceous-pelitic beds (foreland depozones);
- Langhian to Tortonian San Giorgio Fm. and Serra Palazzo Fm., mostly composed of quartzofeldspatic sandstones and calciclastic arenaceous beds (foredeep depozone);
- Tortonian to Early Messinian, quartzose-feldspatic and partly sedimentary carbonatoclastic petrofacies, wedge-top successions (Gorgoglione, Castelvetere, and San Bartolomeo fms.);
- Late Messinian quartzolithic to quartzofeldspatic sandstones (Anzano Molasse and Tufo-Altavilla units; Crotone basin sequence), which can be referred to infilled wedge-top basins;
- Unconformity-bounded Pliocene quartzofeldspatic sandstone strata (wedge-top depozones), characterized by strong synsedimentary tectonic activity.

3. Methods

Our approach includes geological, stratigraphic, structural, and petrological analyses. Stratigraphic studies, detailed geological and structural mapping (at 1:5000 scale) were carried out in about 200 km^2 wide area (Figure 3). The reference lithostratigraphy of the studied geological units and the biostratigraphy data that characterize the studied sections derive from the Sheets n° 450 and 468 of the Geological Map of Italy [20,21]. Stratigraphic sections of geological units defined here provide a detailed stratigraphic framework for the petrographic and tectonostratigraphic analyses.

Subsurface geological data were obtained by detailed stratigraphic and structural surveys made in some key sections of the tunnel, lithological analyses of the excavated materials (i.e., spoil) integrated by petrological analyses of rock samples collected along the tunnel profile, and analysis of the stratigraphy of 20 boreholes up to 350 m in depth.

The arenite-dominated successions have been sampled for compositional analysis along some outcrop and tunnel stratigraphic sections. Twenty-one medium- to coarse-grained sandstones samples were selected for composition analyses. They were impregnated in epoxy resin under vacuum, thin-sectioned, and stained for calcium plagioclase and potassium feldspar identification with hydrofluoric acid and sodium-cobaltonitrite, respectively.

The modal sandstone composition is determined by point-counting on stained thin sections. Counting was performed on all samples using the Gazzi-Dickinson method [73–75]. The framework grain types that are used for discussions of detrital modes are those of Dickinson [76,77], Zuffa [74,75], Critelli and Le Pera [69], Critelli and Ingersoll [78], and Caracciolo et al. [79,80], and comprise:

(a) Quartz grains, including monocrystalline quartz grains (Qm), polycrystalline quartzose lithic fragments (Qp), and total quartzose grains (Qt = Qm + Qp);
(b) Feldspar grains (F), including both plagioclase (P) and potassium feldspar (K);
(c) Aphanitic lithic fragments (L), as the sum of volcanic/metavolcanic (Lv and Lvm), sedimentary (Ls) and metasedimentary (Lm; including Lsm as the sum of Ls and Lm). Carbonate lithic fragments have been reported in Ls (namely, extrabasinal carbonate grains [74,75,81]), because of their importance and occurrence in detrital modes of Apenninic sandstones;
(d) Phaneritic and aphanitic rock/lithic fragments (R), recalculated by point-counting of specific assignment of Lm, Lv, Ls lithic fragments plus quartz, feldspar, micas, and dense minerals in coarse-grained polymineralic fragments in which these minerals individually are larger than the lower limit of the sand range (0.0625 mm), that during counting are summed as quartz (Qm) and feldspar (F) or micas or dense mineral grains (e.g., [69,73–75,78]).

For diagrams, the proportions of quartzose grains, feldspar grains and aphanitic lithic fragments are recalculated to 100 percent, and summary detrital modes are then reported as Qt%–F%–L% and Qm%–F%–Lt%. Description of sandstone petrofacies includes values of QmFLt%. General descriptions of changing nature of detrital budget in terms of calculations of both phaneritic and aphanitic rock/lithic fragments are also included in the petrofacies definition.

4. Results

The Sella di Conza study area is located along the southern Apennines Mts. divide at the borders along the Irpinia, Cilento, and Lucania sectors (Figure 1). The rugged landscape is characterized by high carbonate mountain reliefs, with the Picentini Mts. to the west and the Mt. Marzano massif to the east (Figure 2). They are separated by the smooth and wide upper Sele river and Ofanto river valleys. The former develops in a large tectonic depression bounded by roughly N-S trending structural lineaments, where Liguride and Sicilide units largely crop out. Northward, the Ofanto river valley develops along transversal active seismo-tectonic structures, which caused the 1980 Irpinia earthquake (Figure 2).

4.1. Geological Map: Stratigraphic Units and Structural Elements

The geological surveys performed in the study area allowed to recognize the presence of four main regional tectonic units (Figure 3), referred to different paleogeographic domains, such as the Sicilide unit (Lucanian basin), the Mt. Picentini and Mt. Marzano units (Apennine platform), and the Frigento unit (Lagonegro basin). These units are unconformably overlain by calciclastic and siliciclastic thrust-top basin and foredeep basin fillings, ranging from Early Miocene to Pliocene in age. Quaternary age deposits are mainly represented by debris slope talus, alluvial fan, and terraced and actual alluvial deposits.

The Sicilide Unit includes a Cretaceous to Oligocene-Early Miocene pelagic basin calcareous-pelitic succession, made up of pelitic deposits with a large component of carbonate resedimented deposits, forming the Argille Varicolori Superiori (ALV) and the Monte Sant'Arcangelo Fm. (FMS), belonging to the Argille Variegate Group (AV) [82]. They are overlain by siliciclastic, volcaniclastic, and calciclastic strata, referring to Corleto Arenite and Albanella Arenite (ABA) formations.

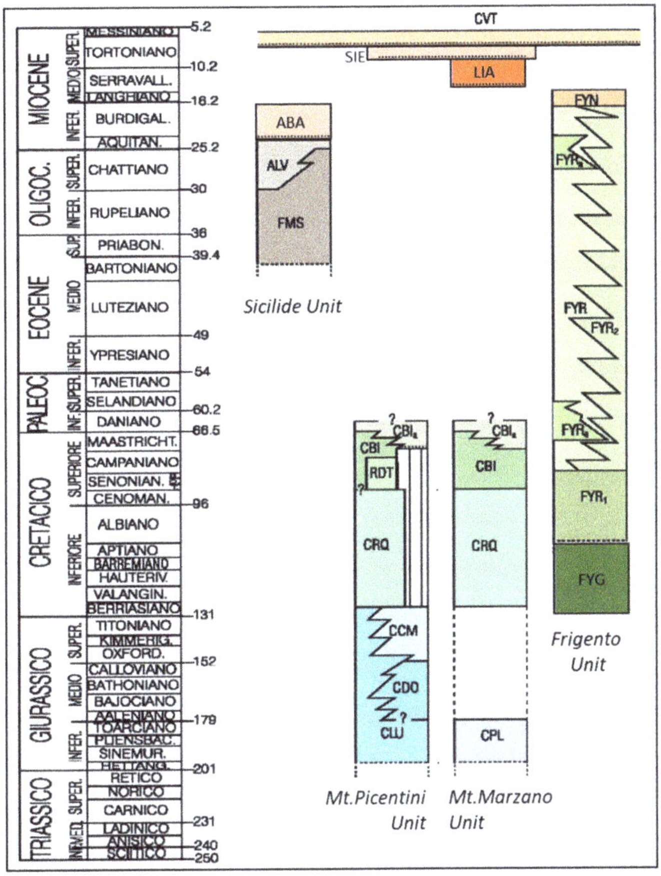

Figure 3. Chronostratigraphic scheme of the geological units cropping out in the study area (modified after [82]). Legend: Monte S. Arcangelo Fm. (**FMS**), Argille Varicolori Superiori Fm. (**ALV**), Albanella/Corleto unit (**ABA**), Palaeodasycladus limestone fm. (**CPL**), limestone and dolomitic limestone fm. (**CLU**), oolitic ed oncolitic limestone fm. (**CDO**), Cladocoropsis and Clypeina limestone fm. (**CCM**), requienia and gasteropods limestone fm. (**CRQ**), Radiolitidae Limestone fm. (**RDT**), bio-litoclastic rudist limestone fm. (**CBI**), "Pseudosaccharoidal" limestone lithofacies (**CBIa**), Flysch Galestrino Fm. (**FYG**), Flysch Rosso Fm. (**FYR**), limestone member (**FYR2**) or lithofacies (**FYRa**), Numidian Flysch (**FYN**), Laviano calcarenites fm. (**LIA**), Monte Sierio Fm. (**SIE**), Castelvetere Formation (**CVT**).

The Monte S. Arcangelo Fm. (FMS) is mainly formed by light brown, yellowish, greenish, and greyish marly limestones, sometimes silicified; subordinately, light brown and greyish graded and laminated calcarenites, dark clayey marls, thin bedded silty marls and micaceous arenites layers are present. The thickness is approximately 350–400 m. The unit is stratigraphically underlying and passing laterally into AVF. The age is referred to Eocene-Upper Oligocene [82].

The Argille Varicolori Superiori Fm. (ALV) is composed by greyish, greenish, and reddish clays interbedded with thin layered whitish limestones and marly limestones, related to a deep marine basin depositional environment. Locally, thin bedded reddish marls and calcareous marls, alternated with clayey marls, pinkish marls, fine grained greyish calcarenites with planar laminations, laminated clayey marls, silicized marls, and radiolarites with typical prismatic fractures are present. Some horizons up to 100 m thick are made of thick bedded whitish limestones, marly limestones and marls, whitish-greyish calcarenites, and calcilutites alternated with arenaceous-calcareous and calcareous-marly turbidites (calcareous lithofacies, ALVc). They are also formed by light gray, green, or reddish calcareous marls in medium layers with conchoid fracture, gray calcarenites with Tb-c Bouma sequences in thin to medium layers. The depositional environment is marine with turbiditic distal inputs. The unit stratigraphically overlies FMS, passes laterally and vertically into ABA. The thickness does not exceed 150 m. The age is Oligocene-Aquitanian [82].

A sequence formed by thin bedded brownish quartz-micaceous sandstones with planar and convolute laminations, poorly cemented yellowish and grey clayey marls, siltstones, blackish shales with calcite veins form the Albanella/Corleto unit (ABA), stratigraphically overlying the AV upper successions. The thickness is approximately 50–100 m. The age is Lower Miocene for the stratigraphic position.

The Apennine Platform units are formed by several lithostratigraphic units ranging in age between Noric and Late Maastrichtian. The lithological description, age and environment interpretation are derived by Torre et al. [82] and Pescatore and Pinto [83]. The sedimentary paleoenvironments of these carbonate successions can be referred to the margin of the carbonate platform, representing the lateral transition to the Lagonegro basin. Some differences in the stratigraphic succession between the Mt. Picentini and Mt. Marzano carbonate tectonic units are present.

The Mt. Picentini unit is Triassic to Lower Cretaceous in age in the lower portion, including several lithostratigraphic units of lagoonal succession, while the upper portion (Late Cretaceous) consists of bioclastic facies related to a high-energy open platform environment.

The Dolomia Superiore fm. (DBS) is formed by light grey layered dolostones, characterized by laminated stromatolites, alternated with massive dolostones. In the area, the Megalodon limestones and dolostones member (DBS4) is present, consisting of thin to thick layered grey limestones, dolomitic limestones (packstone, grainstone), and dolomites, with levels rich in megalodontids valves. The age is Norian–Hettangian p.p. and the depositional environment is peritidal. The total thickness exceeds 1000 m.

The Cladocoropsis and Clypeina limestone fm. (CCM) is formed by medium bedded grey and light brown mudstones, wackestones and packstones interbedded with thin bedded marls rich in *Cladocoropsis mirabilis*. In the upper part, limestones with dasycladacean algae (*Clypeina* sp.) are present. The age is Upper Jurassic–Neocomian and the environment is lagoonal. The thickness is at least 450 m.

The limestone and dolomitic limestone fm. (CLU) is formed by bedded light brown and grey oncolitic limestones of platform environment. The thickness is approximately 750 m. The age is Lower Jurassic p.p.–Neocomian.

The requienia and gasteropods limestone fm. (CRQ) is composed by light brown and grey layered limestones rich in nerineids and requienia, occasionally giving rise to bioclastic rudstones, alternated with mudstones and wackestones. Limestones with oolites, limestones with alveolinids, and laminated dolostones are also present. The total thickness is approximately 450 m, and the age is Barremian–Cenomanian p.p.

The Radiolitidae Limestone fm. (RDT) are formed by thick to medium strata of gray and white dolomitic limestone and limestone, and clastic limestone rich in rudists (*Radiolitidae*, *Hippuritidae*). The thickness is approximately 300 m. The age is Turonian–Campanian.

The bio-litoclastic rudist limestone fm. (CBI) is formed by massive bioclastic limestones (rudstones and floatstones) with abundant large fragments of rudists (radiolitids and hippuritids) and large gastropods and with thin beds of yellow-greenish marly limestones. A lithofacies with thick bedded "pseudosaccharoidal" limestones and graded calcarenites is present. The depositional environment is from open carbonate ramp to upper slope environment. The age is Upper Cenomanian–Paleocene. The thickness is approximately 300 m.

The Mt. Marzano unit crops out along the norther slope of Mt. Marzano massif and consists of the upper part of a discontinuous succession Upper Cretaceous–Lower-Middle Miocene in age, which is composed of the typical "pseudosaccharoidal limestones" units (CBI), made of calcirudites of open carbonate ramp and slope environments, disconformably overlain by the Laviano calcareous-pelitic sequence (LAI) and unconformably overlain by the Mt. Sierio calciclastic Fm. The lower part is formed by DBS, CLU, CRQ, and CPL units [82].

The Dolomia Superiore fm. (DBS) is formed by light colored, massive, and strongly fractured dolostones with stromatolites and bivalve shells, Norian–Hettangian p.p. in age, and about 100 m in thickness. The limestone and dolomitic limestone fm. (CLU) is composed by oncolitic light grey limestones, calcarenites, and light brown calcilutites with rare thin shells of small sized gasteropods. The thickness is approximately 400 m and the age is Lower Jurassic p.p.–Neocomian.

The requienia and gasteropods limestone fm. (CRQ) is formed by limestones rich in requienia, nerineids and other bivalve shells, and by oolitic and intraclasts limestones. The thickness is around 100 m; the age is Barremian–Cenomanian. The Palaeodasycladus limestone fm. (CPL) is made by micritic limestones characterized by peloids, oolites, and oncolites, and with *Palaeodasycladus mediterraneus* and *Orbitopsella* sp.; upwards calcareous-marly levels with spathized valves of *Lithiotis* sp. are locally present. The thickness is approximately 350 m; the age is Upper Hettangian-Pliensbachian.

The bio-litoclastic rudist limestone fm. (CBI) is formed by thick bedded or massive white limestones and calcirudites with abundant rudist fragments, often referred to the "pseudosaccharoidal limestones" lithofacies. The environment ranges from open carbonatic ramp to upper slope. The thickness is several hundreds of meters and the age is Upper Cretaceous-Paleocene.

The Laviano calcarenites fm. (LIA) is formed by a sequence comprising medium to thin layered laminated calcarenites and yellowish-greenish marly limestones, alternated with greyish-greenish marls and marly and silty shales. The basal part is characterized by thin bedded marly limestones and grey-greenish and pinkish marls about 5 m thick. The calcarenites locally contain an arenaceous fraction and quartz grains; upwards thin layered glauconitic bioclastic calcarenites alternated with calcilutites, marls, and whitish calcareous marls, and rare "numidian" quartzarenites are found. The thickness is approximately 100 m. The strata are formed by turbidites related to basin-slope toe, and hemipelagic deposits. The unit overlies in paraconformity on CBI. The age is Serravallian–Lower Tortonian [82].

The Monte Sierio Fm. (SIE) unconformably overlies on CBI. It is formed by calcirudites and calcarenites rich in nummulites and alveolines bioclasts; locally upwards, marly calcilutites, laminated and silicized siltstones, laminated lithic sandstones, calcarenites with nummulites and orbitoids, and blocks and strata composed of "numidian" sandstones are present. The thickness is not less than 150 m and the age is Upper Tortonian.

In the study area, the Frigento Unit consists mainly of Cretaceous to Early Miocene basinal and shelf-margin facies successions (Flysch Galestrino Fm. and Flysch Rosso Fm.), characterized by calcareous clastic and pelitic turbidite associations laterally and vertically passing to marly-calcareous successions; a debris calcareous clastic lithofacies is present at different heights. The Flysch Rosso Fm. pass gradually upward to the upper Burdigalian-upper Langhian p.p. Numidian Flysch with alternations of calcarenites, marls, and quartzarenites.

The Flysch Galestrino Fm. (FYG) is formed by grey-green medium to thick bedded siliciferous marls and clays with intercalations of medium-thin bedded silicified calcarenites and calcilutites of dark grey color; dark grey siliceous shales with intercalations of grey calcilutite in layers up to 30 cm thick are also present. The unit crops out discontinuously with thicknesses around a few tens of meters. It is referred to a bathyal environment with both siliciclastic and carbonate turbiditic inputs. The age is Upper Jurassic p.p.–Cretaceous.

The Flysch Rosso Fm. (FYR) is formed by red and grey-green marls and clays, thin bedded calcilutites, medium bedded commonly graded and laminated calcarenites with Alveolinidae, Nummulites, and Orbitoids, thick bedded litoclastic calcirudites with rudist fragments. The limestone member (FYR2) is formed by grayish and white clastic limestone (calcirudites, crystalline limestone, breccias, calcarenites) in thick and very thick massive layers, with irregular stratification and internal amalgamation, resting with erosive contact on grayish clayey-marly horizons cleavage fractures. In the upper part of the formation, a "pre-numidian sequence," showing the gradual transition to FYN, is formed by grey-green silty clays with intercalations of medium to thick bedded yellowish calcarenites rich in quartz granules. The environment is pelagic ranging from bathyal to carbonate slope toe. The thickness is approximately 200 m. The age is Late Cretaceous p.p.–Early Miocene p.p.

The turbiditic, grain-flow and hemipelagic siliciclastic deposits of the Castelvetere Formation (CVT) and the clastic Pliocene units (SAD and RVM) present an unconformable lower boundary overlying the deformed Sicilide, Mt. Picentini, Mt. Marzano, and Frigento units. The Castelvetere Fm. (CVT) is formed by arenites and granular conglomerates in thick beds with lens-shaped geometry and erosive base, locally amalgamated, containing clay-chips and thin pelitic interlayers, locally reddish and rich in coal fragments, and by quartz-feldspathic arenaceous turbidites and paraconglomerates. The thickness is approximately 250 m. The silty-clayey-marly lithofacies (CVTa) is formed by yellowish quartz-micaceous arenaceous siltstones in thin beds with planar and convolute laminations within bedded clays and grey-green silty marls. The thickness is approximately 150 m. The basal coarse sequences are composed of channeled facies rich in detrital deposits (olistolites) and clayey landslides (olistostromes) that are present at different stratigraphic levels. The upper finer sequence is linked to a turbidite system formed by depositional lobes and overbank deposits. The age is Upper Tortonian–Lower Messinian.

There are two Pliocene units, the Andretta Synthem (SAD) and the Ruvo del Monte Synthem (RVM). The Andretta Synthem (SAD) is formed by continental fan-delta conglomeratic and sandy deposits passing to coastal marine sands and clays. On the northern slope of the Ofanto river valley, the sequence is formed by yellowish silty-marly clays, locally interlayered with fine-grained sands and thin arenite beds. The thickness is of about 150 m, and the age is referred to the Early Pliocene (biozone MPl3/MPl4a).

The Ruvo del Monte Synthem (RVM) is approximately 200 m thick and Early–Middle Pliocene in age (biozone MPl4a); it consists mainly of arenitic-sandy-conglomeratic deposits from paralic to continental environment. The synthem is formed of a fine to medium-grained sandy succession, containing current structures and fragments of shells and valves of ostreids and pectinids, and locally with well-cemented arenitic lenses. Very thick layers and lenses of conglomerates with intercalations of lithic coarse sand and micro-conglomerates with a reddish yellow sandy matrix are also present.

Among the Quaternary units, the Conza della Campania synthem (CZC) crops out with terraced fluvial deposits along the Ofanto river and around the Conza lake. Usually, the succession is formed by gravel lens, pale yellow sands, silts, and gray to brown clayey silts, and is 10 m thick. The age is referred to Middle Pleistocene–Holocene.

More recent deposits linked to the active geomorphic processes are mainly represented by debris slope talus (a3), terraced and actual alluvial deposits and alluvial fan (b), and landslide deposits (a1).

The studied area is characterized by two structurally complex carbonate blocks, forming the morpho-structural heights of the Picentini Mts. to the west and the Mt. Marzano massif to the east (Figure 4). They are separated by the large tectonic depression of the Sele river valley, bounded by roughly N-S trending structural lineaments, and are interrupted northward by the wide E-W trending tectonic depression of the Ofanto river valley developing along transversal active seismo-tectonic structures.

The described tectonic units are strongly deformed by mainly east-verging imbrications and are unconformably covered by Burdigalian to Tortonian sequences. The imbrications are formed by folds related to reverse faults located into north-east verging regional thrust sheets. The tectonic superpositions are complicated by a polyphase structuration, which occurred during the deposition of Late Miocene to Pliocene siliciclastic synorogenic deposits [84]. The morpho-structural setting of the area is controlled by the Plio-Quaternary extensional tectonics, which uplifted the two major carbonate massifs and led to the formation of the Sele and Ofanto river valleys.

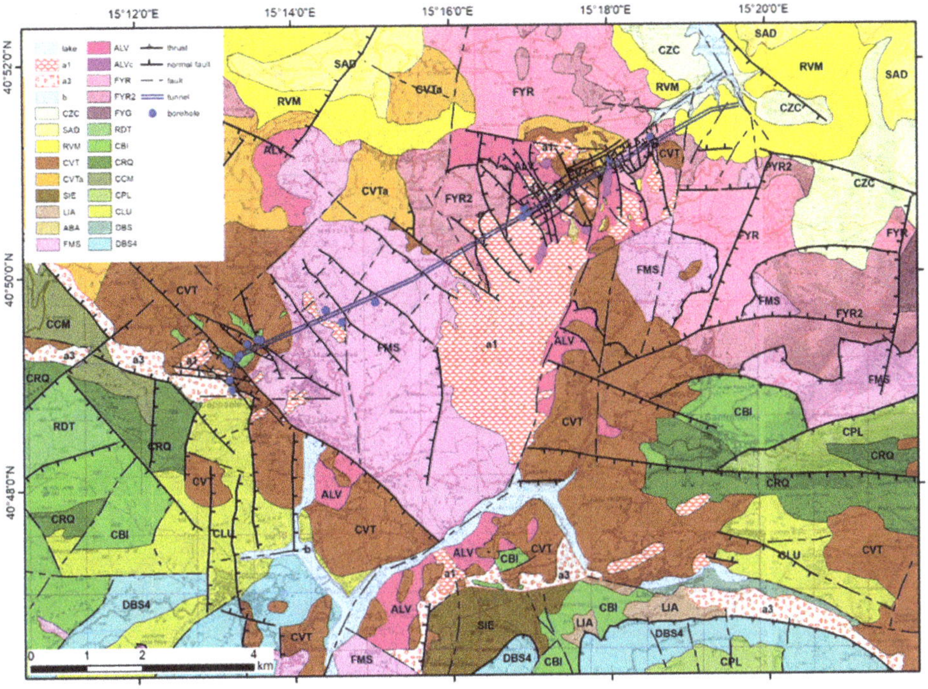

Figure 4. Geological map of the study area, indicating the Pavoncelli-bis tunnel track and the location of the boreholes performed in the different survey campaigns. Legend: (a) Sicilide unit—Monte S. Arcangelo Fm. (FMS), Argille Varicolori Superiori Fm. (ALV) and calcareous lithofacies (ALVc), Albanella/Corleto unit (ABA); (b) Mt. Picentini unit—Dolomia Superiore fm. (DBS) and Megalodon limestones and dolostones member (DBS4), Cladocoropsis and Clypeina limestone fm. (CCM), limestone and dolomitic limestone fm. (CLU), requienia and gasteropods limestone fm. (CRQ), Radiolitidae Limestone fm. (RDT), bio-litoclastic rudist limestone fm. (CBI); (c) Mt. Marzano unit—DBS, CLU, CRQ, Palaeodasycladus limestone fm. (CPL), "Pseudosaccharoidal" limestones lithofacies (CBIa), Laviano calcarenites fm. (LIA); (d) Frigento unit—Flysch Galestrino Fm. (FYG), Flysch Rosso Fm. (FYR) and limestone member (FYR2), pre-numidian sequence (FYN); synorogenic units—Monte Sierio Fm. (SIE), Castelvetere Formation (CVT) with silty-clayey-marly lithofacies (CVTa), Andretta Synthem (SAD), Ruvo del Monte Synthem (RVM); Quaternary units—Conza della Campania synthem (CZC), debris slope talus (a3), terraced and actual alluvial deposits and alluvial fan (b), landslide deposits (a1).

The internal Sicilide units tectonically overly the platform carbonate units, and largely crop out only in the tectonic depressions within the Sele and the Ofanto valleys. Only small scattered remains of the Sicilide stratigraphic succession can be found on the carbonate massif slopes. The carbonate platform units tectonically overly the Frigento unit, and these thrusting structures are probably linked to those observed in the central Picentini area within the Campagna tectonic window [27]. In our study area, these structures are hidden by Quaternary high-angle normal faults bordering the Ofanto River valley tectonic depression. In fact, along the northern mountain front of Picentini and Marzano Massifs, a peculiar structural setting, characterizing the entire southern Apennines axial belt, is well evident. The regional overthrusting of the carbonate platform units onto the Lagonegro units is characterized by a late reimbrication of the older thrust sheets [82]. The reimbrication crops out south of

Laviano, where the Triassic carbonate units (Mt. Picentini unit) thrusted over the "pseudosaccharoidal limestones" and detrital limestones related to the Mt. Marzano unit covered by the Laviano sequence. This reimbrication also involve the tectonic overlapping of the Mt. Marzano unit on Sicilide unit and Frigento unit thrusts.

The border between the Picentini Mts. massif and the Sele valley is bounded by a lateral hanging-wall ramp of the reimbrication thrusts, locally characterized by sub-vertical reverse faults and by a large antiformal fold, with NNE oriented axis parallel to the massif border [82]. The western flank of the Sele valley tectonic depression is defined by this regional fold and associated to later NNE trending high angle normal faults. Indeed, the eastern flank of the Sele valley is essentially controlled by faults produced by mainly SSE and ESE-trending faults with a generally oblique kinematics. Southward the study area, some large carbonate bedrock blocks, bounded by the same transtensional tectonics, crop out also within the valley (see for example Oliveto Citra, Contursi, and Quaglietta carbonate blocks [82]).

4.2. Subsurface Data: Tunnel and Borehole Logs

The subsurface geological data were collected along the tunnel underground line by lithological analyses of the excavated spoil materials, integrated by petrological analyses of rock samples collected along the tunnel excavation surfaces and by detailed stratigraphic and structural surveys made in key sections during some technical stops of the excavation works.

The tunnel crosses three main sectors characterized by very different lithological features (Table 1) separated by first-order tectonic structures. The western sector of the tunnel, from 0 to 655 m, was excavated within massive and fractured carbonate successions belonging to the CBI unit. The central sector of the tunnel, from 655 to 7760 m, was characterized by complex and strongly tectonized lithological units ranging from calcareous-pelitic to arenaceous rocks (FMS, ALV, ABA, CVT). In this sector, based on the lithological description of the excavated spoil materials, some different lithofacies have been recognized within the succession of the FMS unit: FMS-arg (argillaceous lithofacies), FMS-arg-ma (argillaceous-marly lithofacies), FMS-ma-arg (marly-argillaceous lithofacies), and FMS-cal-ma (calcareous-marly lithofacies). Arenitic strata were firstly found at 6100 m nearby a complex fault system. The eastern sector of the tunnel, from 7760 to 8500 m, was excavated in well stratified argillite, marl, and calcarenite sequences (FYR unit), and then up to 10,220 m in Pliocene deposits.

A technical stop during the TBM excavation works allowed to perform detailed stratigraphic and structural surveys along the walls of a small lateral tunnel between 6170–6190 m.

A fining-upward pelitic-arenitic stratified sequence, about 15 m thick, has been recognized (Figure 5). The strata are tilted toward the west, where the initial sector of the tunnel is located; the dipping is usually N240–20° (dip direction-dip angle). It is formed by coarse to medium-grained arenites thick strata with thin layers of dark silt and clays passing to fine grained arenites, dark gray siltstones and argillites thin strata. The thick arenitic strata (20–60 cm) show normal grading structures passing to parallel and cross laminations in the fine-grained arenitic and silty layers (5–20 cm in thickness); in the upper part of the stratigraphic sequence, a pebbly mudstone layer is present.

The rock mass is locally heavily jointed due to several fault systems crossing the bedding. A fault plain, oriented N300–75° with a 10–15 cm thick shear zone, cuts and dislocates the sequence near chainage 6175 m (Figure 6).

The analysis of the stratigraphic logs of n. 17 boreholes (Table 2), extending from 40 m to 350 m in depth, allowed to link the geological units recognized along the tunnel profile to the underground and surface geological units. The boreholes NP1, NP2, NP5, NP6, NP7, NP8, PR1, PB1, and PB2, located in the western sector of the Pavoncelli bis tunnel between Caposele and Materdomini towns (Figure 7), show the thickness variation (in the range of 11 to 120 m) of CVT pelitic-arenaceous sequences laying on the CBI calcareous substratum. Instead, the PR1 borehole show the tectonic superposition of FMS unit on CBI carbonate unit that is sutured by the CVT sequence.

The boreholes PR2, PR3, PR4, S1-CdG, SV1, and SV2, located in the median sector of the Pavoncelli bis tunnel around Cresta del Gallo and Bosco di Contra localities (Figures 7 and 8), show the relationships between FMS and ALV units, while the SV3 and SB boreholes, located in the easternmost tunnel sector (Figure 8), show the stratigraphy details of FYR and CVTa units.

(a) (b) (c) (d)

Figure 5. Arenaceous-pelitic stratigraphic sequences: (**a**) bedding; (**b**) coarse-grained arenite; (**c**) parallel and cross lamination in silty-arenaceous layers; (**d**) dark silty argillite.

Figure 6. Fault plane dislocating the stratigraphic sequence.

Table 1. Lithological units crossed by the Pavoncelli bis tunnel. Geological unit codes: ALV, Argille Varicolori Superiori Fm.; ABA, Albanella/Corleto unit; CBI, bio-litoclastic rudist limestone fm.; FYR, Flysch Rosso Fm.; CVT, Castelvetere Formation; FMS, Monte S. Arcangelo Fm. with four lithofacies: FMS-arg (argillaceous lithofacies), FMS-arg-ma (argillaceous-marly lithofacies), FMS-ma-arg (marly-argillaceous lithofacies), and FMS-cal-ma (calcareous-marly lithofacies).

Chainage Interval	Description of Lithological Intervals	Geological Unit Interpretation
0–655 m	limestones	CBI
655–2000 m	clays, shale, and marls	FMS-arg-ma
2000–4300 m	marls and clays	FMS-ma-arg
4300–5000 m	marly-calcareous turbidites	FMS- cal-ma
5000–5300 m	clay, dark clays, and clayey silts	FMS-arg
5300–5600 m	calcareous-clay turbidites	FMS-cal-ma
5600–6100 m	clayey turbidites, calcareous fault breccia, dark clayey marls, and argillites	ALV
6100–6233 m	well stratified arenites and marly clays, interbedded with thin layers of sandy silt and dark grey argillite, quartzarenites, and dark siltites and argillites	ABA
6233–6357 m	coarse to fine grained quartzofeldspathic sandstones	CVT
6357–7760 m	dark argillites and marls with layers of reddish clays and whitish marly limestone, calcilutites and calcarenites; within 6448 m–6463 m and 6709–6725 m intervals, some thick CVT arenite strata are present	ALV
7760–8500 m	grey argillites and marls with reddish clayey marls, marly limestones, and bioclastic calcirudites	FYR

Table 2. Borehole stratigraphy data available in the study area. For unit codes see Table 1 caption.

Borehole	Chainage (Approx.)	Description of Lithological Intervals	Geological Units (Thickness)
NP6	near Caposele	0–12 m, arenite; 12–21 m, arenite and limestone; 21–40 m, limestone	CVT (0–21 m); CBI (21–40 m)
NP8	near Caposele	0–23 m, arenite and clay; 23–27 m, limestone; 27–42 m, arenite and limestone; 42–50 m, limestone	CVT (0–23 m); CBI (23–50 m)
NP5	10	0–50 m, clays with limestone; 50–58 m, limestone	CVT (0–50 m); CBI (50–58 m)
NP7	20	0–32 m, clay and limestone; 32–60 m, limestone	CVT (0–32 m); CBI (32–60 m)
PB1	150	0–3 m, silt; 3–120 m, limestone	CBI (0–120 m)
PB2	200	0–11 m, silt; 11–120 m, limestone	CVT (0–11m), CBI (11–120 m)
NP2	490	0–4 m, clay; 4–35 m, sand; 35–43 m, clays with limestone; 43–110 m, limestone	CVT (0–43 m); CBI (43–110 m)
NP1	600	0–120 m, limestone	CBI (0–120 m)
PR1	655	0–20 m, no data; 20–80 m, marl, clayey marl, clay; 80–95 m, limestone	CVT (0–20 m), FMS arg-ma (20–80 m), CBI (80–95 m)
PR2	1920	0–95 m, no data; 95–135 m, marl, clayey marl, clay	FMS ma-arg, FMS arg-ma (0–135 m)
SV2	2150	0–39 m, clay and marl; 39–100 m, clayey marl, calcareous marl, marly limestone	FMS ma-arg; FMS cal-ma (0–100 m)
PR3	2810	0–120 m, no data; 120–160 m, marl, clayey marl, clay	FMS ma-arg (0–120 m)
PR4	5850	0–20 m, no data; 20–85 m, marl, clayey marl, clay	ALV; FMS arg (0–85 m)
S1-CdG	5920	0–65 m, marl, clayey marl, clay	ALV (0–65 m)
SV1	5960	0–11 m, silty clay; 11–197 m marl, clayey marl, calcareous marl, limestone, silty marl; 197–199 m clay; 199–350 m, marl, clayey marl, calcareous marl, marly limestone, silty marl	ALV (0–11 m); FMS cal-ma (11–197 m); ALV (197–350 m)
SB	7630	0–50 m, no data; 50–107 m clay and quartz-feldspatic arenite; 107–130 m, arenite, calcarenite, reddish clay	CVT (0–130 m)
SV3	8540	0–100 m, reddish clay and limestone	FYR (0–100 m)

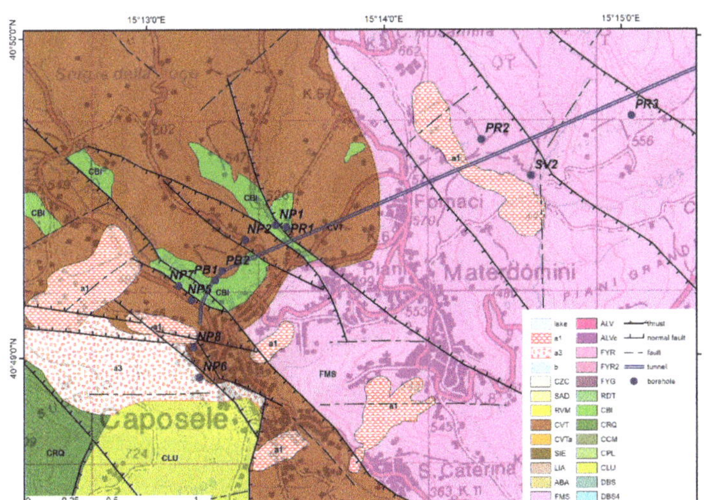

Figure 7. Detail of the Geological map of the Caposele sector of study area with borehole location. For legend unit codes, see Figure 4 caption.

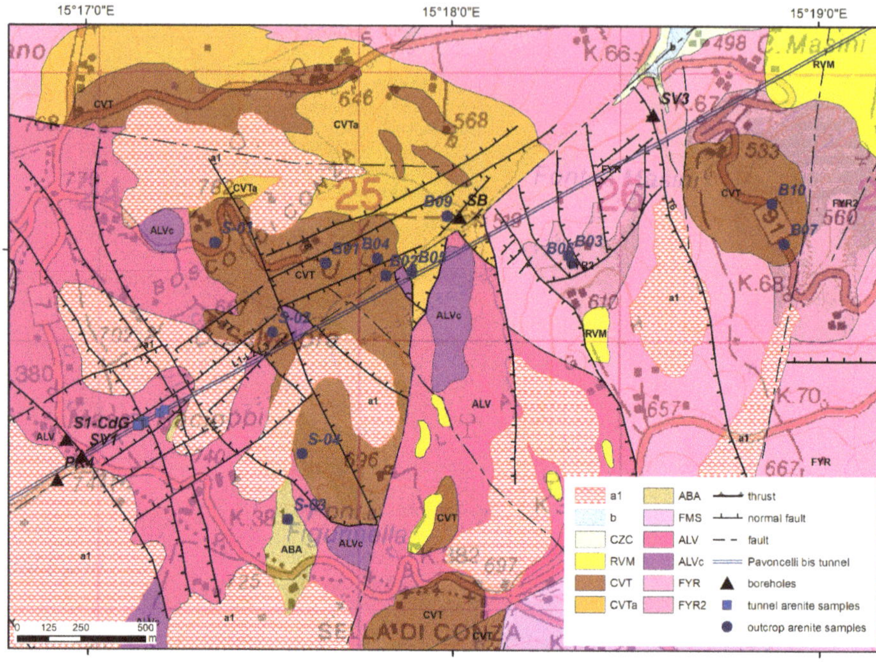

Figure 8. Detail of the Geological map of the Sella di Conza sector of study area with borehole and sample location. For legend unit codes, see Figure 4 caption.

4.3. Sandstone Petrology: Principal Compositional Groupings

Sandstone composition analysis includes 21 sandstone samples. Samples have been collected both along stratigraphic sections cropping out in the area located in correspondence with the tunnel segment comprised within 6000–8600 m chainage interval (Figure 8), and in some tunnel sections between chainage 6100 m and 6300 m (Figure 9).

Table 3 summarizes the raw data of point-counting results of selected sandstones, while Table 4 shows recalculated grain parameters. The analyzed sandstone samples include 6 quartzolithic sandstone of the Albanella-Corleto Formation (ABA), 12 quartzofeldspathic sandstones of the Castelvetere Formation (CVT), and 3 quartzose and hybrid arenites of the pre-Numidian interval of FYR upper part (Figures 10 and 11).

4.3.1. Quartzolithic Sandstone Petrofacies

The petrofacies has abundant quartz and lithic grains (Qm60 F13 Lt26), while feldspar is minor and dominantly plagioclase. Aphanitic lithic fragments (Lm42 Lv31 Ls27) include metavolcanic (serpentinite, Figure 10a to serpentine schist), volcanic (mainly microlithic, Figure 10b and felsitic textures), abundant metasedimentary (phyllite, schist, minor quartzite, and fine-grained schist; Figure 10c–e), and sedimentary (micritic to sparitic limestone grains, and radiolarian chert) grains. The abundant metasedimentary fragments in this petrofacies are derived from the low-to-medium grade metamorphic Paleozoic sections of the Calabrian Terranes [25,33,34,36,37,39,40]. The abundant volcanic lithic fragments are derived from coeval active volcanism located to the west (i.e., Sardinia magmatic arc; [25]) and textures of the volcanic particles suggest an andesitic to dacitic volcanism [68]. Metavolcanic fragments are closely related to accreted oceanic terranes of the Lucanian Oceanic Unit obducted at the time of deposition of the Quartzolithic sandstones of the Corleto Formation [25,31]. Sedimentary detritus can have diverse source terranes from the Mesozoic sedimentary covers of the

Paleozoic metamorphic units, and from sedimentary strata of the accreted oceanic terranes, and from the basinal and carbonate platform units of the internal portions of the Apenninic units [25,33,35,36]. The quartzolithic petrofacies corresponds to the Corleto and Albanella Sandstone Formations (Figure 11).

4.3.2. Quartzofeldspathic Sandstone Petrofacies

The petrofacies has abundant quartz and feldspars (Qm51 F32 Lt15), while lithic fragments are minor. Plagioclase grains are dominant than K-feldspar (P/F = 0.83). Aphanitic lithic grains include abundant metasedimentary and sedimentary, minor volcanic, and ophiolite lithic particles (Lm47Lv7 Ls46). Metasedimentary lithic fragments include abundant phyllite and fine-grained schist. Sedimentary lithic fragments consist of abundant extrabasinal carbonate fragments, mostly Jurassic to Lower Tertiary pelagic and shallow-marine carbonate fragments and minor radiolarian chert, shale, and siltstone fragments. Ophiolitic detritus includes serpentinite and serpentine-schist, whereas volcanic lithic fragments are minor and consist of lithic particles that have a felsitic granular texture. Phaneritic rock fragments (apportioned in Qm, P, K, micas, and dense minerals [69,78,85]) include plutonic and metamorphic detritus; plutonic detritus consists of dominantly quartz-plagioclase-biotite particles of granodioritic to tonalitic compositions (Figure 10g,h), and minor quartz-K-feldspar composite grains of granite composition. Phaneritic metamorphic detritus consists of quartz-plagioclase-sillimanite, and quartz-plagioclase-garnet-sillimanite composite grains of gneiss and micaschist.

The abundant metamorphic and plutonic detritus in the quartzofeldspathic sandstone petrofacies are derived from the metamorphic and plutonic Paleozoic sections of the Calabrian Terranes [25,33,34,36,37,39,40]. Sedimentary detritus can have diverse source terranes from the Mesozoic sedimentary covers of the Paleozoic metamorphic-plutonic terranes, and from sedimentary strata of the basinal and carbonate platform units of the internal portions of the Apenninic units [25,33,35,36]. Ophiolitic fragments are closely related to thrusted oceanic terranes of the Lucanian Oceanic Unit [25,31]. The quartzofeldspathic petrofacies corresponds to the Castelvetere Sandstone Formation (Figure 11).

4.3.3. Hybrid Arenite and Quartzarenite Sandstone Petrofacies

This composite petrofacies has abundant quartz (Qm95 F5 Lt0), while feldspar is minor and dominantly plagioclase. Aphanitic lithic fragments are virtually absent or minor. Quartz grains are dominantly monocrystalline, and they are well rounded. Few dense minerals are ultrastable and consist of zircon, tourmaline, and rutile. Quartzarenite are well cemented by quartz overgrowth and calcite. Hybrid arenites of the pre-Numidian Sandstone Formation include abundant quartz, and extrabasinal carbonate lithic grains. Very abundant intrabasinal carbonate particles of bioclasts (Figure 10f, mainly planktonic forams and sponges), and minor peloids and intrabasinal non-carbonate glauconite grains. Abundance of well-rounded quartzose detritus of the pre-Numidian and Numidian Sandstone Formation has been interpreted as a craton-derived huge arrivals of sand-sized quartz from the African continental margin [25,33,37,38,86,87]. The quartzose and hybrid arenite petrofacies corresponds to the pre-Numidian (hybrid arenite) and Numidian Sandstone Formation (quartzarenite) (Figure 11).

Table 3. Point-count raw results of the analyzed sandstone samples.

Unit	Corleto Perticara Fm.							pre-Numidian seq.			Castelvetere Fm.										
sample	**GAL01**	**GAL02**	**S03**	**B11**	**B13**	**B14**	**B15**	**B03**	**B06**	**B09**	**S01**	**S02**	**S04**	**B01**	**B02**	**B04**	**B05**	**B07**	**B08**	**B10**	**B12**
Petrographic classes																					
Quartz (Qt = Qm + Qp)																					
Quartz (single crystals)	195	114	167	165	116	154	152	45	38	27	173	190	164	153	129	126	137	144	127	149	151
Polycrystalline quartz with tectonic fabric	1		1	2	6	1	3	2	2	1	1	5	5	1	1	1	3	3	1	2	1
Polycrystalline quartz without tectonic fabric													1	1			2				2
Quartz in metamorphic r.f.	17	2	3	16	34	8	3				18	4	2	4	3	6	16		2	1	3
Quartz in plutonic r.f.				1	4						27	3	22	6	5	18	1	1	13	1	6
Quartz in plutonic or gneissic r.f.											3		4								
Calcite replacement on quartz												6									
Feldspars (F = K + P)																					
K-feldspar (single crystals)	5		3	1	5	1	2				20	38	23	10	8	14	2	8	16	7	23
K-feldspar in plutonic r.f.	5			3	5						16	13	19	6	5	4	2		10	5	9
Calcite replacement in k-feldspar			1									1									
Plagioclase (single crystals)	51	18	37	29	29	25	19		1	1	89	83	89	68	63	66	36	57	69	57	85
Plagioclase in metamorphic r.f.	8			7	8	2					5	4	5		2	1	6		2	1	1
Plagioclase in plutonic r.f.				9	12	3					38	21	44	21	19	39	11	15	25	12	33
Plagioclase in plutonic or gneissic r.f.	2												6								
Plagioclasio in volcanic r.f.											1										
Calcite replacement in Plagioclase												5	1								
Micas																					
Micas and clorite (single crystals)	58	40	73	18	15	30	25	1	2	1	2	7	3	4	3	2	31	2	6	1	3
Micas in plutonic r.f.	1				1		1														
Micas in metamorphic r.f.	2				5						3		1								
Lithic fragments (L = Lm + Lv + Ls)																					
Volcanic lithic with microlithic fabric	37	1		19	25	8	7										4		1		2
Volcanic lithic with felsitic granular texture	2			9	9	6	4				4	2	1	2	5	3	5	1	1	1	
Volcanic lithic with felsitic seriate texture	3			5	8	3											1				
Volcanic lithic with lathwork texture	1																				
Volcanic lithic with vitric texture				1													1			1	
Serpentinite	13		3	3	8	4	2					1									
Serpentine-schist	4	1		4	2	1	2										1			1	
Phyllite	53	7	31	27	27	19	15				19	12	15	16	11	23	36	15	10	14	8
Fine-grained Schist	4			16	6	1					13	3	11		1	1	6	2	1	1	5
Fine-grained Gneiss											7										
Siltstone				2								2	2	3	1	2				2	1
Impure Chert	3			4	3	3	1					2		2	3		3	1		3	1
Shale												1		3	3	2		1		3	1

Table 3. *Cont.*

Unit	Corleto Perticara Fm.							pre-Numidian seq.			Castelvetere Fm.										
sample	**GAL01**	**GAL02**	**S03**	**B11**	**B13**	**B14**	**B15**	**B03**	**B06**	**B09**	**S01**	**S02**	**S04**	**B01**	**B02**	**B04**	**B05**	**B07**	**B08**	**B10**	**B12**
Dense minerals																					
Dense mineral (single crystal)	10	5	2	10	8	13	12	2	1	3	1	2	2	1	2		4	3	2	2	1
Dense mineral in plutonic r.f.													1								
Dense mineral in metamorphic r.f.	4	3			1																
Opaque minerals			4		1	3	3				1		1	1			2	1			1
Extrabasinal Carbonates (EC)																					
Dolostone					6							2					1				1
Micritic Limestone	7	3	11	9	6	8	7	3	4	4		4	4	7	6	7	4	10	4	4	1
Sparitic Limestone			1	5	3	2	4					3	4	2	2	2	8	3		3	2
Microsparitic Limestone	6			4	4	1						2	3		1			4	1	1	
Biomicritic Limestone	4		4	4	7	2	2	1	1	1		8	3	7	6	8	2	11	3	10	6
Biosparitic Limestone					3	2						3	1	1						1	1
Fossil (single skeleton)				1	1	1		6	9	4			1		1	1	1	1			
Fossil in Limestone-Dolostone					2	1		3	4	2											
Single spar (calcite)	1			1	4							5	4	2	1						
Single spar (dolomite)																			2		1
Intrabasinal Carbonates (IC) and noncarbonates (INC)																					
Bioclast	3		1					187	180	190			4							5	
Peloids								2	4	3											
Glauconite		2	3	1		1		3	2	3		1		1					1		
Rip-Up clasts							2					2	3	18	16	9	2	5	3	7	5
Interstitial components (matrix and cements)																					
Siliciclastic matrix	24	21	13	12	8	8	16				6	4	8	27	17	6	31	15	12	23	13
Carbonate matrix (micrite)								54	50	47						2	2	3	2	3	
Carbonate cement (pore-filling)	1	7	14	2	4	5	2	65	55	57		20	9	19	56	66	6	63	13	52	8
Carbonate cement (patchy calcite)	18	48	91	11	26	38	32	14	22	12	1	52	28	38	16	11	16	7	14	21	21
Calcite replacement on underterm. grain	2		9	15	5	17						15	5	11	21	24	5	21	7	17	
Siliceous cement				1			2												1		
Phyllosilicate cement				2	2	1	4							3			5	1		1	4
Oxid-Fe cement						2		12	8	8					1						
Alterites (indeterminate alterite grain)	1				1							1	1					2			
total counted points	**546**	**272**	**472**	**419**	**420**	**374**	**322**	**400**	**383**	**364**	**448**	**527**	**500**	**438**	**408**	**444**	**393**	**400**	**349**	**412**	**400**

Table 4. Recalculated modal point for collected samples (X = mean SD = standard deviation). Grain parameters: Qm = monocrystalline quartz; Qp = polycristalline quartz; Qt = Qm þ Qp; K = K-feldspar; P = Plagioclase; F = P þ K; L = aphanitic lithic grains; Lt = L þ Qp þ CE; Lvm = volcanic and metavolcanic; Lsm = sedimentary and metasedimentary; Lm = metamorphic; Lv = volcanic; Ls = sedimentary lithics grains; Rg = plutonic; Rv = volcanic; Rm = metamorphic rock fragments; NCE = non carbonate extrabasinal grains; CE = carbonate extrabasinal; CI = carbonate intrabasinal grains.

	Corleto Perticara Fm.							pre-Numidian			Castelvetere Fm.									
Sample	**GAL-01**	**GAL-02**	**S-03**	**B-11**	**B-13**	**B-14**	**B-15**	**B-03**	**B-06**	**B-09**	**S-02**	**S-04**	**B-01**	**B-02**	**B-04**	**B-05**	**B-07**	**B-08**	**B-10**	**B-12**
Qm	50	80	65	52	46	63	70	75	64	68	48	44	52	50	47	53	52	49	54	46
F	17	12	16	14	17	12	9	0	2	3	39	43	33	35	38	20	29	42	29	44
Lt	33	8	19	34	37	25	21	25	34	29	13	13	15	15	15	27	19	9	17	10
Qt	51	80	65	54	48	65	72	78	68	70	49	46	53	51	47	56	54	50	56	48
F	17	12	16	14	17	12	9	0	2	3	39	43	33	35	38	20	29	42	29	44
L	32	8	19	32	35	23	19	22	32	27	12	11	14	14	15	24	17	8	15	8
Qm	74	87	80	79	72	83	88	100	97	96	55	51	61	58	54	73	64	53	65	52
K	4	0	2	2	5	1	1	0	0	0	14	11	6	6	7	2	4	10	5	10
P	22	13	18	19	23	16	11	0	3	4	31	38	33	36	39	25	32	37	30	38
Qp	3	0	2	5	7	6	9	13	11	8	13	11	9	10	2	10	8	8	11	12
Lvm	43	17	6	35	41	35	32	0	0	0	5	2	4	12	6	15	2	8	6	6
Lsm	54	83	92	60	52	59	59	87	89	92	82	87	87	78	92	75	90	84	83	82
Lm	44	67	62	42	29	34	38	0	0	0	30	53	36	29	50	59	35	46	36	43
Lv	41	8	6	32	42	34	30	0	0	0	6	2	4	12	6	15	2	8	4	7
Ls	15	25	32	26	29	32	32	100	100	100	64	45	60	59	44	26	63	46	60	50

Table 5. Mean detrital modes of selected sandstone suites of the southern Apennines correlated with well data of studied successions. Data source from Critelli (2018).

Description and Location of Unit	N	QtFL	QmFLt	QmKP	QpLvmLsm	LmLvLs	RgRsRm
Quartzolithic suite							
Corleto-Perticara Formation	55	66 (± 4)–18 (± 6)–16 (± 7)	62 (± 4)–18 (± 6)–20 (± 7)	71 (± 5)–8 (± 3)–16 (± 4)	8 (± 3)–16 (± 5)–76 (± 7)	64 (± 11)–16 (± 3)–20 (± 6)	-
Albanella Fm.	15	66 (± 4)–16 (± 6)–18 (± 7)	62 (± 4)–16 (± 6)–22 (± 7)	74 (± 6)–5 (± 3)–21 (± 5)	3 (± 3)–10 (± 6)–87 (± 6)	84 (± 6)–6 (± 5)–10 (± 5)	6 (± 3)–6 (± 6)–88 (± 11)
Colle Cappella Formation	24	64 (± 4)–16 (± 6)–20 (± 7)	60 (± 4)–16 (± 6)–24 (± 7)	73 (± 3)–6 (± 2)–16 (± 2)	11 (± 6)–2 (± 3)–87 (± 7)	72 (± 14)–4 (± 3)–24 (± 6)	6 (± 3)–15 (± 6)–79 (± 11)
Tufiti di Tusa Formation (quartzolithic suite)	16	50 (± 18)–18 (± 3)–32 (± 16)	46 (± 18)–18 (± 3)–36 (± 16)	70 (± 5)–3 (± 2)–27 (± 5)	8 (± 3)–26 (± 5)–66 (± 8)	47 (± 7)–29 (± 7)–24 (± 3)	-
Tufiti di Tusa Formation (volcanolithic suite)	22	14 (± 6)–28 (± 15) 58 (± 13)	12 (± 6)–28 (± 15)–60 (± 13)	28 (± 6)–2 (± 15)–70 (± 13)	4 (± 3)–90 (± 6)–6 (± 4)	8 (± 3)–88 (± 7)–4 (± 3)	-
Corleto Formation in wells	7	62 (± 11)–14 (± 3)–24 (± 9)	61 (± 11)–14 (± 3)–25 (± 9)	80 (± 6)–2 (± 2)–18 (± 4)	4 (± 3)–30 (± 12)–66 (± 14)	45 (± 13)–28 (± 14)–27 (± 6)	6 (± 3)–15 (± 6)–79 (± 11)
Quartzofeldspathic suite							
Castelvetere Formation	68	46 (± 4)–42 (± 6)–12 (± 5)	45 (± 4)–42 (± 6)–13 (± 5)	52 (± 6)–15 (± 5)–33 (± 4)	9 (± 4)–6 (± 7)–85 (± 9)	41 (± 14)–5 (± 5)–54 (± 16)	54 (± 16)–20 (± 9)–26 (± 12)
Castelvetere Formation in wells	**11**	**50 (± 4)–42 (± 6)–8 (± 5)**	**48 (± 4)–42 (± 6)–10 (± 5)**	**52 (± 6)–15 (± 5)–33 (± 4)**	**9 (± 4)–6 (± 7)–85 (± 9)**	**41 (± 14)–5 (± 5)–54 (± 16)**	**54 (± 16)–20 (± 9)–26 (± 12)**
Sorrento Sandstone Formation	30	55 (± 5)–40 (± 6)–5 (± 2)	53 (± 5)–40 (± 6)–7 (± 2)	56 (± 5)–22 (± 4)–22 (± 4)	22 (± 12)–4 (± 2)–74 (± 12)	41 (± 14)–5 (± 5)–54 (± 16)	54 (± 7)–13 (± 8)–33 (± 8)
San Bartolomeo Formation	46	55 (± 7)–34 (± 7)–11 (± 7)	52 (± 7)–34 (± 7)–14 (± 7)	60 (± 7)–17 (± 3)–23 (± 5)	25 (± 12)–0 (± 0)–75 (± 12)	45 (± 18)–0 (± 0)–55 (± 18)	12 (± 8)–39 (± 16)–49 (± 17)
Quartzarenite suite							
pre-Numidian Hybrid arenites	3	**72 (± 3)–2 (± 2)–26 (± 3)**	**69 (± 3)–2 (± 2)–29 (± 3)**	**98 (± 4)–0 (± 0)–2 (± 1)**	-	-	-
Numidian Sandstone (southern Apennines)	39	95 (± 5)–5 (± 5)–0 (± 0)	95 (±5)–5 (±5)–0 (±0)	97 (± 5)–0 (± 0)–3 (± 1)	-	-	-

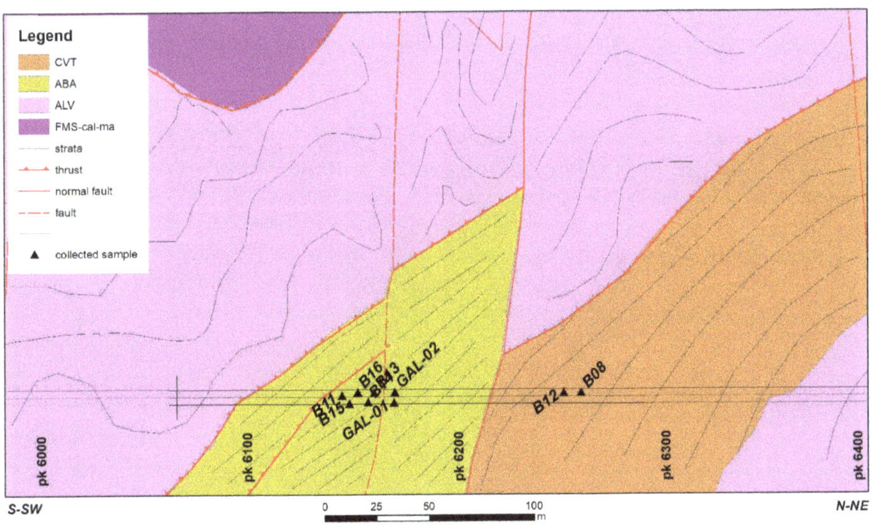

Figure 9. Location of sample collected along tunnel section from 6100 m to 6300 m chainage (pk). Geological unit codes: ALV, Argille Varicolori Superiori Fm.; ABA, Albanella/Corleto unit; CBI, bio-litoclastic rudist limestone fm.; CVT, Castelvetere Formation; FMS, Monte S. Arcangelo Fm. with calcareous-marly lithofacies (FMS-cal-ma).

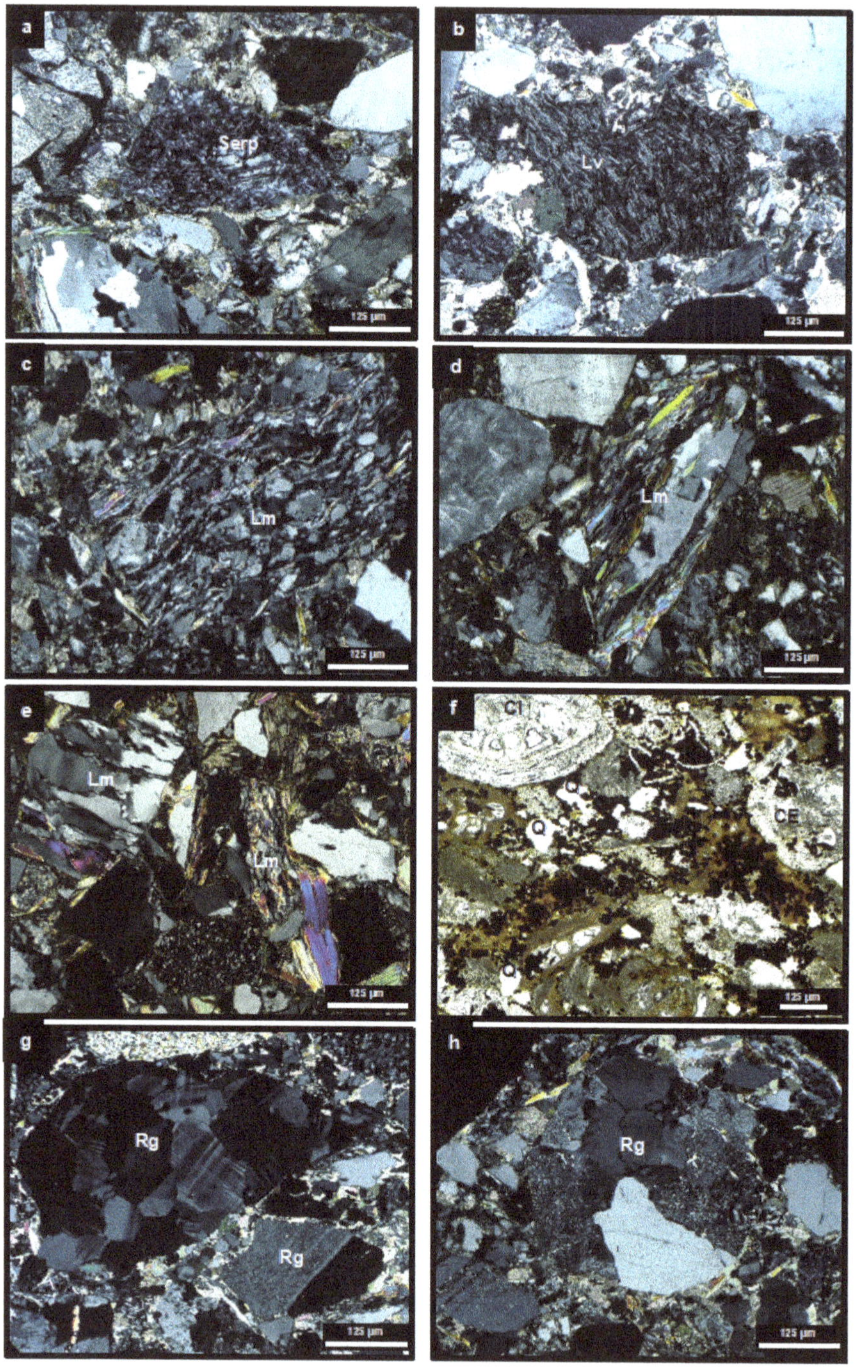

Figure 10. Photomicrographs of the main detrital fragments for quartzolithic, quartz-feldspathic sandstone petrofacies and hybrid arenites of the derived from the borehole stratigraphy of the study area. Quartzolithic petrofacies(**a–e**): (**a**) serpentinite having cellular texture; (**b**) volcanic lithic fragments with microlithic texture; (**c**) low-grade metamorphic lithic fragment (Lm; phyllite); (**d**,**e**) low-to-medium

grade metamorphic lithic fragments (Lm, phyllite to micaschist). Hybrid Arenites of the pre-Numidian Sandstone (**f**): (**f**) hybrid arenite having abundant bioclast tests (Cl) and quartz (Q), Quartzofeldspathic Petrofacies (**g**,**h**): (**g**) phaneritic rock fragments of plagioclase-bearing tonalite-to-granodiorite fragments; (**h**) phaneritic rock fragment of K-feldspar-Quartz granite fragment. Scale bars = 125 μm; (**f**) plane-polarized light; (**a**–**h**) crossed nicols.

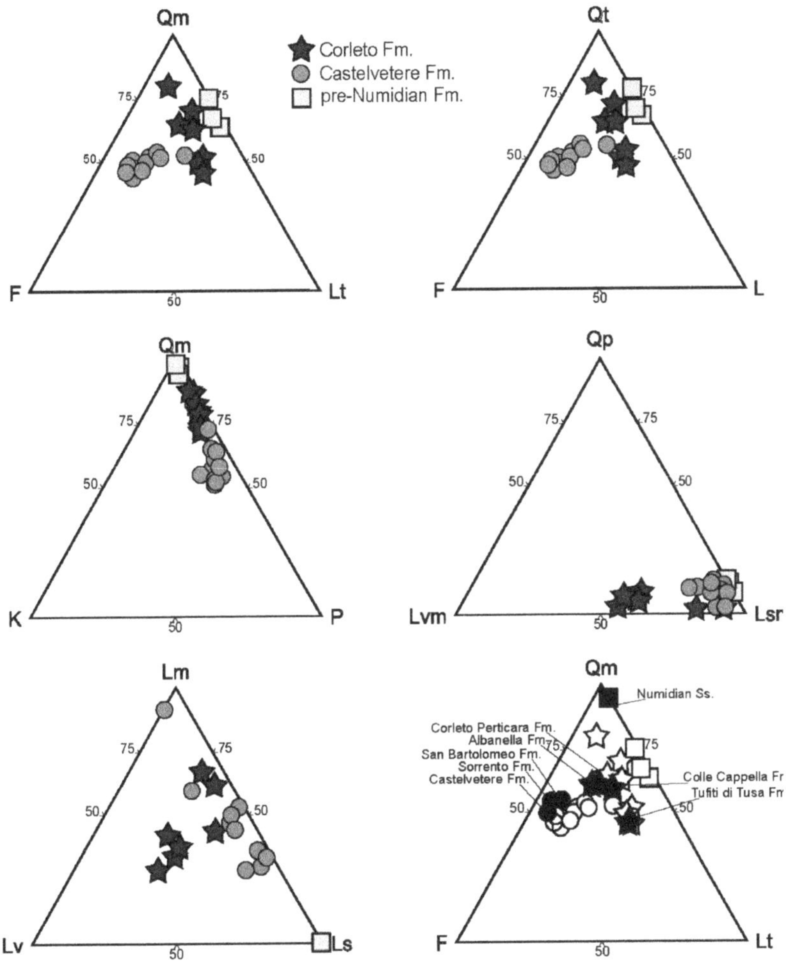

Figure 11. Ternary compositional plots of the Corleto/Albanella, Castelvetere, and pre-Numidian sandstones and their relations with similar sandstone suites of the southern Apennines foreland basin system. Qm, monocrystalline quartz; Qp, polycrystalline quartz, Qt, (Qm + Qp), F, feldspars (K + P); K, K-feldspar; P, plagioclase; Lt, aphanitic lithic fragments; Lm, aphanitic metamorphic lithic fragments; Lv and Lvm, aphanitic volcanic (Lv) and metavolcanic (Lvm) lithic fragments; Ls, aphanitic sedimentary lithic fragments, Lsm, metasedimentary and sedimentary lithic fragments. Latter ternary compositional plot Qm-F-L documents the relations of the Corleto, Castelvetere, and pre-Numidian sandstones and with related sandstone suites of the southern Apennines foreland basin system (Regional dataset of the southern Apennines foreland sandstone suites is in Table 5).

5. Discussion

5.1. Geological Profile along Tunnel Pavoncelli Bis

All data obtained by the geological surveys performed on the field in the study area and the subsurface geological data derived by the tunnel excavations and the boreholes, integrated by petrological analyses of arenitic samples, allowed to obtain a detailed geological profile along the tunnel Pavoncelli bis track. The geological profile (Figure 12a) shows complex stratigraphic structures and tectonic patterns and three main different structural sectors can be recognized.

In the western sectors, located between Caposele springs and chainage 655 m of the tunnel (Figure 12a), a Late Cretaceous formation (CBI fm.) belonging to the Mt. Lattari carbonate tectonic unit is unconformably covered by the Late Miocene arenaceous-clayey successions (CVT Fm.). The stratigraphic sequence is faulted with horst and graben structures originated by Quaternary NW-SE trending normal faults.

The large central sector is showed in the geological profile encompasses from chainages 655 m to 7700 m, and it is characterized by the presence of complex structures involving mainly the Sicilide Unit. In the initial part, ranging from 665 m to 5400 m (Figure 12a), different lithofacies of the Monte Sant'Arcangelo Fm. (FMS) are present and are locally covered by the unconformable deposits of the Castelvetere Fm. (CVT). The stratigraphic succession if strongly folded and faulted by normal faults, NW-SE, NE-SW, and N-S trending. The NW-SE oriented set is formed by high-angle normal faults, dipping toward SE up to the chainage 4200 m and toward NW after that sector. This change in the tectonic pattern allows to observe the back-thrusting of the Frigento Unit (Lagonegro basin) onto the Sicilide Unit in the Cresta del Gallo area, corresponding to the tunnel interval ranging from 4200 m and 5400 m. In fact, in this area an approximately 150 m thick calcareous-clastic sequence of the calcareous member (FYR2) of the Flysch Rosso Fm. forms a small calcareous ridge tectonically laying over the calcareous-pelitic poly-folded sequences of the Monte Sant'Arcangelo Fm. (FMS), recognized along the tunnel. The thrust surface is cut off by NE-dipping high angle normal faults; in this sector, the large and complex Buoninventre landslide [88] was reactivated after the 1980 earthquake and is still partly active. In the final part of this central sector, between 5400 m and 7760 m (Figure 12a), the tectonic structures are more complex. Two thrusting surfaces, crossing the tunnel at about 5400 m and 6200 m, involve the whole stratigraphic succession of the Sicilide Unit. The first thrust fault is characterized by a folded strata sequence of FMS and ALV units in the hanging-wall and a sequence of ALV and ABA units in the foot-wall, where a minor reverse fault separates the ALV and ABA units. This thrust plane is cut and displaced by two sub-vertical normal faults at 5900 m and 6100 m and is interrupted by a major normal fault at 6200 m. The Early Miocene arenite (ABA) strata, about 100 m thick, are entrapped within these normal faults at the tunnel depth. The latter fault divides the described tectonic structure by a different structural sub-sector, characterized by the thrusting of a folded strata sequence of FMS, ALV, and CVT units (hanging-wall) onto a sequence of ALV and CVT units (foot-wall). Thick sequence (200–300 m) of Late Miocene arenites (CVT) characterizes this area. Several normal faults cut the stratigraphic succession and further dislocate the thrust plane.

The eastern sector of the geological profile encompasses from chainage 7700 m to 10,220 m (Figure 12a) and is characterized by more simple tectonic structures and completely different stratigraphic units. An important normal fault cut the thrust surface of the Frigento unit onto the Sicilide unit, crossed by the tunnel at 7800 m. The Frigento unit is formed by the thick argillaceous-calcareous sequence of the Flysch Rosso Fm. (FYR, FYR2), comprising pre-numidian strata in the uppermost part. Several high-angle normal faults, mainly NNW-SSE oriented, have originated horst and graben structures. At about 8925 m, the unconformity basal contact of the Pliocene stratigraphic sequences (RVM synthem) on the Frigento unit is preserved. The RVM sandy-silty succession is little deformed.

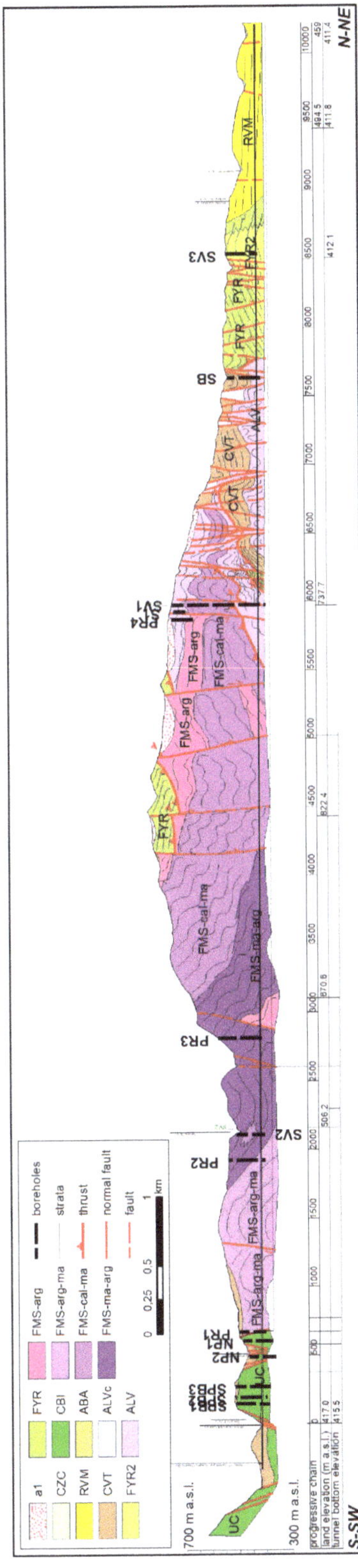
S-SW
N-NE
700 m a.s.l.
300 m a.s.l.
a1
CZC
RVM
CVT
FYR2
FYR
CBI
ABA
ALVc
ALV
FMS-arg
FMS-arg-ma
FMS-cal-ma
FMS-ma-arg
boreholes
strata
thrust
normal fault
fault
0 0.25 0.5 1 km
progressive chain
land elevation (m a.s.l.)
tunnel bottom elevation
SV2
PR2
PR3
SB
SV3
RVM
FYR
CVT
FMS-cal-ma
FMS-ma-arg
FMS-arg-ma
FMS-arg
UC

(a)

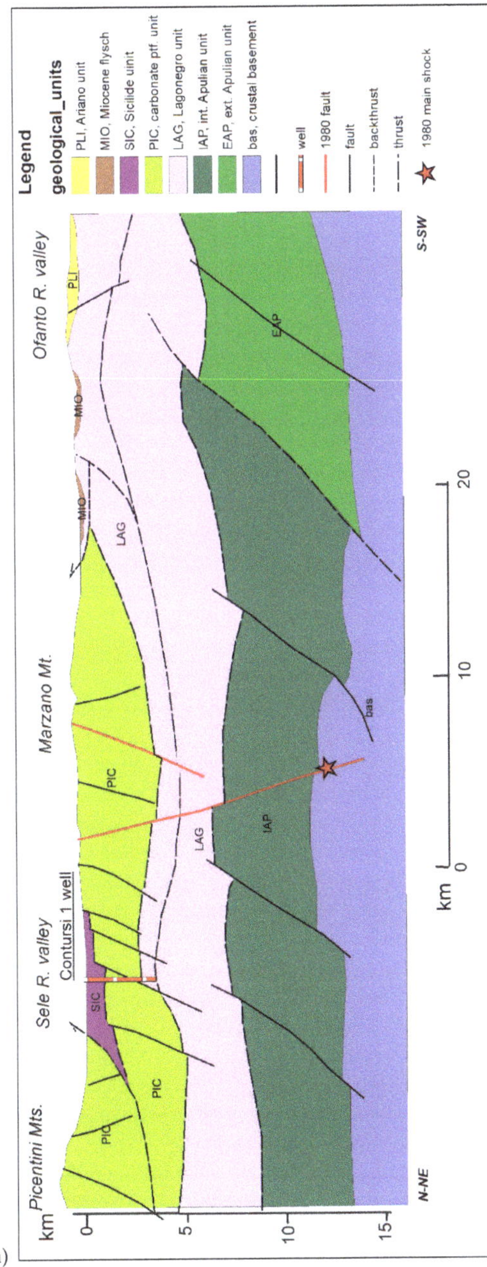
Legend
geological_units
PLI, Ariano unit
MIO, Miocene flysch
SIC, Sicilide uinit
PIC, carbonate ptf. unit
LAG, Lagonegro unit
IAP, int. Apulian unit
EAP, ext. Apulian unit
bas, crustal basement
well
1980 fault
fault
backthrust
thrust
1980 main shock
Picentini Mts.
Sele R. valley
Contursi 1 well
Marzano Mt.
Ofanto R. valley
N-NE
S-SW
km
0
5
10
15
km 0 10 20
PIC
SIC
LAG
IAP
MIO
PLI
EAP
bas

(b)

Figure 12. Geological sections. (**a**) Detailed geological section along the Pavoncelli-bis tunnel profile (for tunnel trace, see Figures 2 and 4). The vertical scale is double respect to horizontal scale. Legend: (a) Sicilide unit—Monte S. Arcangelo Fm. (FMS) with four lithofacies: FMS-arg (argillaceous lithofacies), FMS-arg-ma (argillaceous-marly lithofacies), FMS-ma-arg (marly-argillaceous lithofacies), and FMS-cal-ma (calcareous-marly lithofacies). Argille Varicolori Superiori Fm. (ALV) and calcareous lithofacies (ALVc), Albanella/Corleto unit (ABA); (**b**) Mt. Picentini unit—bio-litoclastic rudist limestone fm. (CBI); (c) Frigento unit—Flysch Rosso Fm. (FYR) and limestone member (FYR2); synorogenic units—Castelvetere Formation (CVT), Ruvo del Monte Synthem (RVM); Quaternary units—Conza della Campania synthem (CZC), landslide deposits (a1). (**b**) Geological sketch section (for trace, see Figures 2 and 4). Modified after [27,28,89].

5.2. Geology and Tectonic Evolution

The studied region is located at the intersection of the Irpinia, Sele Valley, and Lucania sectors in the axial-eastern margin of the Southern Apennines (Figure 1). This sector represents a key area for understanding the paleogeographic and tectonic evolution. Trias to Early Miocene basin to slope successions and Middle Miocene to Pliocene foreland clastic successions widely occur. The pre-orogenic successions are arranged into several regional tectonic units, such as the Sicilide, Picentini Mt., Marzano Mt., and Frigento units [27,84,90]. The Sicilide Unit consists of basinal facies ranging in age from Late Cretaceous to Early Miocene; the unit overthrusts both Picentini Mts. Unit and Frigento Unit. The Picentini Mts. Unit is a fragment of the Southern Apenninic platform (namely "Campano-Lucana" platform) and is formed by pre-orogenic carbonate platform successions ranging from Late Jurassic to Paleogene; it is tectonically lying on the Frigento Unit. The Mt. Marzano unit is formed by upper Trias-Langhian pre-orogenic carbonate platform to slope successions, followed by Serravallian foredeep deposits (Laviano sequence. LAI). The Frigento Unit [27] includes lower Cretaceous shales with resedimented limestones (Flysch Galestrino) and Upper Cretaceous–Lower Miocene succession of hemipelagic mudstones and with resedimented limestones (i.e., 'Flysch Rosso' Fm.), conformably overlain by Numidian quartzarenites. These tectonic units are unconformable overlain by thrust-top basin and foredeep basin fillings of mixed to siliciclastic successions, ranging from Early Miocene to Pliocene in age (ABA, SIE, CVT, RVM). These successions are separated by regional unconformities [27].

The described units are strongly deformed and thrusted eastward, with minor westward back-thrust structures, and testify a poly-phase orogenic evolution during Early Miocene, Middle–Late Miocene, and latest Miocene–Pliocene times. These phases are characterized by the subduction of the oceanic crust along the Adria margin producing an accretionary prism during Paleogene–Early Miocene, Adria lithospheric flexing and tectonic contraction of Meso-Cenozoic platforms and basins during Tortonian and Messinian and by the reimbrication, thrusting, and refolding of older thrust-sheets [27,33,84].

During Early Miocene tectonic phases, the final closure of the southern and eastern Tethyan realm and onset of accretionary processes of the Mesomediterranean microplate (e.g., Calabrian Terranes [25]) is responsible of huge volumes of clastic sedimentary sequences, dominantly deep-marine turbidite systems, which were deposited at the front of the accretionary orogenic terranes in foreland basin systems [25]. The latest Oligocene-to earliest Miocene correspond with the final closure of the Lucanian Oceanic realm, where sandstone turbidite successions having dominantly a quartzolithic composition (i.e., Corleto sandstone Formation), interbedded with volcanolithic sandstones (i.e., Tufiti di Tusa Formation), were accommodated. After deformation of the oceanic terranes, at the end of Burdigalian, the Langhian to Tortonian foreland sequences unconformably covered the oceanic units and correspond with deposition of the Cilento Group.

The Tortonian phase was characterized by a north-east direction of the tectonic transport, causing the sequential superimposition of the Sicilide Unit onto the Apennine Platform units and of these units on the Frigento Unit. The phase produced geometric relationships of hanginwall-flat on footwall-flat type [27]. These thrusts are suturated by the Late Tortonian–Early Messinian Castelvetere Formation.

This latter unit, a sand-rich turbidite system, has dominantly quartzofeldspathic sandstones, which has an abundance of coarse grained plutonic and high-grade metamorphic phaneritic rock fragments, derived from weathered [25,33,81,91] highest tectonic units of the northern Calabrian terranes.

During Early Messinian–Middle Pliocene, a stage of collisional tectonics with sequential effects of break-back-thrust type developed [27]. During the Early-Late Messinian phase, new geometric relationships of the previously assembled units were produced by out of sequence thrusting structures (breaching, [43]), which were sutured by the Late Messinian successions (Altavilla unit and Anzano Molasse Fm. [23]). The Early–Middle Pliocene phase was characterized by the development of severe rearrangements within the orogenic wedge. Since Early Pliocene, the breaching effects determined high angle faults, namely fault cut-off, involving the older faults within the tectonic wedge [27]. A back-regressive evolution of the fault cut-off from the current marginal areas to that internal to the chain (from east to west) occurred at regional scale. Middle Pliocene unit (Ruvo del Monte Synthem) sutured the fault cut-off in the Ofanto river valley. This tectonic evolution can be attributed to the structuring processes of the buried Apulian Unit as recorded in the more external Irpinia sector [92,93]. The uplift generated by the buried duplexing of the Apulian continental crust blocked the sequential kinematics progression towards the Adriatic sectors. The resistance to progression, during Africa-Adria plates convergence, generated surface dislocations towards the Tyrrhenian due to a gravity processes of re-adjustments in the tectonic wedge behind the outer edge, causing the back-thrusting of the Lagonegro unit onto the Sicilide unit [27].

The regional geological section (Figure 12b) shows the relationships among assembled tectonic units, Quaternary normal faults and 1980 seismic faults. The studied axial areas of southern Apennines are also characterized by high heat flux data suggesting the presence of melt intruded into the crust along lithospheric faults in the zone [94,95]. A relationship between the geodynamic evolution, the melt intrusions, and the genesis of the large earthquakes in the region is suggested [94].

The few volcanic rocks found close to the 1980 epicentral area are those of Tertiary age crossed by deep oil drillings [28], and the allochthonous diabase within Sicilide unit cropping out near Frigento [92,93,95], but there is no evidence of relation between these magmatic rocks and the analyzed volcaniclastic rocks.

6. Conclusions

The geological study carried out in the southern Apennine divide area, where the new Caposele-Conza hydraulic tunnel called Pavoncelli bis was built, allowed to obtain relevant new elements on the tectonic and stratigraphic evolution of the Apennine thrust belt.

The excavation of the Pavoncelli bis gallery has given us a new opportunity for obtaining a detailed knowledge of the geological structures that are present in the subsurface of the studied sector of the Apennine chain, where the tectonic events have almost completely hidden the original geometric relationships existing between the various geological units. The complexity of the geological structures and the mainly pelitic nature of the lithologies here present made the reconstruction of the geological structures crossed by the excavation very difficult, especially in correspondence with the tunnel sections with high thickness coverage. However, the collection of original subsoil data and the results of detailed geological surveys, supported by in-depth analysis of the petrographic characteristics of the arenitic successions, allowed to elaborate a geological detailed model of the area crossed by the tunnel excavations, which has been reported in the geological profile (Figure 12).

For the first time for this area, the presence of quartzolithic arenaceous successions with volcanoclastic composition, related to the Early Miocene arenaceous successions of the Corleto/Albanella Sandstone Formations, and a quartz-feldspathic arenaceous successions, related to the late Miocene Castelvetere Formation, have been documented in an exhaustive way both in the outcrops and in the tunnel sectors.

The stratigraphic and petrological data allowed to recognize the presence in the study area of the arenitic sequence of quartzolithic petrofacies corresponding to the lower Miocene Corleto/Albanella

Sandstone Formation. These units usually unconformably cover the Argille Varicolori Superiori Fm., typical of the upper part of the Sicilide Unit, and were originally deposited in the Lucanian basin. These data may allow conforming the presence of the Sicilide unit in the axial sector of the Campania mountain belt eastward to the Apennine carbonate units.

Author Contributions: Methodology, S.d.N., F.M., and S.C. (Salvatore Critelli); petrological analysis, S.C. (Sara Criniti) and S.C. (Salvatore Critelli); field investigation, S.d.N. and F.M.; writing, F.M. and S.C. (Salvatore Critelli); project administration, S.d.N. All authors have read and agreed to the published version of the manuscript.

Funding: This research received no external funding.

Acknowledgments: We acknowledge the support given in the field surveys by geol. Antonio Riviello, and in the tunnel surveys by Eng. Roberto De Angelis. We also acknowledge for the technical support (access to excavated rock and soil sample, assistance during tunnel visits, and logistic support) given by Caposele scarl.

Conflicts of Interest: The authors declare no conflict of interest.

References

1. Guidoboni, E.; Ferrari, G.; Mariotti, D.; Comastri, A.; Tarabusi, G.; Sgattoni, G.; Valensise, G. *CFTI5Med, Catalogo dei Forti Terremoti in Italia (461 a.C.-1997) e nell'area Mediterranea (760 a.C.-1500)*; Istituto Nazionale di Geofisica e Vulcanologia: Roma, Italy, 2018. Available online: storing.ingv.it/cfti/cfti5/ (accessed on 18 June 2020).
2. Locati, M.; Camassi, R.; Rovida, A.; Ercolani, E.; Bernardini, F.; Castelli, V.; Caracciolo, C.H.; Tertulliani, A.; Rossi, A.; Azzaro, R.; et al. *Database Macrosismico Italiano (DBMI15), versione 2.0*; Istituto Nazionale di Geofisica e Vulcanologia: Roma, Italy, 2019. Available online: emidius.mi.ingv.it/CPTI15-DBMI15/description_DBMI15.htm (accessed on 18 June 2020).
3. Branno, A.; Esposito, E.; Luongo, G.; Marturano, A.; Porfido, S.; Rinaldis, V. The largest earthquakes of the Apennines, southern Italy. In Proceedings of the AIGI-IAEG Proceedings, Bari, Italy, 13–19 April 1986; Volume 4, pp. 3–14.
4. Bernard, P.; Zollo, A. The Irpinia (Italy) 1980 earthquake: Detailed analysis of a complex normal faulting. *J. Geophys. Res.* **1989**, *94*, 1631–1648. [CrossRef]
5. Valensise, G. Summary of Contributions on the 23 November 1980, Irpinia earthquake. *Ann. di Geofis.* **1993**, *36*, 345–351.
6. Westaway, R.W.C.; Jackson, J.A. Surface faulting in the southern Italian Campania—Basilicata earthquake of 23 November 1980. *Nature* **1984**, *312*, 436–438. [CrossRef]
7. Pingue, F.; De Natale, G. Fault mechanism of the 40 seconds subevent of the 1980 Irpinia (Southern Italy) earthquake from levelling data. *Geophys. Res. Lett.* **1993**, *20*, 911–914. [CrossRef]
8. Porfido, S.; Alessio, G.; Gaudiosi, G.; Nappi, R.; Spiga, E. The resilience of some villages 36 years after the Irpinia-Basilicata (Southern Italy) 1980 earthquake. In *Proc. 4th WLF2017, Advancing Culture of Living with Landslides*; Mikošet, M., Casagli, N., Yin, Y., Sassa, K., Eds.; Springer: Cham, Switzerland, 2017; Volume 15.
9. Cotecchia, V. Ground deformations and slope instability produced by the earthquake of 23 November 1980 in Campania and Basilicata. *Geol. Appl. Idrogeol.* **1986**, *21*, 31–100.
10. Cotecchia, V.; Salvemini, A.; Ventrella, N.A. Interpretazione degli abbassamenti territoriali indotti dal terremoto del 23 Novembre 1980 e correlazioni con i danni osservati su talune strutture ingegneristiche dell'area epicentrale irpina. *Riv. Ital. Geotec.* **1990**, *4*, 145–158.
11. Esposito, E.; Pece, R.; Porfido, S.; Tranfaglia, G. Hydrological anomalies connected to earthquakes in Southern Apennines (Italy). *Nat. Hazards Earth Syst. Sci.* **2001**, *1*, 137–144. [CrossRef]
12. Porfido, S.; Esposito, E.; Michetti, A.M.; Blumetti, A.M.; Vittori, E.; Tranfaglia, G.; Guerrieri, L.; Ferreli, L.; Serva, L. Areal distribution of ground effects induced by strong earthquakes in the Southern Apennines (Italy). *Surv. Geophys.* **2002**, *23*, 529–562. [CrossRef]
13. Gizzi, F.T.; Potenza, M.R.; Zotta, C. 23 November 1980 Irpinia–Basilicata earthquake (Southern Italy): Towards a full knowledge of the seismic effects. *Bull. Earthq. Eng.* **2012**, *10*, 1109–1131. [CrossRef]
14. Galli, P.; Peronace, E. New paleoseismic data from the Irpinia fault. A different seismogenic perspective for southern Apennines (Italy). *Earth Sci. Rev.* **2014**, *136*, 175–201. [CrossRef]

15. Galli, P.A.; Peronace, E.; Quadrio, B.; Esposito, G. Earthquake fingerprints along fault scarps: A case study of the Irpinia 1980 earthquake fault (southern Apennines). *Geomorphology* **2014**, *206*, 97–106. [CrossRef]
16. Engineering Office. The Apulian Acqueduct, Southern Italy. *Engineering* **1929**, 35–36.
17. Servizio Geologico d'Italia. *Carta Geologica d'Italia alla Scala 1:100.000, F° 186 «S. Angelo dè Lombardi»*, 1st ed.; Servizio Geologico d'Italia: Roma, Italy, 1970. Available online: http://193.206.192.231/carta_geologica_italia/tavoletta.php?foglio=186 (accessed on 9 May 2020).
18. Servizio Geologico d'Italia. *Carta Geologica d'Italia alla Scala 1:100.000, F° 198 «Eboli»*; Servizio Geologico d'Italia: Roma, Italy, 1970. Available online: http://193.206.192.231/carta_geologica_italia/tavoletta.php?foglio=198 (accessed on 9 May 2020).
19. Bonardi, G.; D'argenio, B.; Perrone, B. Carta geologica dell'Appennino meridionale. *Mem. Soc. Geol. Ital.* **1988**, *41*, 1341.
20. ISPRA—Servizio Geologico d'Italia. Carta Geologica d'Italia Alla Scala 1:50000, Foglio 468 Eboli. Available online: www.isprambiente.gov.it/Media/carg/468_EBOLI/Foglio.html (accessed on 9 May 2020).
21. ISPRA—Servizio Geologico d'Italia. Carta Geologica d'Italia Alla Scala 1:50000, Foglio 450 Sant'Angelo dei Lombardi. Available online: www.isprambiente.gov.it/Media/carg/450_SANTANGELOLOMBARDI/Foglio.html (accessed on 9 May 2020).
22. Bonardi, G.; Ciarcia, S.; Di Nocera, S.; Matano, F.; Sgrosso, I.; Torre, M. Carta delle principali Unità Cinematiche dell'Appennino Meridionale. Nota illustrativa. *Ital. J. Geosci.* **2009**, *128*, 47–60.
23. Matano, F.; Critelli, S.; Barone, M.; Muto, F.; Di Nocera, S. Stratigraphic and provenance evolution of the southern Apennines foreland basin system during the Middle Miocene to Pliocene (Irpinia-Sannio successions, Italy). *Mar. Pet. Geol.* **2014**, *57*, 652–670. [CrossRef]
24. Postpischl, D.; Branno, A.; Esposito, E.; Ferrari, G.; Maturano, A.; Porfido, S.; Rinaldis, V.; Stucchi, M. The Irpinia earthquake of November 23; 1980. *Atlas Isoseismal Maps Ital. Earthq. CNR-PFG* **1985**, *114*, 152–157.
25. Critelli, S. Provenance of Mesozoic to Cenozoic Circum-Mediterranean sandstones in relation to tectonic setting. *Earth-Sci. Rev.* **2018**, *185*, 624–648. [CrossRef]
26. Di Nocera, S.; Matano, F.; Torre, M. Le unità "sannnitiche" Auct. (Appennino centromeridionale) rassegna delle correnti interpretazioni stratigrafiche e paleogeografiche e nuove ipotesi con l'introduzione dell'Unità di Frigento. *St. Geol. Camerti* **2002**, *1*, 87–102.
27. Di Nocera, S.; Matano, F.; Pescatore, T.; Pinto, F.; Quarantiello, R.; Senatore, M.; Torre, M. Schema geologico del transetto Monti Picentini orientali—Monti della Daunia meridonali: Unità stratigrafiche ed evoluzione tettonica del settore esterno dell'Appennino meridionale. *Boll. Soc. Geol. Ital.* **2006**, *125*, 1–20.
28. Mostardini, F.; Merlini, S. Appennino Centro-meridionale: Sezioni geologiche e proposta di modello strutturale. *Mem. Della Soc. Geol. Ital.* **1986**, *35*, 177–202.
29. Pescatore, T.; Renda, P.; Tramutoli, M. Rapporti tra le Unità lagonegresi e l'Unità sicilide nella media valle del Basento. *Mem. Soc. Geol. Ital.* **1988**, *41*, 353–361.
30. Ogniben, L. Schema introduttivo alla geologia del confine calabro-lucano. *Mem. Soc. Geol. Ital.* **1969**, *8*, 453–763.
31. Critelli, S. Sandstone detrital modes in the Paleogene Liguride Complex, accretionary wedge of the Southern Apennines (Italy). *J. Sediment. Petrol.* **1993**, *63*, 464–476.
32. Critelli, S. The interplay of lithospheric flexure and thrust accomodation in forming stratigraphic sequences in the southern Apennines foreland basin system, Italy. *Mem. Dell'Accad. Naz. Dei Lincei* **1999**, *10*, 257–326. [CrossRef]
33. Critelli, S.; Muto, F.; Perri, F.; Tripodi, V. Interpreting Provenance Relations from Sandstone Detrital Modes, Southern Italy Foreland Region: Stratigraphic record of Miocene tectonic evolution. *Mar. Pet. Geol.* **2017**, *87*, 47–59. [CrossRef]
34. Barone, M.; Critelli, S.; Le Pera, E.; Di Nocera, S.; Matano, F.; Torre, M. Stratigraphy and detrital modes of Upper Messinian Post-Evaporitic sandstones of the Southern Apennines, Italy: Evidence of foreland-basin evolution during the Messinian Mediterranean salinity crisis. *Int. Geol. Rev.* **2006**, *48*, 702–724. [CrossRef]
35. Critelli, S.; Muto, F.; Tripodi, V.; Perri, F. Relationships between lithospheric flexure, thrust tectonics and stratigraphic sequences in foreland setting: The Southern Apennines foreland basin system, Italy. In *New Frontiers in Tectonic Research at the Midst of Plate Convergence*; Schattner, U., Ed.; Intech Open: London, UK, 2011; pp. 121–170.

36. Critelli, S.; Muto, F.; Tripodi, V.; Perri, F. Link between thrust tectonics and sedimentation processes of stratigraphic sequences from the southern Apennines foreland basin system, Italy. *Rend. Online Soc. Geol. Ital.* **2013**, *25*, 21–42.
37. Critelli, S.; Le Pera, E. Tectonic evolution of the Southern Apennines thrust-belt (Italy) as reflected in modal compositions of Cenozoic sandstone. *J. Geol.* **1995**, *103*, 95–105. [CrossRef]
38. Critelli, S.; Le Pera, E. Post-Oligocene sediment dispersal systems and unroofing history of the Calabrian Microplate, Italy. *Int. Geol. Rev.* **1998**, *48*, 609–637. [CrossRef]
39. Doglioni, C.; Harabaglia, P.; Martinelli, G.; Mongelli, F.; Zito, G. A geodynamic model of the southern Apennines accretionary prism. *Terra Nova* **1996**, *8*, 540–547. [CrossRef]
40. Malinverno, A.; Ryan, W.B.F. Extension in the Tyrrhenian Sea and shortening in the Apennines as result of arc migration driven by sinking of the lithosphere. *Tectonics* **1986**, *5*, 227–245. [CrossRef]
41. Patacca, E.; Scandone, P. Post-Tortonian mountain building in the Apennines. The role of the passive sinking of a relict lithospheric slab. In *The Lithosphere in Italy. Advances in Earth Science Research*; Boriani, A., Bonafede, M., Piccardo, G.B., Vai, G.B., Eds.; Accademia nazionale dei Lincei: Rome, Italy, 1989; Volume 80, pp. 157–176, Atti Conv. Lincei.
42. Doglioni, C. A proposal for the kinematic modelling of W-dipping subduction possible applications to the Tyrrhenian-Apennines system. *Terra Nova* **1991**, *3*, 423–434. [CrossRef]
43. Butler, R.W. Thrust sequences. *J. Geol. Soc.* **1987**, *144*, 619–634. [CrossRef]
44. Selli, R. Sulla trasgressione del Miocene nell'Italia Meridionale. *G. Geol.* **1957**, *26*, 1–54.
45. Selli, R. Il Paleogene nel quadro della geologia dell'Italia centro meridionale. *Mem. Soc. Geol. Ital.* **1962**, *3*, 737–789.
46. D'Argenio, B.; Pescatore, T.; Scandone, P. Structural pattern of the Campania-Lucania Apennines. In *Structural model of Italy*; Ogniben, L., Parotto, M., Praturlon, A., Eds.; Quaderni de La ricerca scientifica: Roma, Italy, 1975; Volume 90, pp. 313–327.
47. D'Argenio, B.; Pescatore, T.; Scandone, P. Schema geologico dell'Appennini meridionale (Canpania, Lucania). *Acc. Naz. Lincei Atti del Convegno Mod. Vedute Sulla Geol. Dell'Appennino.* **1973**, *182*, 49–72.
48. Sgrosso, I. Nuovi elementi per un più articolato modello paleogeografico nell'Appennino centro-meridionale. *Mem. Soc. Geol. Ital.* **1988**, *41*, 225–252.
49. Sgrosso, I. Possibile evoluzione cinematica miocenica nell'orogene centro-sud appenninico. *Boll. Soc. Geol. Ital.* **1998**, *117*, 679–724.
50. Patacca, E.; Scandone, P. Late thrust propagation and sedimentary response in the thrust-belt-foredeep system of the Southern Apennines (Pliocene-Pleistocene). In *Anatomy of an Orogen: The Apennines and adjacent Mediterranean Basins*; Vai, G.B., Martini, I.P., Eds.; Kluwer Academic Publ.: Berlin, Germany, 2001; pp. 401–440.
51. Mazzoli, S.; Aldega, L.; Corrado, S.; Invernizzi, C.; Zattin, M. Pliocene-quaternary thrusting, syn-orogenic extension and tectonic exhumation in the Southern Apennines (Italy): Insights from the Monte Alpi area. *Geol. Soc. Am. Sp. Pap.* **2006**, *414*, 55–77.
52. Mazzoli, S.; D'Errico, M.; Aldega, L.; Corrado, S.; Invernizzi, C.; Shiner, P.; Zattin, M. Tectonic burial and "young" (<10 Ma) exhumation in the southern Apennines fold and thrust belt (Italy). *Geology* **2008**, *36*, 243–246. [CrossRef]
53. Mazzoli, S.; Ascione, A.; Buscher, J.T.; Pignalosa, A.; Valente, E.; Zattin, M. Low-angle normal faulting and focused exhumation associated with late Pliocene change in tectonic style in the southern Apennines (Italy). *Tectonics* **2014**, *33*, 1802–1818. [CrossRef]
54. Pescatore, T.; Di Nocera, S.; Matano, F.; Pinto, F. L'Unità del Fortore nel quadro della geologia del settore orientale dei Monti del Sannio (Appennino meridionale). *Boll. Soc. Geol. Ital.* **2000**, *119*, 587–601.
55. Pescatore, T.; Pinto, F.; Renda, P.; Senatore, M.R.; Tramutoli, M.; Valente, A. Avanfosse mioceniche dell'Appennino meridionale (Italia). *Rend. Acc. Sc. Fis. Mat. Napoli* **1996**, *63*, 85–121.
56. Bonardi, G.; Amore, F.O.; Ciampo, G.; De Capoa, P.; Miconnet, P.; Perrone, V. Il complesso Liguride auct.: Stato delle conoscenze e problemi aperti sulla sua evoluzione pre-appenninica ed i suoi rapporti con l'Arco Calabro. *Mem. Soc. Geol. Ital.* **1988**, *41*, 17–35.
57. Carannante, G.; Severi, C.; Simone, L. Off-shelf carbonate transport along foramol (temperate-type) open shelf margins: An example from the Miocene of the central-southern Apennines (Italy). *Mem. Soc. Geol. Fr.* **1994**, *169*, 277–288.

58. D'Argenio, B. Zone isopiche e faglie trascorrenti nell'Appennino meridionale. *Mem. Soc. Geol. Ital.* **1966**, *4*, 279–299.
59. D'Argenio, B. Le piattaforme carbonatiche periadriatiche. Una rassegna di problemi nel quadro geodinamico mesozoico dell'area mediterranea. *Mem. Soc. Geol. Ital.* **1974**, *13*, 137–160.
60. Dazzaro, L.; Di Nocera, S.; Pescatore, T.; Rapisardi, L.; Romeo, M.; Russo, B.; Senatore, M.R.; Torre, M. Geologia del margine della catena appenninica tra il F. Fortore ed il T.Calaggio (Monti della Daunia—App. Merid.). *Mem. Soc. Geol. Ital.* **1988**, *41*, 411–422.
61. Scandone, P. Studi di geologia lucana: La serie calcareo-silico-marnosa e i suoi rapporti con l'Appennino calcareo. *Boll. Soc. Nat. Napoli* **1967**, *76*, 1–175.
62. Scandone, P. Note illustrative della Carta Geologica d'Italia alla scala 1:100,000 dei Fogli 199 e 210, Potenza e Lauria. *Serv. Geol. d'Ital.* **1971**, 71. Available online: http://sgi.isprambiente.it/geologia100k/mostra_pdf.aspx?pdffile=199.pdf (accessed on 9 May 2020).
63. Basso, C.; Ciampo, G.; Ciarcia, S.; Di Nocera, S.; Matano, F.; Staiti, D.; Torre, M. Geologia del settore irpino-dauno dell'Appennino meridionale: Implicazioni sui domini paleogeografici delle unità bacinali meso-cenozoiche e nuovi vincoli stratigrafici nell'evoluzione tettonica mio-pliocenica del settore esterno della catena. *Stud. Geol. Camerti* **2002**, *1*, 7–26.
64. Basso, C.; Di Nocera, S.; Matano, F.; Esposito, P.; Russo, B.; Torre, M. Stratigrafia delle successioni sedimentarie evaporitiche e post-evaporitiche del Messiniano superiore in Irpinia settentrionale (Appennino meridionale). *Boll. Della Soc. Geol. Ital.* **2001**, *120*, 211–231.
65. Pescatore, T.; Senatore, M.R. A comparison between a present-day (Taranto Gulf) and Miocene (Irpinian basin) foredeep of the Southern Apennines (Italy). *Spec. Publs Int. Ass. Sediment.* **1986**, *8*, 169–182.
66. Pavan, G.; Pirini, C. Stratigrafia del Foglio 157 «Monte, S. Angelo». *Boll. Serv. Geol. d'Ital.* **1966**, *86*, 123–189.
67. Ricchetti, G.; Ciaranfi, N.; Luperto Sinni, E.; Mongelli, F.; Pieri, P. Geodinamica ed evoluzione sedimentaria e tettonica dell'avampaese apulo. *Mem. Soc. Geol. Ital.* **1988**, *41*, 57–82.
68. Perri, F.; Critelli, S.; Cavalcante, F.; Mongelli, G.; Sonnino, M.; Dominici, R.; De Rosa, R. Provenance signatures for the Miocene volcaniclastic succession of the Tufiti di Tusa Formation, southern Apennines, Italy. *Geol. Mag.* **2012**, *149*, 423–442. [CrossRef]
69. Critelli, S.; Le Pera, E. Detrital modes and Provenance of Miocene sandstones and Modern sands of the Southern Apennines thrust-top basins (Italy). *J. Sediment. Rese.* **1994**, *A64*, 824–835. [CrossRef]
70. Cavuoto, G.; Valente, A.; Nardi, G.; Martelli, L. Turbidite Depositional Systems and Architectures, Cilento, Italy. In *The Atlas of Deep-Water Outcrops of the World*; AAPG Special Publication, Studies in Geology: Tulsa, OK, USA, 2008; Volume 56.
71. Zecchin, M.; Civile, D.; Caffau, M.; Muto, F.; Di Stefano, A.; Maniscalco, R.; Critelli, S. The Messinian succession of the Crotone Basin (southern Italy) I: Stratigraphic architecture reconstructed by seismic and well data. *Mar. Pet. Geol.* **2013**, *48*, 455–473. [CrossRef]
72. Zecchin, M.; Caffau, M.; Di Stefano, A.; Maniscalco, R.; Lenaz, D.; Civile, D.; Muto, F.; Critelli, S. The Messinian succession of the Crotone Basin (southern Italy) II: Facies architecture and stratal surfaces across the Miocene-Pliocene boundary. *Mar. Pet. Geol.* **2013**, *48*, 474–492. [CrossRef]
73. Ingersoll, R.V.; Bullard, T.F.; Ford, R.L.; Grimm, J.P.; Pickle, J.D.; Sares, S.W. The effect of grain size on detrital modes: A test of the Gazzi-Dickinson point-counting method. *J. Sediment. Petrol.* **1984**, *54*, 103–116.
74. Zuffa, G.G. Optical analysis of arenites: Influence of methodology on compositional results. In *Provenance of Arenites*; Zuffa, G.G., Ed.; D. Reidel: Dordrecht, The Netherlands, 1985; pp. 165–189.
75. Zuffa, G.G. Unravelling hinterland and offshore paleogeography from deep water arenites. In *Deep-Marine Clastic Sedimentology: Concepts and Case Studies*; Leggett, J.K., Zuffa, G.G., Eds.; Graham and Trotman: London, UK, 1987; pp. 39–61.
76. Dickinson, W.R. Interpreting detrital modes of graywacke and arkose. *J. Sediment. Petrol.* **1970**, *40*, 695–707.
77. Dickinson, W.R. Interpreting provenance relations from detrital modes of sandstones. In *Provenance of Arenites*; Zuffa, G.G., Ed.; D. Reidel: Dordrecht, The Netherlands, 1985; Volume 148, pp. 333–361.
78. Critelli, S.; Ingersoll, R.V. Interpretation of neovolcanic versus palaeovolcanic sand grains: An example from Miocene deep-marine sandstone of the Topanga Group (southern California). *Sedimentology* **1995**, *42*, 783–804. [CrossRef]
79. Caracciolo, L.; Critelli, S.; Innocenti, F.; Kolios, N.; Manetti, P. Unravelling provenance from Eocene-Oligocene sandstones of the Thrace Basin, North-east Greece. *Sedimentology* **2011**, *58*, 1988–2011. [CrossRef]

80. Caracciolo, L.; Von Eynatten, H.; Tolosana-Delgado, R.; Critelli, S.; Manetti, P.; Marchev, P. Petrological, geochemical, and statistical analysis of Eocene-Oligocene sandstones of the Western Thrace Basin, Greece and Bulgaria. *J. Sediment. Res.* **2012**, *82*, 482–498. [CrossRef]
81. Critelli, S.; Le Pera, E.; Galluzzo, F.; Milli, S.; Moscatelli, M.; Perrotta, S.; Santantonio, M. Interpreting siliciclastic-carbonate detrital modes in Foreland Basin Systems: An example from Upper Miocene arenites of the Central Apennines, Italy. In *Sedimentary Provenance: Petrographic and Geochemical Perspectives*; Arribas, J., Critelli, S., Johnsson, M., Eds.; Geological Society of America Special Paper: Boulder, CO, USA, 2007; Volume 420, pp. 107–133.
82. Di Nocera, S.; Iannace, A.; Torre, M.; Basso, C.; Caiazzo, C.; Ciarcia, S.; Cinque, A.; Gasparrini, M.; Matano, F.; Mazzoli, S.; et al. Note Illustrative del Foglio 468 «Eboli». ISPRA—Servizio Geologico d'Italia, Note Illustrative della Carta Geologica d'Italia alla scala 1:50,000. Available online: www.isprambiente.gov.it/Media/carg/note_illustrative/468_Eboli.pdf (accessed on 9 May 2020).
83. Pescatore, T.S.; Pinto, F. Note Illustrative del Foglio 450 «S. Angelo dei Lombardi». ISPRA—Servizio Geologico d'Italia, Note Illustrative della Carta Geologica d'Italia alla scala 1:50,000; 2016. Available online: www.isprambiente.gov.it/Media/carg/note_illustrative/450_SAngeloLombardi.pdf (accessed on 9 May 2020).
84. Pescatore, T.S.; Di Nocera, S.; Matano, F.; Pinto, F.; Boiano, U.; Civile, D.; Martino, C.; Quarantiello, R. Prime considerazioni sulla geologia del settore centrale dei monti del Sannio. *Mem. Descr. Carta Geol. Ital.* **2008**, *77*, 77–94.
85. Critelli, S.; Le Pera, E. La formazione di Catelvetere nell'evoluzione petrostratigrafica dell'avanfossa del Tortoniano-Messiniano dell'Appennino meridionale. *Boll. Soc. Geol. Ital.* **1995**, *114*, 615–634.
86. Fornelli, A.; Micheletti, F.; Langone, A.; Perrone, V. First U-Pb detrital zircon ages from Numidian sandstone in southern Apennines: Evidences of African provenance. *Sediment. Geol.* **2015**, *320*, 19–29. [CrossRef]
87. Thomas, M.F.H.; Bodin, S.; Redfern, J.; Irving, D.H.B. Constrained African craton source for the Cenozoic Numidian Flysch: Implications for the palaeogeography of the western Mediterranean basin. *Earth Sci. Rev.* **2010**, *101*, 1–23. [CrossRef]
88. Cotecchia, V.; Lenti, V.; Salvemini, A.; Spilotro, G. Reactivation of the large "Buoninventre" slide by the Irpinia earthquake of 23 November 1980. *Geol. Appl. Idrogeol.* **1986**, *21*, 217–253.
89. Ascione, A.; Mazzoli, S.; Petrosino, P.; Valente, E. A decoupled kinematic model for active normal faults: Insights from the 1980, MS = 6.9 Irpinia earthquake, southern Italy. *GSA Bull.* **2013**, *125*, 1239–1259. [CrossRef]
90. Di Nocera, S.; Matano, F.; Pescatore, T.; Pinto, F.; Torre, M. Geological characteristics of the external sector of the Campania-Lucania Apennines in the CARG maps. *Rend. Online Soc. Geol. Ital.* **2011**, *12*, 39–43.
91. Scarciglia, F.; Le Pera, E.; Critelli, S. The onset of sedimentary cycle in a mid-latitude upland environment: Weathering, pedogenesis and geomorphic processes on plutonic rocks (Sila Massif, Calabria). In *Sedimentary Provenance: Petrographic and Geochemical Perspectives*; Arribas, J., Critelli, S., Johnsson, M., Eds.; Geological Society of America Special Paper; Geological Society of America: Boulder, CO, USA, 2007; Volume 420, pp. 149–166.
92. Di Nocera, S.; Imperato, M.; Matano, F.; Stanzione, D.; Valentino, G.M. Caratteri geologici ed idrogeochimici della valle di Ansanto (Irpinia Centrale, Appennino Campano-Lucano). *Boll. Soc. Geol. Ital.* **1999**, *118*, 395–406.
93. Matano, F.; Di Nocera, S. Geologia del settore centrale dell'Irpinia (Appennino meridionale): Nuovi dati e interpretazioni. *Boll. Soc. Geol. Ital.* **2001**, *120*, 3–14.
94. Sinno, R. I minerali della valle di Ansanto. *Atti Accad. Sci.Fis. Matem.* **1969**, *7*, 219–258.
95. Italiano, F.; Martelli, M.; Martinelli, G.; Nuccio, P.M. Geochemical evidence of melt intrusions along lithospheric faults of the Southern Apennines, Italy: Geodynamic and seismogenic implications. *J. Geophys. Res. Solid Earth* **2000**, *105*, 13569–13578. [CrossRef]

geosciences

MDPI

Review

Roman to Middle Age Earthquakes Sourced by the 1980 Irpinia Fault: Historical, Archaeoseismological, and Paleoseismological Hints

Paolo Galli [1,2]

1 Dipartimento della Protezione Civile, Via Vitorchiano 4, 00189 Rome, Italy; paolo.galli@protezionecivile.it
2 CNR-IGAG, Via Salaria km 29,300, 00015 Monterotondo Stazione (Rome), Italy

Received: 11 June 2020; Accepted: 22 July 2020; Published: 27 July 2020

Abstract: The Italian seismic compilations are among the most complete and back-in time extended worldwide, with earthquakes on record even before the Common Era. However, we have surely lost the memory of dozen strong events of the historical period, mostly in the first millennium CE. Given the lack of certain or conclusive written sources, besides paleoseismological investigations, a complementary way to infer the occurrence of lost earthquakes is to cross-check archaeoseismic evidence from ancient settlements. This usually happens by investigating collapses/restorations/reconstructions of buildings, the general re-organization of the urban texture, or even the abrupt abandonment of the settlement. Exceptionally, epigraphs mentioning more or less explicitly the effects of the earthquake strengthened the field working hypothesis. Here, I deal with both paleoseismological clues from the Monte Marzano Fault System (the structure responsible for the catastrophic, Mw 6.9 1980 earthquake) and archaeoseismological evidence of settlements founded in its surroundings to cast light on two poorly known earthquakes that occurred at the onset and at the end of the first millennium CE, likely in 62 and in 989 CE. Both should share the same seismogenic structure and the size of the 1980 event (Mw 6.9).

Keywords: Irpinia fault; historical earthquakes; archaeoseismology; paleoseismology

1. Introduction

As widely demonstrated by several works, the Irpinia fault—which is the popular name of the Monte Marzano Fault System (MMFS, [1])—was responsible in November 1980 for a devastating earthquake (Mw 6.9), which was accompanied by the longest surface faulting ever observed in Italy (>30 km), at least before the recent central Apennines event (October 2016, Mw 6.6). Early paleoseismological studies [2] claimed that the recurrence time for 1980-like characteristic earthquakes during the Holocene was approximately 2 kyr, which is a value clashing with both the frequent destructive seismicity of the area affected by the fault [3] and the high GPS-derived strain rate of the region (at least 2.9 mm/year [4,5]). This inconsistency was fixed through high-precision topographic leveling run across the compound fault scarp [6], and new paleoseismological trenches and pits opened on different fault segments of the 43 km long MMFS [7]. Results revealed that most of the destructive earthquakes that hit the upper Ofanto valley in the past two millennia (e.g., events of 1980, Mw 6.9; 1694, Mw 7.0; 1466, Mw 6.3, and many others; as shown in Figure 1) might be ascribed to the MMFS periodical activation, without the necessity of evoking other, unknown faults.

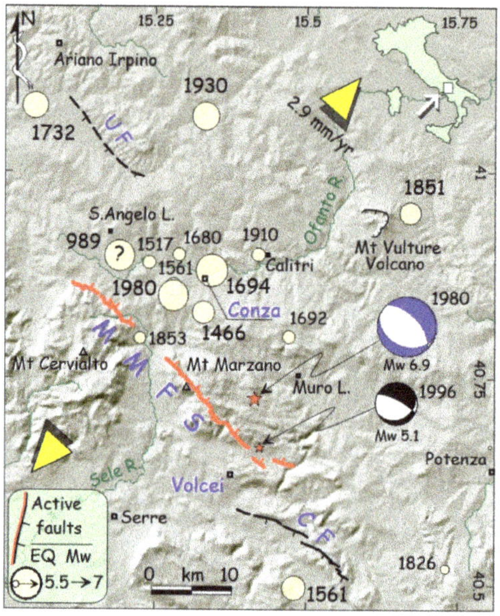

Figure 1. Macroseismic epicenters of Mw > 5.5 earthquakes in southern Irpinia (modified from [7]). Notwithstanding possible approximate location, earthquakes cluster mainly in the hanging wall of the Monte Marzano Fault System (MMFS). CF, Caggiano fault; UF, Ufita fault. Stars are the instrumental epicenters of 1980 (Mw 6.9; epicenter from [8]) and 1996 (Mw 5.1) events. Diverging triangles indicate crustal extension, as deduced by GPS analyses published by [4,5].

Macroseismic studies of recent normal faulting earthquakes in Italy showed that the spatial distribution of damage is strongly influenced by rupture directivity effects [9–14], explaining why the same seismogenic source might generate different shaking scenarios time after time. For instance, the 1980 rupture, nucleating from the SE tip of the MMFS toward NW [15], induced the northwestward damage distribution depicted by [16] (Figure 2D). Conversely, in the stronger 1694 earthquake, the rupture nucleated from the NW tip toward SE [7], as proven by the huge damage suffered by many villages located on the eastern side of the highest-intensity data points distribution (from now: HIDD; see Figure 2C).

This paper deals with two little-known earthquakes that were likely generated by the recurring rupture of the Irpinia fault. The first one occurred in the second half of the 1st century CE (Common Era), whereas the second occurred at the end of the 10th century (e.g., 989 CE). Both have left their destructive imprint in some archaeological settlement of the region, and both are mentioned in a few primary historical sources. Moreover, their signature can be read across the stratigraphy exposed within the paleoseismological trenches opened in the past years along the Irpinia fault or scanning the stepped profile of its compound scarp. By joining historical/epigraphical sources, archaeological evidence, and paleoseismic data, I have tried to provide more reliable seismogenic hypothesis and robust parameters for these events.

Here, I will refer to the MCS scale, which is the Mercalli–Cancani–Sieberg Macroseismic scale [17,18] adopted in all the seismic catalogues and works in Italy. Io and Is are epicentral and site intensity, respectively.

2. Historical Seismicity of the MMFS Region

The historical seismicity of the investigated area is among the strongest of the Apennines, both in terms of frequency and of maximum magnitude, with epicenters mainly concentrated in the MMFS hanging wall (Figure 1). As observed for the longest normal fault systems dissecting the Apennines [19], the largest events of Irpinia (Mw ≥ 6.9) were sourced by the contemporary rupture of the entire MMFS, whereas minor earthquakes occurred on single or grouped segments of this system [7].

Among the latter, the strongest was on January 14, 1466 (Figure 2B), striking the upper Ofanto and Sele Valleys, destroying the town of Conza (Is 9–10 MCS) and causing damage, collapses, and casualties in many other villages in the area, as witnessed by both archaeoseismic indications [20] and contemporary sources [21]. Although we have only 30 intensity points, the 1466 HIDD clearly falls in the hanging wall of the MMFS, roughly mimicking the 1694 and 1980 effects distribution (Figure 2C–D). Nonetheless, as the 1466 event had a lesser destructive impact, it was not sourced by the entire rupture of the MMFS, as happened in the two forthcoming, catastrophic earthquakes. By inverting its HIDD with the Boxer algorithm [22], its equivalent magnitude was rated as Mw 6.3 [7].

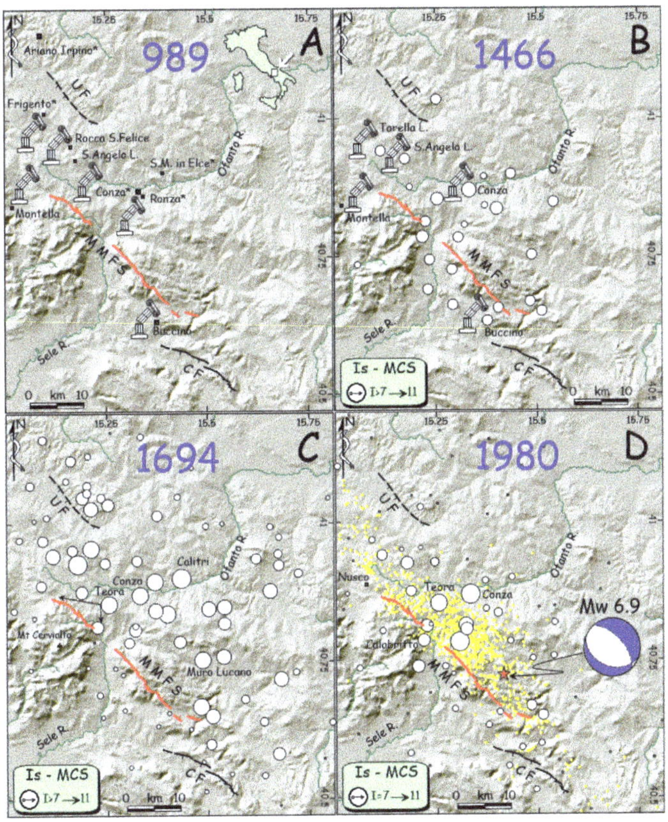

Figure 2. Distribution of effects induced by the 989, 1466, 1694, and 1980 earthquakes (MCS intensities from [7]). (**A**), 989 event; collapsing columns indicate coeval archaeoseismic evidence (see text). Asterisks, damaged villages quoted by contemporary sources. (**B**), 1466 event (data from [20,23]. (**C**), 1694 event; the arrows near Teora indicate the sector of the slope where contemporary accounts describe the opening of a long chasm, which was probably due to surface faulting [7]. (**D**), 1980 event; yellow circles in the background are 1980 aftershocks. Focal mechanism from [24]. MCS: Mercalli–Cancani–Sieberg Macroseismic scale.

During the 16th century, two damaging earthquakes had again their epicenters in the MMFS hanging wall. On March 29, 1517, the first one caused collapses and victims in Conza, and it was strongly felt from Ariano Irpino to Serre [20]. The second is part of a complex seismic sequence, which included three mainshocks occurring between July 30 and August 19, 1561. Castelli et al. [3] located one of these in the upper Ofanto Valley (Figure 1), where it caused heavy damage in several settlements. However, both the 1517 and 1561 events lack conclusive parameters, as their poor or confusing historical descriptions make it difficult to assess robust epicentral coordinates or magnitude.

Starting from the end of the 17th century, the region suffered the strongest seismic period of its long history. Two strong long-term foreshocks in 1680 (9 November) and 1692 (4 March; Figure 1) paved the way to the catastrophic September 8, 1694 earthquake (Mw 7.0), which caused ca. 6000 deaths, razing to the ground (i.e., Is 10–11 MCS) dozens of villages in the upper Ofanto Valley [25]. Even in 1694, the HIDD focuses on the MMFS hanging wall, with an abrupt intensity decrease in the footwall (Figure 2C). Coeval accounts described an impressive ground rupture formed in the Teora mountains, with a length of 10 Neapolitan miles (18.5 km), suggesting the occurrence of surface faulting along the northern sections of the MMFS, as confirmed by paleoseismological investigations [7].

Leaving aside the catastrophic earthquake of 29 November 1732, which also struck towns located in the upper Ofanto Valley, but with an epicentral area shifted further northwest (Figure 1; Ufita Fault), in 1853 and 1910, the MMFS hanging wall was affected by two other moderate earthquakes. The former (Mw 5.5) caused severe damage in the upper Sele Valley villages (e.g., Caposele and Calabritto [26]), whereas the latter (Mw 5.7) induced collapses and victims mainly in Calitri. The 1853 earthquake can be directly associated with the rupture of a segment of the MMFS, whereas the 1910 was likely sourced by an antithetic fault bounding the hanging wall to the north [7].

At the end, the 23 November 1980 earthquake (Mw 6.9; Io 10) hit the same villages already devastated in 1466 and 1694, destroying 75,000 buildings and severely damaging another 275,000 (Figure 2D). The death toll was 3000, the highest in Italy in the 20th century, after the 1908 (Messina) and 1915 (Fucino) events, with the total destruction of some towns such as Conza (30% of the victims; Is 11), which was abandoned and rebuilt in another place [16]. The physics of this earthquake have been investigated in several studies [27–30] that reconstructed the complex rupture process along both the main fault (mainshock, and 20-s sub-event) and the antithetic, SW-dipping fault (40-s sub-event). The rupture nucleated at ca. 10 km in depth [29] and caused impressive surface faulting across the Mount Marzano massif, especially east of the Sele Valley [7,24,31–34].

3. The Little Known Earthquake of the 1st Century CE

3.1. Archaeoseismic Evidence from Volcei

Robust proofs of multiple destructive events have been found in this ancient town, starting from the late 3rd century BCE [35]. The Roman municipium of Volcei (today Buccino) developed over a settlement existing since the Iron Age close to the Apennine watershed, between the Campania and Basilicata regions (Figure 1). The modern Buccino was heavily damaged (Is 8 MCS) by the 1980 Irpinia earthquake (Mw 6.9), with the successive reconstruction works providing an exceptional opportunity to rediscover the buried remains of the Roman town. As Buccino shares the same ruinous destiny as the other villages struck by the 1980, 1694, 1561, and 1466 events [3,23], the recognition of widespread, seismically induced effects to the ancient structures was not an unexpected discovery, and it allowed extending our information well behind the memory of the written Modern sources.

During the post-1980 reconstruction works, many indications of ancient collapses, including *butti* (i.e., stacks of archeological debris), fills, and the leveling of destroyed buildings were found everywhere in the Buccino underground. There was also evidence of restoring and/or rebuilding several Roman houses with architectural elements from previous buildings, such as architraves and epigraphs, which were recycled in the new constructions. Besides all these findings, a coeval epigraph explicitly mentions the restoration made after a collapse due to an earthquake. In the whole, all the indications point to a destructive event falling in the second half of the 1st century CE, as summarized in the forthcoming points (see Figure 3 for sites location), which account mostly for unpublished data collected and discussed with A. Lagi, who was formerly responsible for the Salerno Superintendence for the Buccino excavations, and with many other archaeologists who followed one another in the project.

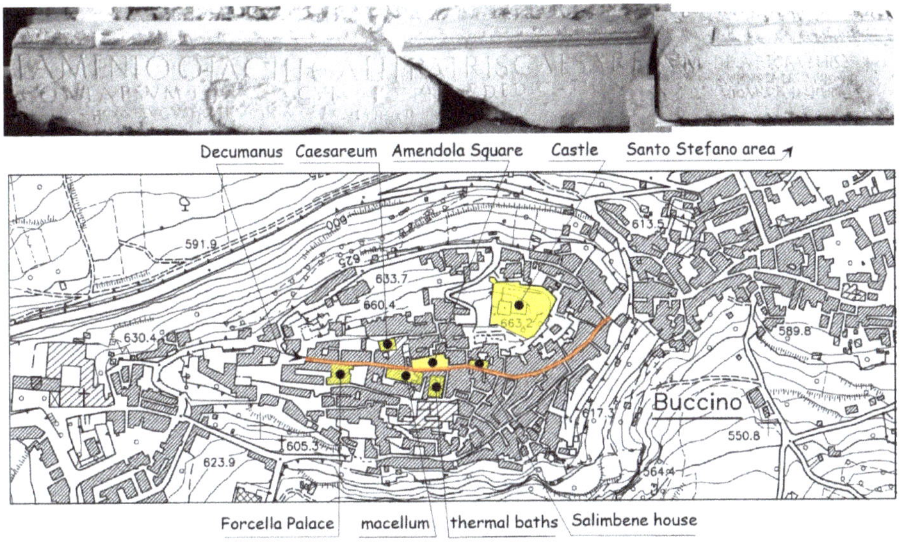

Figure 3. Map of Buccino showing the main archaeological sites attesting the 1st century CE earthquake. Upper panel is the photomosaic of the epigraph of Otacilius Gallus, a far ancestor of mine, attesting the collapse of the *Caesareum* (photo by P.G.).

3.1.1. Forcella Palace

Below this palace, which was built along the Roman Decumanus, the indications consist of a dumping grave containing domestic pottery (lamps and dishes) datable within the first half of the 1st century CE. As no sigillata chiara A pottery (early second half of 1st century CE) was found within the grave, the dumping age must fall at the onset of the second half of the 1st century CE.

3.1.2. Castle

Over the southern side of the main 12th century Norman tower (i.e., the donjon, resting over the basement of a Roman temple), the excavations unearthed a broad, rubble fill, which was supported by a wall. The fill was rich in domestic material, bricks, tiles, and limestone masonry blocks, and it was lined upward by a raw concrete pouring. More importantly, it also contains sigillata italica and Africana chiara A pottery, the latter datable to the second half of the 1st century CE (maximum 60–70 CE). At the bottom of the fill, a coin of Emperor Tiberius (23–30 CE) provides a certain post quem term for the rubbles mass, which was thus leveled between the late 1st and the beginning of the 2nd century.

3.1.3. Sotto San Nicola Street

In the southeastern slope of Buccino, another dumping grave of both building and domestic rubble has been found. It contains vases and lamps datable within the first half of the 1st century CE (i.e., sigillata italica, pareti sottili pottery), being hence coeval with the castle area fill.

3.1.4. Amendola Square

In this place, along the Decumanus, three different dumping graves have been overlaid and sealed with a restoration floor; the infill material contains pottery shards with pareti sottili (early 1st century CE). Amongst all the pottery, the archaeologist found the relic of a pot (glilarium or vivarium in doliis) containing the skeleton of a dormouse (Glis glis), which was ready to be cooked. At least one of the buildings facing the Decumanus was restored in the 2nd century CE, when also a porticus with four pillars was added to the house. One of the pillars supported an Osco-Latin epigraph, likely recalling the restoration of a nearby vicum venerlum (a brothel?) in the 2nd century (G. Camodeca and A. La Regina, personal communications, 2006). Here, I sampled and dated some charred materials belonging to the wooden structure of the porticus, which was buried by the subsequent collapse of this building during the Early-High Middle Age (see next section). The calibrated age (110–330 CE, 2σ cal.; Table 1) fits the period of general restoration of the town, providing the ante quem term for the collapse.

Table 1. Radiocarbon ages of samples collected in the investigated area (AMS, Accelerator Mass Spectrometry and Radiometric ages by Beta Analytic Inc., Miami, FL, USA). 2σ calibration with software Calib 7.1 [36].

Site	Sample	Laboratory	Dated Material	Dating	d13C	Measured Age (BP)	2s cal. 95%
Volcei	BucAm1c	BETA-175146	Charred material	AMS	−26.7	1810 ± 40	110–330 AD
	BucAm4b	BETA-173042	Bone Collagene	AMS	−21.1	900 ± 40	1030–1230 AD
Compsa	Conz01	BETA-180232	Charred material	Radiometric	–	1240 ± 70	660–970 AD

3.1.5. Thermal Baths

Between the 1st and the 2nd century CE, the thermal baths were restored, and their orientation was changed, whereas the floors were completely renewed with different mosaics.

3.1.6. Macellum

In the same period (1st–2nd century CE), in the area of the macellum, the two tholoi were dismantled, and the macellum itself was abandoned, whereas its remains were leveled and occupied by new workshops.

3.1.7. Salimbene House

Below this house facing the Decumanus, the archaeologists found the remains of a 1st century BCE room ceiling that abruptly collapsed and was successively buried by other structures. By removing the fallen material, it was possible to observe that the incannucciato ceiling collapsed directly over the mortar floor of the room, where it also buried pottery shards of the 1st century CE (Figure 4). This clearly means that the room was in use when the collapse happened.

3.1.8. Caesareum Temple

Here, in the same period (1st century CE), an opus caementicium cistern was built with the aim of supporting the damaged retaining wall of the temple. The severe damage suffered by this temple is also testified by the epigraph of Otacilius Gallus that will be hereafter described.

In the whole, the archaeological data evidence a general, abrupt discontinuity within the urban texture of the Roman Volcei, followed by a reconstruction phase focused between the 1st and 2nd century CE. The great abundance of domestic pottery, tiles, bricks, and stones in the dumping graves, summed to the existence of leveled rubble fills, are the proof of contemporary, extensive building collapses in the town. Moreover, the discovery of the pot with the dormouse, ready to be cooked when it was buried under the rubble, and the Otacilius' epigraph, attesting the collapse of the *Caesareum*, are conclusive proofs concerning the occurrence of this event. Last but not least, a further clue evocating the tragedy caused by this earthquake comes from the funerary monument of Gresia Tertia, which is located only 10 km SE to Volcei. Here, archaeological investigations unearthed an epigraph datable within the 1st century CE, where an *infelix mater* (a desperate mother) cries over the death of her family, namely all the four sons and the father. Even if the cause is not declared, the simultaneous decease of five persons in the same family could really be related to the collapse of their house.

Figure 4. Simultaneous collapse of the *incannucciato* ceiling (unit 24) and of the plaster (25) over the mortar floor (30) of the inhabited room. Unit 21 is instead the foundation of a medieval wall that was carved within the Roman rubble. For the record, this medieval wall collapsed due to a further earthquake, and its relics were found inside an adjacent room (photo by P.G.).

3.2. *Archaeoseismic Evidence from Compsa*

The municipium of Compsa (today Conza, Figure 1), similar to the nearby Volcei, starting from the 1st century BCE, flourished throughout the whole Imperial period, surviving to the Late Antiquity decline without dramatic urbanistic breaks [35]. However, despite the monumental buildings of the 1st century BCE being renewed under Emperor Augustus at the beginning of the 1st century CE [37,38], the archaeologists have found sparse indications of further works made just after a little interval, consisting of the reworking of the *Capitolium*, its *podium*, and its staircase. Moreover, they noticed the restoration of other new buildings surrounding the *forum*, with the demolition of structures built only a few years before, as the columns of the *porticus* or the wall of the *sacellum* (F. Soriano, personal communications, 2020). Thus, whereas the architectural improvement of the monumental buildings during the Augustus period is typical in almost all the Roman towns of the time, the rebuilding of the same structure only a few years after might suggest the occurrence of a destructive event, such as an earthquake. In this case, the archaeologist placed it just after the first half of the 1st century CE.

3.3. *Paleoseismic Evidence from MMFS Trenching*

Besides the archaeological proofs of this earthquake, there exists also geological evidence of its occurrence and size. As aforementioned, the authors of [6,7] performed both high-precision topographic leveling across the MMFS scarp and new paleoseismic trenches across different segments of the master fault. About 60 topographic profiles revealed the presence of at least 4 clusters of retreated scarp edges (e.g., up-slope: migrating inflection points: [6]), representing as many surface faulting events (Figure 5). These match 4 paleoseismic episodes recognized within trenches, the age of which was successively defined by the AMS dating of numerous key samples collected there.

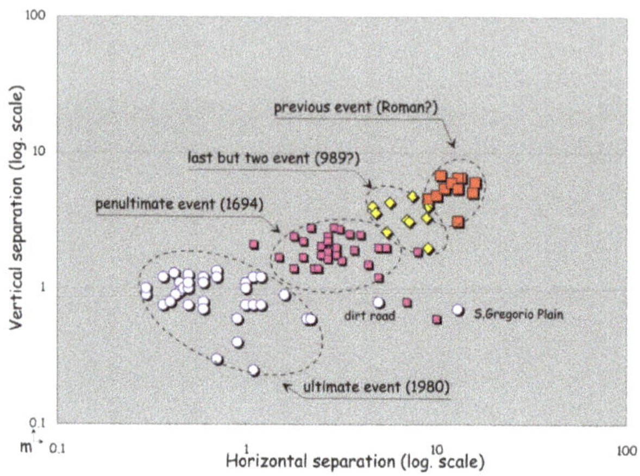

Figure 5. Horizontal versus vertical separation values measured through high-precision topographic profiles across the compound fault scarp of Monte Marzano faults (vertical values are cumulated). The different clusters provide evidence for at least four major surface-faulting events. The older the rupture, the higher the vertical and horizontal separation (mod. from [6]).

In detail, the fourth cluster was dated in the trenches T1 and T2 of [7]. These excavations (Figure 6) provided robust evidence of slope debris offset that occurred just before 80–310 CE, which is a time matching the oldest event recognized also in T3, where this was dated between 540–390 BCE and 540–650 CE. The event also fits event 2 in [2], which occurred between 620 and 230 CE. Considering all the above, the preferred age of [7] was bracketed within the early Imperial Period (1st–2nd century CE).

Figure 6. View of trenches T1–T2 during the excavation along the >30° dipping slope of Monte Valva-Marzano (summer 2010). Note, in the right panel, the 1980 surface offset, still visible after 30 years. The blow-up of T2 (left panel) focuses on the tectonic wedges formed in the 989 and 1694 surface faulting (see details in [7]) (photos by P.G.).

3.4. Historical Sources

In a strict sense, historical accounts concerning the effects of this earthquake in Irpinia do not exist. Leaving aside the intriguing epigraph of Gresia Tertia and the ambiguous Osco-Latin epigraph in Amendola square, the only certain information is provided by the epigraph of Otacilius Gallus. This was carved on an architrave (Figure 3), and it recalls the collapse of the *Caesareum*, which is a temple built only one century before, around 50–60 BCE. The text is:

OTACILIVS EX TESTAMENTO OTACILI GALLI PATRIS CAESAREVM/[TERRAE MOTV] CONLAPSVM P(ecVnia) [S(Va) R(estitVit)]. CVIVS OPER[IS] DEDICATIONE/ [DEDIT DECVRIONIBVS] (sestertios) XXX, AVGVSTA[L]IBVS (sestertios) XX, VICANIS (sestertios) XII, VX[ORIBVS]/ DECVRIONVM (sestertios) XVI, AVGVSTALIVM (sestertios) VIII, VICANORVM (sestertios) IIII

which roughly means: "Otacilius, according to the will of his father, rebuilt by its own the Cesareum which was destroyed by an earthquake … etc.". If the integration of the missing text (square-brackets: [39]) is correct, the *Caesareum* was destroyed by an earthquake that occurred before the end of the 1st century CE, and it was then restored by Otacilius in the 2nd century (G. Camodeca, personal communication, 2006).

4. The 989 CE Earthquake

4.1. Historical Sources

This is the first earthquake explicitly quoted in the Irpinia area, and it was already reported by the first Italian editors of seismic catalogues, as Giannozzo Manetti, Colanello Pacca, and Marcello Bonito [40–42]. Manetti mentions the collapse of several buildings in Benevento and Capua, as in Ariano, Ronza, and Conza, where many people and the Bishop died. In turn, Pacca places the earthquake in 982 in the Naples Kingdom, describing the effects in Benevento, Capua, Ariano, Frigento, Conza, and Ronza. Bonito places it in 989, reporting texts and sources that he painstakingly collected. In short, the primary sources to which these three authors refer are the Annales Beneventani [43], the Chronica Monasterii Casinensis [44], and the Chronicon of Romualdo Guarna [45]. The exact year of the earthquake has been fixed to 989 by [46], who cross-checked other certain historical episodes to which the event can be chronologically referred. Nevertheless, on the basis of the only historical descriptions, its epicentral parameters were not univocally defined. The most interesting information derives from Leone Ostiense, the author of [44], who was writing in the Montecassino Monastery and likely used first-hand sources, which are lost today. In few words, he informs us that the earthquake struck Vipera (this toponym still exists in the Istituto Geografico Militare maps near Benevento) and Benevento, where 150 died and 15 towers collapsed, whereas in Capua, many houses fell down, and the bells rung. Ariano Irpino and Frigento were almost completely destroyed, as was Conza, where many inhabitants died together with the Bishop, while Ronza (a settlement close to Conza) was totally razed to the ground. Leaving aside the inconsistent news from the very far town of Capua (where bell towers just ring while houses are collapsing), on the basis of all the news, [20] assigned Is 10–11 MCS to Ronza, 10 MCS to Conza, and 9–10 MCS to Ariano and Frigento, while in Benevento, the most important and populated town of the region, the number of casualties may suggest at least an Is 8–9 MCS intensity.

4.2. Archaeoseismic Evidence from Compsa

Here, the archaeological excavation performed after the 1980 earthquake by the former Superintendency of Salerno, Avellino, and Benevento unearthed vast areas of the Roman and Medieval town. As a matter of facts, buildings and tombs of Byzantine age (6th century) before, and Langobardic (7th century) then, occupied the Forum area, always resting in phase with the Roman town levels. A proto-cathedral was also built over the Roman Capitolium, with a staircase connecting the church to the forum pavement [47–49]. The archaeoseismic survey conducted with the archaeologists during the excavation in 2003 allowed collecting indications of a strong break in the Early Middle Age history of Compsa, which was namely between the 10th and 11th century. After this break, for the first time after its foundation, the town was not rebuilt, respecting the inherited urbanistic Roman texture. An extended level of rubble, containing pottery shards and coins of the 10th century, buried the limestone slabs of the *Forum* and all its inscriptions [50]. The first cathedral, erected on the *Capitolium*, collapsed and was later rebuilt close to the *Forum*, although it was rotated with respect to the axes of the Roman–Early Middle Age town. The new apse occupied a part of the previous *Forum*, covering the leveled strata of destruction. Here, I made a radiocarbon dating of a huge charred fragment sampled within the collapse levels, which provided an age of 660–970 CE (2σ cal. Age; Table 1), representing the *post quem* term of the earthquake destruction (Figure 7). Unfortunately, this new *basilica*, consecrated in 1122 [51], will be soon destroyed by the forthcoming 1466 earthquake, and then again in 1694, 1732, and definitely 1980 [20].

Figure 7. Rubble and abandonment layers excavated in the Forum area of Compsa, over the limestone slabs (below the booklet). The radiocarbon dating of a charred wood (660–970 CE; 2σ cal. age) provided the *post quem* term for the destruction of the Early Middle Age town in the 989 earthquake (photo by P.G.).

4.3. *Archaeoseismic Evidence in Volcei*

Here, the evidence of the 989 event is represented by the synchronous and total collapse of the buildings excavated below Amendola Square and in other neighboring *insulae*. Moreover, it is witnessed by the general abandonment of the surviving Late Antiquity buildings, which were still inhabited during the Langobardic period, and by the new urban topography that, as in Conza, drifts apart from the Roman imprint, assuming a concentric path around the new castle. Actually, in Volcei, it is difficult to provide univocal age brackets for this earthquake, mainly because of the paucity of the Early Middle Age pottery. However, archaeoseismic clues are constrained between the 7th–8th and the 12th century, even if I cannot exclude the occurrence of multiple events within this time span. The crude set of indications can be summarized as follows (Figure 3 for location).

4.3.1. Amendola Square

The excavations have revealed the synchronous collapse of all the buildings surviving since the Late Roman times. Below the rubble, it has been possible to read the history of these houses, with the different redistricting of each room during times, the wall restorations, the floors overlapping, and the doorstep reutilization. The collapse affected all the masonry walls, the pillars of the *porticus*, and the roofs, which have been found all directly overlaying the floors (Figure 8). This catastrophic collapse also definitely buried the *Decumanus*, which was still in use and well maintained at least during the 7th century, as testified by the materials found in the ditches. At first glance, the collapse also killed a small sheep, which was hit and buried by the rubble on the road *basoli* (see the arrow in Figure 8). I sampled and obtained an AMS collagen dating of 1034–1214 AD (2σ cal.) from the sheep bones. Inside the porticated building, beside the collapse of the wall plasters of the *incannucciato* ceiling, tiles, and masonry, it was possible to observe some walls and the four bricks/pillars that fell away from the road, burying *bande rosse* pottery (used all along the Early Middle Age).

Figure 8. Amendola Square insula. View looking west of the amazing, total, and simultaneous collapse of the porticated house over the *Decumanus* before the removal of the roof (reddish material). The arrow points to the limestone *basoli*, where the sheep was found (photo by P.G.).

4.3.2. Salimbene House

An Early Middle Age cobble wall was founded inside the fill burying the Roman buildings (Figure 4). The pottery shards inside the foundation trench are the same as in Amendola Square, i.e., it contains *bande rosse* pottery. Therefore, this wall—which successively collapsed over a nearby room—might suggest the onset of the reconstruction of the Early Middle Age Buccino after and over the earthquake rubble.

4.4. Other Settlements

In order to enrich the framework of effects distribution related to this earthquake, I collected further indications concerning damage to ancient buildings (see Figure 2A) with the assistance of M. Rotili (Naples University; personal Communication, 2002).

4.4.1. Frigento

Here, [52] described the burying of the right apse of the Langobardic Mother Church, which collapsed and was successively leveled prior to the early 11th century. The rebuilding of the church is dated between the 11th and 12th century, when the new structures were founded over the walls of the collapsed church, or directly over the infill of its ruins.

4.4.2. Montella

In the Angevin Park, [53] found the 13th–14th century boundary wall built over the leveling of the collapsed 9th century wall. In turn, other remains of the 9th century wall were still laying outside the newer wall, burying tombs of the 7th–8th century.

4.4.3. Rocca San Felice

Here, [53] hypothesizes the destruction and successive reconstruction of the 7th–8th fortified structure because of the 989 earthquake. This is also attested by the 10th–12th century *donjon* that was built over the leveling of the Langobardic structures.

4.4.4. Sant'Angelo dei Lombardi

The same author [53] claims for the 989 earthquake shaking in order to explain the destruction of the Langobardic fortified settlement, which was rebuilt in the 11th–12th century. As well as the restoration and reinforcing of the previous damaged wall, the Norman *donjon* was erected over the leveling of the Langobardic structures, as in Rocca San Felice. Moreover, it is worth noting that the right apse of the 11th–12th century cathedral was founded over the leveling of the old, collapsed boundary wall.

4.5. Paleoseismic Evidence from MMFS Trenching

In the trenches opened by [7] and along the fault-scarp profiles discussed in [6], between the signatures of the 1st century CE and the 1694 earthquakes, these authors have found the clear evidence of a Middle Age surface faulting (Figure 5). In trenches T1–T2 (Figure 6), they placed it contemporary or slightly after 720–970 CE (age of the infilling colluvium at the bottom of a coseismic tectonic wedge), and well before 1190–1270 CE and 1260–1390 CE (age of the units sealing offset levels). The same event was also found in T8, where it occurred after 685–892 CE.

5. Discussion and Conclusions

An unexpected and tragic event, such as the Mw 6.9, 1980 Irpinia earthquake, allowed rediscovering the buried relics of the Roman *municipium* of Volcei (Buccino) and of Compsa (Conza). The amazing discovery is that below the ruins of the 1980 earthquake, the archaeologists have found a palimpsest of constructions/collapses/reconstructions attributable to as many seismic events that, each time, have partly allowed the freezing of the buildings history below their own rubble. Although sometimes, the unraveling of this tangled skein of construction/destruction events is a very hard task, with the guide and help of the archaeologists, it has been possible to collect several indications from different sectors of the towns that suggest the occurrence of different earthquakes striking Volcei/Buccino and

Compsa/Conza in the past, and in particular, one in the 1st century CE and one in the 10th century, i.e., the 989 earthquake.

1st century CE earthquake—In Volcei, many archaeological evidence indicate a strong, destructive event that occurred in the second half of the 1st century AD, as constrained well by the age of the huge amount of pottery materials found by the archaeologists, which is supported also by the AMS dating of wooden structures. Conversely, in Compsa, this event was not clearly identified due to the lack of the Roman stratigraphy, which was mostly removed by the early post-1980 works. However, here archaeologists agree that the [1]st century BCE *Forum* was newly rebuilt in the early 1st century CE [39,50], whereas it was again restored and reworked only a few years after (F. Soriano, personal communication, 2020), i.e., contemporary to the well documented Volcei reconstruction. This earthquake, which is unknown to the seismological compilations in this region, can be related to the one found by means of paleoseismological analyses across the Mount Marzano Fault System [7], where it was dated a time before 80–310 CE. Alongside the archaeoseismic and paleoseismic indications, this event is recorded by an epigraph that, although incomplete, clearly mentions the collapse of the *Caesareum* of Volcei. Given the lack of classical literary sources for the Irpinia area, a very speculative hypothesis concerning its age may be suggested by the concurrence of the so-called Pompei earthquake in 62 CE. Bearing in mind that the strongest earthquakes sourced by the MMFS, as well as by many others Apennines faults [54], always induced high-intensity effects in the surrounding of Naples (e.g., Is 7 MCS in both 1694 and 1980 events, with isolated collapses and casualties; see [55]), it could be possible that the damage reported by the historical sources in ancient Pompei, Naples, Ercolano, and Nocera might represent the far-field effects of the same earthquake that razed to the ground Compsa and Volcei. It is worth remembering that Seneca in the *Naturales quaestiones* [56], besides describing the effects in the aforementioned towns, also stresses that the earthquake "*Campaniam … magna strage vastavi*" (destroyed the Region Campania with many casualties) and "*non desiit enim assidue tremere Campania*" (Campania continued trembling), likely alluding that damage was spread over a vast area. If this were true, according to the theory of elastic stress transfer, one could also tentatively hypothesize that this Apennine earthquake was the one that triggered the forthcoming 79 CE Vesuvius eruption. This is suggested by some of the case histories presented and discussed in [57], although more robust hypotheses were discussed in [58].

989 CE earthquake—Differently from the 1st century earthquake, Compsa was explicitly quoted among the towns destroyed by the 989 event. In turn, even in this case, its archaeological evidence is less detailed than in Volcei, as most of the first excavations after the 1980 earthquake were made without a systematic description, documentation, and collection of archaeological materials. However, the available data undoubtedly suggest that an earthquake occurred at the end of the 10th century (i.e., post 660–970 CE, before 1122) causing the almost total destruction of Compsa, which was successively abandoned and then slowly rebuilt and repopulated. Besides Compsa, other settlements of the region show archaeoseismic evidence of contemporary destruction and reconstruction, such as Frigento (also mentioned by the historical sources), Montella, Rocca San Felice, and Sant'Angelo dei Lombardi, which were all heavily struck by the 1980 event as well.

In Volcei, there are many indications of coeval destruction around the end of the 10th century, although the univocal dating of collapses is problematic. Indeed, whereas the pottery shards involved and buried by the collapses predate the onset of the second millennium, the AMS dating of the sheep buried under the rubble would indicate a slightly later age. Considering the uncertainty that often arises from collagen dating, the simplest hypothesis is that an event occurred between the uppermost and lowermost boundary of the two terms, i.e., around 1000 AD, a time fitting the October 25, 989 earthquake. However, this framework is complicated by the presence of a 16th century historical source who mentions that an earthquake occurred at the times of Pope Callistus II (1119–1124), the destructive effects of which were still visible in Buccino in the "*horti...nella parrocchia de S Maria Sollitta...*" (… in the gardens of St Mary Sollitta church [59]). The period of Pope Callistus fully matches the AMS age of the sheep squashed under the rubble, although it is not consistent with the time span suggested by the pottery. Nevertheless, considering the high frequency of earthquakes occurrence in this region [3,20], we cannot exclude that more than one event hit Buccino just before and after 1000 CE, cumulating damage and favoring the collapse of the highly vulnerable Middle Age buildings. As a matter of fact, in one of the houses below Amendola Square, the archaeologists unearthed a small lime furnace that was operating at the time of the collapse. The furnace overlays a thin abandonment level on the floor, suggesting that men were working inside an uninhabited house [33]. Thus, an attractive hypothesis, which makes no claims to being conclusive, is that while works were in progress for repairing the damage of the 989 earthquake, another event caused the complete collapse of the buildings around 1120 CE and the definitive burying of the ancient *Decumanus*.

As well as in the case of the Roman event, paleoseismic results along the MMFS support the presence of a Middle Age surface faulting. This was dated as contemporary or slightly after 720–970 CE, and well before 1190–1270 CE. Last but not least, the rupture of the MMFS in this period matches the distribution of the settlements destroyed by the 989 earthquake, which all fall in the hanging wall or in the surroundings of the Irpinia fault.

As a concluding remark, I would like to stress that the results summarized in this short review represent some intriguing case histories in the investigation of ancient earthquakes. The concurrent presence of historical sources, archaeoseismic indications, and paleoseismic evidence of the same earthquake have been rarely observed not just in Italy, but also worldwide. The contribution of each discipline strengthens the parametrization of the earthquake, both in terms of age, location, and energy released. The probable association with the seismogenic source of the 1980 earthquake suggests that both the 1st century CE and 989 CE events had a comparable magnitude (Mw 6.9), whereas the age brackets provided by the archaeoseismic and paleoseismic results suggest the challenging hypothesis that the older event might be the one that in 62 CE damaged Pompei, anticipating and, dubiously triggering, the famous Vesuvius eruption in 79 CE.

Funding: This research received no external funding.

Acknowledgments: I am grateful to all the archaeologists who shared with me all their findings, discussing with me the hypotheses presented in this paper. In particular, I am indebted to Adele Lagi, who spent a lot of time explaining me the secrets of all the stones of Volcei. I wish to thank also Gabriella Colucci Pescatori, Enzo di Giovanni, Pierfrancesco Talamo, Fiammetta Soriano and Marcello Rotili.

Conflicts of Interest: The author declares no conflict of interest.

References

1. Galli, P.; Galadini, F.; Pantosti, D. Twenty years of paleoseismology in Italy. *Earth-Sci. Rev.* **2008**, *88*, 89–117. [CrossRef]
2. Pantosti, D.; Schwartz, D.P.; Valensise, G. Paleoseismology along the 1980 surface rupture of the Irpinia Fault: Implications for earthquake recurrence in the southern Apennines, Italy. *J. Geophys. Res. Space Phys.* **1993**, *98*, 6561–6577. [CrossRef]
3. Castelli, V.; Galli, P.; Camassi, R.; Caracciolo, C. The 1561 Earthquake(s) in Southern Italy: New Insights into a Complex Seismic Sequence. *J. Earthq. Eng.* **2008**, *12*, 1054–1077. [CrossRef]
4. Palano, M.; Cannavo, F.; Ferranti, L.; Mattia, M.; Mazzella, M. Strain and stress fields in the Southern Apennines (Italy) constrained by geodetic, seismological and borehole data. *Geophys. J. Int.* **2011**, *187*, 1270–1282. [CrossRef]
5. D'Agostino, N. Complete seismic release of tectonic strain and earthquake recurrence in the Apennines (Italy). *Geophys. Res. Lett.* **2014**, *41*, 1155–1162. [CrossRef]
6. Galli, P.; Peronace, E.; Quadrio, B.; Esposito, G. Earthquake fingerprints along fault scarps: A case study of the Irpinia 1980 earthquake fault (southern Apennines). *Geomorphology* **2014**, *206*, 97–106. [CrossRef]
7. Galli, P.; Peronace, E. New paleoseismic data from the Irpinia Fault. A different seismogenic perspective for southern Apennines (Italy). *Earth-Sci. Rev.* **2014**, *136*, 175–201. [CrossRef]
8. Westaway, R. Seismic moment summation for historical earthquakes in Italy: Tectonic implications. *J. Geophys. Res. Space Phys.* **1992**, *97*, 15437. [CrossRef]

9. Berardi, R.; Contri, P.; Galli, P.; Mendez, A.; Pacor, F. Modellazione degli effetti di amplificazione locale nelle città di Avezzano, Ortucchio e Sora. In *13 Gennaio del 1915. Il Terremoto Nella Marsica*; Castenetto, S., Galadini, F., Eds.; Servizio Sismico Nazionale, Istituto Poligrafico dello Stato: Rome, Italy, 1999; pp. 349–371.
10. Ameri, G.; Bindi, D.; Pacor, F.; Galadini, F. The 2009 April 6, Mw 6.3, L'Aquila (central Italy) earthquake: Finite-fault effects on intensity data. *Geophys. J. Int.* **2011**, *186*, 837–851. [CrossRef]
11. Galli, P.; Bosi, V.; Galadini, F.; Meghraoui, M.; Messina, P.; Basili, R.; Moro, M.; Sposato, A. Fratturazione superficiale connessa ai terremoti umbro-marchigiani del settembre-ottobre 1997. *Il Quaternario* **1997**, *10*, 255–262.
12. Galli, P.; Camassi, R.; Azzaro, R.; Bernardini, F.; Castenetto, S.; Molin, D.; Tertulliani, A. Il terremoto aquilano del 6 aprile 2009: Rilievo macrosismico, effetti di superficie ed implicazioni sismotettoniche. *Il Quaternario* **2009**, *22*, 235–246.
13. Galli, P.; Peronace, E.; Bramerini, F.; Castenetto, S.; Naso, G.; Cassone, F.; Pallone, F. The MCS intensity distribution of the devastating 24 August 2016 earthquake in central Italy (Mw 6.2). *Ann. Geophys.* **2016**, *59*, 13.
14. Galli, P.; Castenetto, S.; Peronace, E. The Macroseismic Intensity Distribution of the 30 October 2016 Earthquake in Central Italy (Mw 6.6): Seismotectonic Implications. *Tectonics* **2017**, *36*, 2179–2191. [CrossRef]
15. Cocco, M.; Pacor, F. The rupture process of the 1980 Irpinia, Italy, earthquake from the inversion of strong motion waveforms. *Tectonophysics* **1993**, *218*, 157–177. [CrossRef]
16. Postpischl, D. Atlas of Isoseismal Maps of Italian Earthquakes. *Quad. Ric. Sci.* **1985**, *114*, 2A.
17. Sieberg, A. Geologie der Erdbeben. *Handb. Geophys.* **1930**, *2*, 552–554.
18. Molin, D. Rilievo Macrosismico in Emergenza. Rapporto Interno del Dipartimento della Protezione Civile, Ufficio Valutazione, Prevenzione e Mitigazione del Rischio Sismico. Unpublished work, 2009; 13p.
19. Galli, P. Recurrence times of central-southern Apennine faults (Italy): Hints from palaeoseismology. *Terra Nova* **2020**. [CrossRef]
20. Galli, P. La storia sismica di Conza. In *Conza Storia Arte Fede*; Ricciardi, E., Ed.; Grafiche Pannisco: Calitri, Italy, 2010; pp. 23–70.
21. Figliuolo, B.; Marturano, A. Il terremoto del 1466. *Rass. Stor. Salernitana* **1996**, *25*, 93–109.
22. Gasperini, P.; Vannucci, G.; Tripone, D.; Boschi, E. The Location and Sizing of Historical Earthquakes Using the Attenuation of Macroseismic Intensity with Distance. *Bull. Seism. Soc. Am.* **2010**, *100*, 2035–2066. [CrossRef]
23. Galli, P.; Bosi, V.; Piscitelli, S.; Giocoli, A.; Scionti, V. Late Holocene earthquakes in southern Apennine: Paleoseismology of the Caggiano fault. *Acta Diabetol.* **2006**, *95*, 855–870. [CrossRef]
24. Westaway, R.; Jackson, J. The earthquake of 1980 November 23 in Campania—Basilicata (southern Italy). *Geophys. J. Int.* **1987**, *90*, 375–443. [CrossRef]
25. Serva, L. Il terremoto del 1694 in Irpinia e Basilicata. In *Contributo alla Caratterizzazione della Sismicità del Terremoto Italiano*; CNEN: Udine, Italy, 1981; 67p.
26. Porfido, S.; Esposito, E.; Luongo, G.; Marturano, A. I terremoti del XIX secolo dell'Appennino Campano-Lucano. *Mem. Soc. Geol. Ital.* **1988**, *41*, 1105–1116.
27. Bernard, P.; Zollo, A. The Irpinia (Italy) 1980 earthquake: Detailed analysis of a complex normal faulting. *J. Geophys. Res. Space Phys.* **1989**, *94*, 1631. [CrossRef]
28. Boschi, E.; Pantosti, D.; Slejko, D.; Stucchi, M.; Valensise, G. Special issue on the meeting "Irpinia dieci anni dopo", Sorrento, 19–24 November 1990. *Ann. Geofis.* **1993**, *36*, 376p.
29. Westaway, R. Fault rupture geometry for the 1980 Irpinia earthquake: A working hypothesis. *Ann. Geofis.* **1996**, *36*, 51–69.
30. Amato, A.; Selvaggi, G. Aftershock location and P-velocity structure in the epicentral region of the 1980 Irpinia earthquake. *Ann. Geofis.* **1993**, *36*, 3–15.
31. Cinque, A.; Lambiase, S.; Sgrosso, I. Su due faglie nell'alta valle del Sele legate al terremoto del 23.11.1980. *Rend. Soc. Geol. Ital.* **1981**, *4*, 127–129.
32. Bollettinari, G.; Panizza, M. Una "faglia di superficie" presso San Gregorio Magno in occasione del sisma de! 23/11/1980 in Irpinia. *Rend. Soc. Geol. Ital.* **1981**, *4*, 135–136.
33. Funiciello, R.; Pantosti, D.; Valensise, G. Fagliazione superficiale associata al terremoto irpino del 23/11/1980. *Mem. Soc. Geol. Ital.* **1998**, *41*, 1139–1144.

34. Blumetti, A.M.; Esposito, E.; Ferreli, L.; Michetti, A.M.; Porfido, S.; Serva, L.; Vittori, E. Ground effects of the 1980 Irpinia earthquake revisited: Evidence for surface faulting near Muro Lucano. In *Large-Scale Vertical Movements and Related Gravitational Processes*; Dramis, F., Farabollini, P., Molin, P., Eds.; Studi Geologici Camerti, 2002; pp. 19–27.
35. Galli, P.; Lagi, A. *Lost Earthquakes Between Antiquity and Middle Age. Archeological Indication from Volcei (Southern Apennines)*; 34° Convegno Nazionale GNGTS: Trieste, Italy, 2015; pp. 39–46.
36. Stuiver, M.; Reimer, P.J.; Reimer, R. CALIB Radiocarbon Calibration, Execute Version 7.1. 2010. Available online: http://calib.qub.ac.uk/calib/ (accessed on 28 May 2020).
37. Soriano, F. Compsa Romana. Sviluppo Storico e Architettonico. Ph.D. Thesis, University of Verona, Verona, Italy, 2013; 325p.
38. Buonopane, A. L'iscrizione con litterae caelatae del foro di Compsa (Italia, Regio II). *Sylloge Epigr. Barc.* **2015**, *13*, 163–178.
39. Bracco, V. Inscriptiones Italiae, Civitates Vallum Silari et Tanagri, Regio III, I, Roma. *Poligr. Stato* **1974**, 208.
40. Manetti, G. 1457. De Terraemotu Libri tres. In *Collana Commissione ENEA/ENEL per lo Studio dei Problemi Sismici Connessi con la Realizzazione di Impianti Nucleari*; Molin, D., Scopelliti, C., Eds.; Romana Editrice: Rome, Italy, 1983.
41. Pacca, C. *Discorso del Terremoto*; MS 7/A3; Biblioteca della Società Napoletana di Storia Patria, Fondo sismico: Naples, Italy, N.A. 16th century.
42. Bonito, M. *Terra Tremante, Overo Continuatione de' Terremoti dalla Creatione del Mondo Fino al Tempo Presente*; Biblioteca di sismologia, Catalogo dei terremoti: Naples, Italy, 1691; 822p.
43. Bertolini, O. Annales Beneventani, 12th Century. *BISI* **1923**, *42*, 1–159.
44. Hoffmann, H. Chronica Monasterii Casinensis, 11th century. *MGH* **1980**, *1.II c.11*, 189.
45. Guarna, R. 12th century Chronicon. In *RIS, VII/1, Citta di Castello-Bologna 1903–35*; Garufi, C.A., Ed.; pp. 170–171.
46. Figliuolo, B.; Marturano, A. Terremoti in Italia Meridionale dal IX all'XI secolo. In *Contributi per la storia dei terremoti nel Bacino del Mediterraneo*; Marturano, A., Ed.; Osservatorio Vesuviano, Istituto Italiano Studi Filosofici, Laveglia, Nocera Inferiore: Naples, Italy, 2002; pp. 33–68.
47. Pescatori, G. Citta dell'Hirpinia. In *Le Città Campane Fra Tarda Antichità e Alto Medioevo*; Vitolo, G., Ed.; Laveglia Editore: Salerno, Italy, 2005; pp. 283–311.
48. Sgobbo, I. Conza: Ricerche topografiche sull'antica Compsa. *Not. Scavi* **1938**, 98–103.
49. Johannowsky, W. *Conza alla Ricerca del Passato, Atti dell'incontro Dibattito (Conza, 11 Agosto 1979)*; Calitri, Italy, 1982; pp. 7–40.
50. Di Giovanni, V. Compsa tra Irpini e Romani. Ph.D. Thesis, University Federico II, Naples, Italy, 2015; 204p.
51. Gargano, G. *Ricerche Storiche su Conza Antica*; Calitri: Avellino, Italy, 1934; p. 57.
52. Rotili, M. Ricerche archeologiche in Alta Irpinia testimonianze di età romanobarbarica. *Romanobarbarica* **1995**, *13*, 297–324.
53. Rotili, M. Archeologia del donjon di Montella. Memorie dell'Accademia di Archeologia Lettere e Belle Arti. **1999**, *XIII*, 25.
54. Galli, P.; Naso, J.A. Unmasking the 1349 earthquake source (southern Italy): Paleoseismological and archaeoseismological indications from the Aquae Iuliae fault. *J. Struct. Geol.* **2009**, *31*, 128–149. [CrossRef]
55. Esposito, E.; Porfido, S.; Luongo, G.; Petrazzuoli, S.M. Damage scenarios induced by the major seismic events from XV to XIX century in Naples city with partcular reference to the seismic response. In *Earthquake Engineering. Tenth World Conference*; Balkema: Rotterdam, The Netherlands, 1992.
56. Seneca, L. *Naturales Quaestiones*; Vottero, D., Ed.; Torino, Italy, 1989.
57. Nostro, C.; Stein, R.S.; Cocco, M.; Belardinelli, M.E.; Marzocchi, W. Two-way coupling between Vesuvius eruptions and southern Apennine earthquakes, Italy, by elastic stress transfer. *J. Geophys. Res. Space Phys.* **1998**, *103*, 24487–24504. [CrossRef]
58. Marturano, A. Sources of ground movement at Vesuvius before the AD 79 eruption: Evidence from contemporary accounts and archaeological studies. *J. Volcanol. Geotherm. Res.* **2008**, *177*, 959–970. [CrossRef]
59. Bardario, B. *Bucino, Vulcinum, vel Vulceiana Civitas*; Archivio Generale degli Agostiniani: Roma, Italy, 1581.

geosciences

Article

Insights into Mechanical Properties of the 1980 Irpinia Fault System from the Analysis of a Seismic Sequence

Gaetano Festa *, Guido Maria Adinolfi, Alessandro Caruso, Simona Colombelli, Grazia De Landro, Luca Elia, Antonio Emolo, Matteo Picozzi, Antonio Scala, Francesco Carotenuto, Sergio Gammaldi, Antonio Giovanni Iaccarino, Sahar Nazeri, Rosario Riccio, Guido Russo, Stefania Tarantino and Aldo Zollo

Physics Department "Ettore Pancini", University of Naples Federico II, 80126 Napoli, Italy; guidomaria.adinolfi@unina.it (G.M.A.); alessandro.caruso@unina.it (A.C.); simona.colombelli@unina.it (S.C.); grazia.delandro@unina.it (G.D.L.); luca.elia@unina.it (L.E.); antonio.emolo@unina.it (A.E.); matteo.picozzi@unina.it (M.P.); antonio.scala@unina.it (A.S.); francesco.carotenuto2@unina.it (F.C.); sergio.gammaldi@unina.it (S.G.); antoniogiovanni.iaccarino@unina.it (A.G.I.); sahar.nazeri@unina.it (S.N.); rosario.riccio@unina.it (R.R.); guido.russo2@unina.it (G.R.); stefania.tarantino@unina.it (S.T.); aldo.zollo@unina.it (A.Z.)

* Correspondence: gaetano.festa@unina.it; Tel.: +39-081-675248

Abstract: Seismic sequences are a powerful tool to locally infer geometrical and mechanical properties of faults and fault systems. In this study, we provided detailed location and characterization of events of the 3–7 July 2020 Irpinia sequence (southern Italy) that occurred at the northern tip of the main segment that ruptured during the 1980 Irpinia earthquake. Using an autocorrelation technique, we detected more than 340 events within the sequence, with local magnitude ranging between −0.5 and 3.0. We thus provided double difference locations, source parameter estimation, and focal mechanisms determination for the largest quality events. We found that the sequence ruptured an asperity with a size of about 800 m, along a fault structure having a strike compatible with the one of the main segments of the 1980 Irpinia earthquake, and a dip of 50–55° at depth of 10.5–12 km and 60–65° at shallower depths (7.5–9 km). Low stress drop release (average of 0.64 MPa) indicates a fluid-driven initiation mechanism of the sequence. We also evaluated the performance of the earthquake early warning systems running in real-time during the sequence, retrieving a minimum size for the blind zone in the area of about 15 km.

Keywords: earthquake seismology; microseismicity; seismic techniques; seismotectonics

Citation: Festa, G.; Adinolfi, G.M.; Caruso, A.; Colombelli, S.; De Landro, G.; Elia, L.; Emolo, A.; Picozzi, M.; Scala, A.; Carotenuto, F.; et al. Insights into Mechanical Properties of the 1980 Irpinia Fault System from the Analysis of a Seismic Sequence. *Geosciences* **2021**, *11*, 28. https://doi.org/10.3390/geosciences11010028

Received: 16 November 2020
Accepted: 3 January 2021
Published: 5 January 2021

Publisher's Note: MDPI stays neutral with regard to jurisdictional claims in published maps and institutional affiliations.

1. Introduction

Seismic sequences are a useful tool to shed light on fault mechanics and geometry, on the chemical and physical processes occurring on faults and ultimately to illuminate the preparatory phase of large earthquakes. For the goal, diverse dense multi-disciplinary observation infrastructures were created around faults that can potentially host large destructive earthquakes, to understand how those faults slip, what is the seismic and aseismic balance during strain release, and what is the role of fluids in earthquake production.

The study of microseismicity can provide mechanical constraints on the faults, through accurate earthquake location and source parameter computation. Earthquake location can benefit from measurements of P and S wave arrival times and back-azimuth estimation. In dense networks, absolute locations of microearthquakes can reach a sub-kilometric accuracy, using global-search techniques in 3D velocity media tailored for the area [1,2]. When using relative location techniques, such as double differences [3], the accuracy can be pushed down to a decametric scale [4], when relative arrival time determination is performed through cross-correlation [5]. With this resolution, earthquake location can delineate the geometry of faults [3,6] and following event space and time evolution, it is possible to infer the role of the fluids in seismicity migration [7,8].

Source parameters are usually inferred through the fit of a specific rupture model with a simplified description of the observations, such as the displacement amplitude spectrum [9]. Rupture evolution is synthesized in few macroscopic parameters, such as the seismic moment, the source size, and the average stress drop. Large epistemic uncertainties in source parameter estimation come from the medium description, which is usually assumed as homogeneous [10]. This problem can be mitigated either using smaller magnitude earthquakes as empirical Green's functions [11] or computing a non-parametric, data-driven attenuation function [12]. Source parameters provide insights on the stress released during earthquakes and on the earthquake size, enabling the understanding of the role of fluids in earthquake production [13] and the recognition of possible forcing aseismic mechanisms in sequence generation [14].

Characterization of the seismic activity during sequences can constrain fault geometry and mechanics in the vicinity of the swarm location. This information can be crossed with tomographic models in velocity and attenuation, to get a picture of fluid-filled domains and infer the role of the pore pressure in the stress distribution and release at depth [15].

While in most cases seismic sequences end into background seismicity, sometimes acceleration of the cumulative seismic moment release during sequences can yield a large magnitude event, especially for interplate events [16]. Capability to recognize foreshocks before the occurrence of a large event is fundamental for risk monitoring and reduction. There is a large debate in the seismological community on precursory seismicity before a large event and contrasting results have been obtained in indicating the occurrence of foreshock sequences as a persistent feature of moderate to large events (see for California [17,18]). Nevertheless, when a sequence clearly evolves to a large event, foreshocks emerge only when high-resolution seismic catalogs are available, with a magnitude of completeness much smaller than the one presently available for standard monitoring networks. Thus, advanced techniques for event detection are required to improve the catalog, such as the template matching [19] or autocorrelation algorithms [20].

In this study, we have applied some of these advanced techniques to analyze a seismic sequence that occurred in the area of the 1980, M 6.9, Irpinia earthquake. Since 2005, in this region, INFO (Irpinia Near Fault Observatory) operates ISNet—the Irpinia Seismic Network—with the goal of monitoring the seismicity evolution and its relationship with the underlying fault system [21]. ISNet is a dense, high-dynamic range seismic network of 31 stations, covering an area of about 100×70 km^2 along the Campania-Lucania Apennine chain, surrounding the fault system that generated the 1980 Irpinia earthquake [22]. The seismic stations are deployed within two imaginary concentric ellipses, with the major axis parallel to the Apennine chain. The mean distance between stations in the outer ellipse is about 20 km, while the distance between the two ellipses is about 10 km. The average inter-station distance within the inner ellipse is less than 10 km. All stations are equipped with a strong-motion accelerometer and a weak motion sensor, the latter being a short-period, a broadband velocimeter or an accelerometer with lower full-scale. Stations of ISNet provide real-time data with controlled delay at the ISNet control center, enabling the network to be the backbone infrastructure for earthquake early warning systems [23] and for near real-time computation of regional ground-shaking maps.

We used ISNet data to investigate the mechanical properties of the Rocca San Felice seismic sequence, that started on 3 July 2020 and lasted for 4 days. In this study, we reviewed the seismotectonic setting of the area, then we applied advanced techniques for event detection and characterization, to enlarge the automatic catalog and to provide accurate locations, based on double-difference techniques. We thus computed source parameters and focal mechanisms to infer the stress release and its main directions, the size of the events, and to discriminate the rupture plane. Finally, we analyze the ground motion and the performances of earthquake early warning systems, as real-time risk reduction tools.

2. Seismotectonic Setting

2.1. Structural Setting of Southern Apennines

The structure of the Southern Apennines is associated with the Meso-Cenozoic tectonic processes that involved the African and European plates and, in accordance with different tectonic phases of rifting, drifting, and shortening, deformed the Corsica-Sardinia and Adriatic-Apulian forelands [24,25]. The backbone of the mountain belt is characterized by E-NE verging duplexes geometries and out-of-sequence thrusting due to orogenic contraction that was active since upper Eocene-Oligocene Miocene up to late Pliocene [26]. Meso-Cenozoic successions of Adriatic-Apulian-African passive margin, deposited as carbonate platforms with interposed pelagic basins, were involved in folding and thrusting according to imbricated structural units detached from their crystalline basement, as documented by seismic profiles obtained since 1980s during hydrocarbon exploration [27]. Migrating eastward over the Apulian foreland, these sheets were tectonically overthrusted by internal units, deposited originally above the oceanic crust of Tethys. Considering the lithostratigraphic relationships, from top to bottom, units are grouped in: (1) post-orogenic intramontane basin units of marine, terrestrial, and volcanic origin, deposited during Plio-Pleistocene or Holocene in the Adriatic-Bradanic foredeep; (2) syntectonic top-thrust basin successions formed during the progressive shortening toward east; (3) orogenic wedge tectonic units involved in the NE-verging overthrusting from upper, internal domains (Tethyan oceanic crust or Adriatic-Apulian continental crust) to the lower, external domains (Apennine carbonate platforms with inter-basins pelagic units); (4) Apulian carbonates, buried, deformed, and overthrusted in the inner belt and undeformed in the outcropping foreland [25,28]. During the Quaternary, the Southern Apennine thrust belt was dissected by NW-SE oriented normal faults that accommodated an extensional tectonic phase, according to a stress field with the axis of maximum extension coaxial to the axis of maximum compression of Apennines belt (SW-NE trend [24,29,30]). In fact, contractional and extensional deformations took place simultaneously in different sectors of the Apennines with both E-migrating fronts at a similar rate of ~4 cm/year.

2.2. Seismogenic Normal Faults

In the Southern Apennines, normal faulting corresponds to a significant deformation process in terms of regional size of structures, morphotectonics, active displacement, and seismogenesis. Normal faults with NW-SE striking regulate the active tectonics in the thrust belt, accommodating an extension of 3–5 mm/year as evidenced by surface geology, borehole breakout, and available fault plane solutions of earthquakes [31,32]. Earthquakes up to X-XI MCS intensity struck the Southern Apennine chain, indicating that this sector is one of the highest seismic hazard areas of the Mediterranean region, with segmented, seismogenic structures capable of generating up to M 7 earthquakes [33,34] (Figure 1). These earthquakes occur principally in the axial sector of the Apennine chain with recurrence periods $\geq$1000 years and depths down to 10–15 km in the upper crust, in a 30–50 km wide belt that follows the orographic divide [35]. As proposed in different models [36–38], normal faults that dissect Apennines show an evolutionary trend from young, high-angle planar faults, seated in the upper crust and characterized by small extensional strains, to mature, listric faults reaching the crystalline basal detachment with high amount of extensional strain. A consequence of progressive E-migration of the extensional front is that normal faults show a different aging and degree of evolution from mature faults of the inner sector of the Apennine chain (to W, toward the Tyrrenian margin) to young faults superimposed on the accretionary wedge along its axial sector (to E, gradually decreasing toward the foreland). Large seismic events are believed to be associated to younger normal faults connected to the extensional front and, in the eastern sector, to fluids reaching an over-pressure condition at shallow depths due to the low permeability of the crust that inhibits their circulation [34].

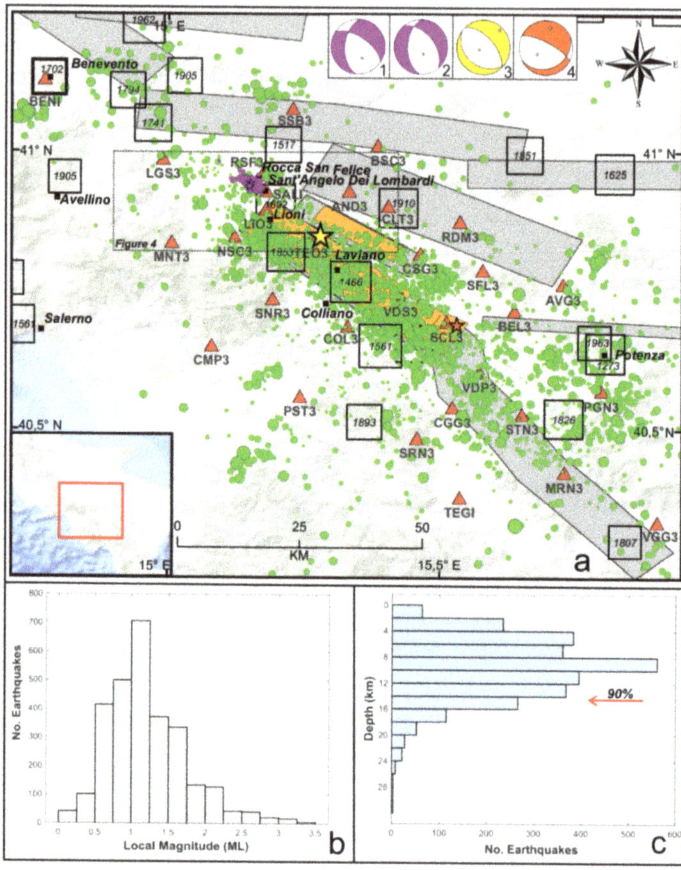

Figure 1. (**a**) Epicentral map of the earthquakes (green circles) recorded by Irpinia Seismic Network (ISNet, red triangles) from 2008 to 2020 (http://isnet-bulletin.fisica.unina.it/cgi-bin/isnet-events/isnet.cgi). The analyzed seismic sequence is highlighted with violet circles. The yellow and orange stars refer to the epicentral location of the 1980, M 6.9, and of the 1996, M 4.9 earthquakes. Historical seismicity is shown with black squares ($I_0 \geq$ 6–7 MCS). Seismogenic sources related to the Irpinia fault system are indicated by orange rectangles; potential sources for earthquakes larger than M 5.5 in surrounding areas are indicated in grey (Database of Individual Seismogenic Sources, DISS, Version 3.2.1). Focal mechanism solutions for four instrumental earthquakes are reported (1) Ml 3.0, 3 July 2020, 16:14, Rocca San Felice sequence, from this study; (2) Ml 3.0, 3 July 2020, 16:19, Rocca San Felice sequence, from this study; (3) 1980, M 6.9, Irpinia earthquake [40]; (4) 1996, M 4.9 earthquake [57]. The area plotted in Figure 4 is identified by a dashed rectangle. (**b**) Histograms of magnitude and (**c**) depth for the microseismicity inside ISNet. The red arrow points to the depth above which 90% of earthquakes occur.

2.3. *1980 Irpinia Earthquake*

The 1980, M 6.9, Irpinia earthquake was the most destructive, instrumental earthquake of the Southern Apennines that occurred along NW-SE trending normal faults. This event is characterized by a complex rupture process involving multiple fault segments according to (at least) three different nucleation episodes delayed each other of 20 s. Several models, characterized by different geometries and locations of activated fault segments, have been proposed to explain its complex rupture [39–42]. Retrieved seismic moment ranges between 2.4 × 10^{19} Nm and 3.0 × 10^{19} Nm, the fault dip activated by the mainshock ranges between 53° and 63° while the strike ranges between 305° and 33° [43]. Location of the event epicenter and surface projection of the three main fault segments are shown in Figure 1.

Kinematic models [41] have proposed the nucleation of the mainshock (0 s) along 30-km-long, NE-dipping, Mount Marzano and Picentini fault segments, with the first subevent activated at 18 s toward a SW, low angle, NE-dipping fault of 20 km length and a second subevent nucleated at 39 s in the Ofanto basin area, along a SW-dipping, antithetic fault. The mainshock initially ruptured at the north tip of the Mount Marzano segment and propagated bilaterally along two NW-SE striking faults. Toward south, it completed the rupture of the Marzano segment and toward north it progressed along the Mount Picentini segment.

The 1980 Irpinia faulting model has been also constrained on the evidence of coseismic surface ruptures [42], and shows that the event activated three main fault segments, along a 38-km-long, NE-dipping scarp with an average N128° direction and one main antithetic blind fault, SW-dipping, that was not capable to reach the surface due to its smaller size, leaving a ground deformation anomaly at NE of mainshock, near the Ofanto river. The NE-dipping fault rupture corresponds to three main strands separated by gaps and anomalies in the continuity of surface faulting identified as: (1) the Cervialto scarp (Mount Picentini, strike of 125°); (2) the Mount Marzano-Valva scarp (strike of 135°) connected southward with the Mount Carpineta scarp (strike ranging between 110° and 135°), and (3) the San Gregorio Magno scarp and related scarplets in and close to the Pantano basin (strike of 120–130°). The Mount Cervialto and Mount Marzano–Mount Carpineta segments have been correlated to the 0 s event (mainshock), the Pantano of San Gregorio Magno segment has been correlated to the 20 s subevent; and the antithetic Ofanto fault segment has been associated with the 40 s subevent. The three segments are represented on the geological map of Figure 2.

Finally, relocated aftershocks and 3D velocity models have provided new constraints about the 1980 Irpinia source model [44,45]: (1) the main rupture started at 10 km depth, on a 60° dipping plane, in a high velocity region, associated with stiff Apulian carbonates and propagated upward along the softer Meso-Cenozoic succession; (2) most of the aftershocks are spread in a volume delimited by the NE-dipping, Marzano-Cervialto normal fault segment and the SW-dipping, antithetic normal fault, associated to the 40 s event; (3) seismicity depth is confined in the first 12 km of the upper crust and mainly concentrated beneath the Marzano-Valva fault segment; (4) largest magnitude aftershocks clustered between the fault segments activated at 0 s and 20 s, near their tips; (5) the Marzano segment is bounded by two clusters of seismicity likely separated by a lithological discontinuity, acting as a seismic barrier, where the stress of the main ruptures was concentrated (the northern Sele Valley and the southern tip of the Carpineta fault segment). In particular, the Marzano-Cervialto normal fault segments have been interpreted as two asperities separated by a low-strength zone, referred to as the Sele barrier [41,42]. New tomographic studies [34] confirm that the mainshock nucleated in basement, below the Apulian carbonates, and propagated in the high Vp/high-Vp/Vs region associated with fractured, water-saturated carbonates of the Apulian platform domain, in agreement with well-data (S. Gregorio Magno 1, [25]) and magnetotelluric surveys [46].

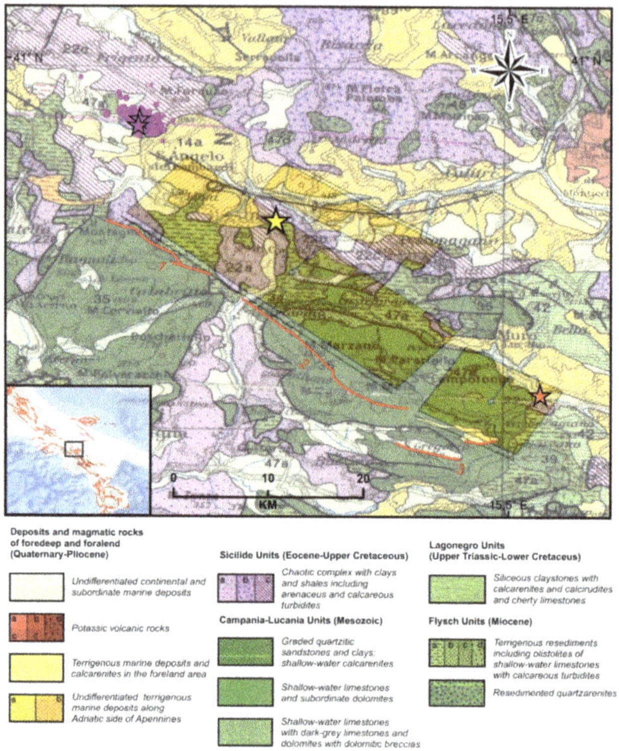

Figure 2. Geological and structural map of the Irpinia region. Here, surface traces (red lines) and sources (orange boxes, Database of Individual Seismogenic Sources, DISS, Version 3.2.1) of faults activated during the 23 November 1980, M 6.9, Irpinia earthquake (yellow star) are shown. Key: 1 = Cervialto fault segment; 2 = Marzano-Carpineta fault segment; 3 = San Gregorio fault segment. The seismic sequence is highlighted with violet circles and stars; the orange star refers to the epicentral location of the 3 April 1996, Ml 4.9 earthquake.

2.4. Recent Seismicity of the Irpinia Region

After the 1980 event, no large earthquakes have struck the area. In 1996, a seismic sequence with a mainshock of M 4.9 took place inside the epicentral area of the 1980 earthquake. Present-day low-magnitude seismicity (Ml < 3.5) occurs mainly in the first 15 km of the crust showing fault plane solutions with normal and normal-strike slip kinematics, indicating a dominant SW–NE extensional regime in agreement with the inversion of fault-slip data of the Irpinia region [32,47–49]. The background low magnitude seismicity appears to be spread into a large volume, and the related stress field is closely linked with the major fault segments activated during the 1980 Irpinia earthquake. In addition, microseismicity seems to be controlled by high pore pressure of water-saturated Apulian carbonates within a fault-bounded crustal volume [15,50]. Location of microseismicity epicentres is shown in Figure 1.

Several studies pointed out a strong relationship between seismicity and high-fluid pressure evidenced by the same location of crustal seismicity and CO_2 degassing areas along the Apennines [38,51]. In the Irpinia area, underneath Mount Forcuso, tomographic images reveal a low-Vp/Vs dome-shaped body, 20 km long and 15 km wide, located between 6 km and 11 km depth [34]. This spot is interpreted as a pressurized CO_2-rich rock volume, filled below the Apulian platform carbonates by fluid-rich mantle melts intruded into the crust. This anomaly correlates with high heat flow values (100–215 mW/m^2) observed along the Mount Forcuso antiform and with geochemical data. This area is characterized by a huge amount of nonvolcanic CO_2-rich gas emission [52,53], the main expression of which occurs in the Mefite d'Ansanto degassing site, where deep, mantle-related fluids are released through the active faults, as indicated by geochemical data.

Deep CO_2-rich fluids may play a key role in the seismogenesis in the Southern Apennines [15,34,38]. Fluids migrating upward along lithospheric faults can be stored in reservoirs under the Apulian carbonates, sealed by Triassic anhydrites, reaching overpressure conditions that can periodically trigger large normal faulting earthquakes. The influence of high-pressure CO_2 in the nucleation of large earthquakes has been also observed in the Central Apennine sector, for the 1997 Colfiorito and the 2009 L'Aquila events [54,55]. The same mechanism has been invoked for the nucleation of the 1980 Irpinia earthquake, being the initiation patch located in the basement, under the Apulian carbonate, at the top of a pressurized CO_2 reservoir (low-Vp/Vs anomaly, [34]).

3. Data

The data used for the sequence analysis have been provided by ISNet. The available stations are shown in Figure 1, together with the epicenters of the seismicity covering the period 2008–2020. We used automatic event detections performed by the software Earthworm [56] to identify the seismic sequence. Specifically, when more than three events occur within 10 km and a couple of hours in time, an indicator light is turned on at the network control center and few days of continuous data before and after the declaration of the sequence are extracted for automatic and visual inspection, followed by machine and human controlled procedures to eventually extract additional events. For this sequence, the automatic system identified 43 events, starting from a Ml 1.9 event, declared on 3 July 2020, at 09:31 and ending with a Ml 1.2 event, that occurred on 6 July 2020, at 16:55. The two largest events in the sequence (Ml 3.0) occurred on 3 July 2020, at 16:14 and 16:19. All these events have been manually inspected, providing revised phase pickings (372 P arrival times and 208 S arrival times), locations, and local magnitude estimations.

During the manual revision of data, the near real-time software INERTIA [22] also provided moment magnitude estimations and ground motion maps. This information is made promptly available at the ISNet bulletin webpage (http://isnet.unina.it/). Continuous waveforms, event data and related information (picking, location, magnitude) were also used as starting point for further analysis as detailed in the rest of this paper. In Figure 3, we show an example of waveforms for a Ml 2.8 event of the sequence, that occurred on 5 July 2020, at 15:18. In the section, stations are ordered for increasing distance from the event epicenter. We can recognize the complexity of the waveforms as the distance increases and the diverse duration and frequency content at some stations, owing to site effects and instrumental filters.

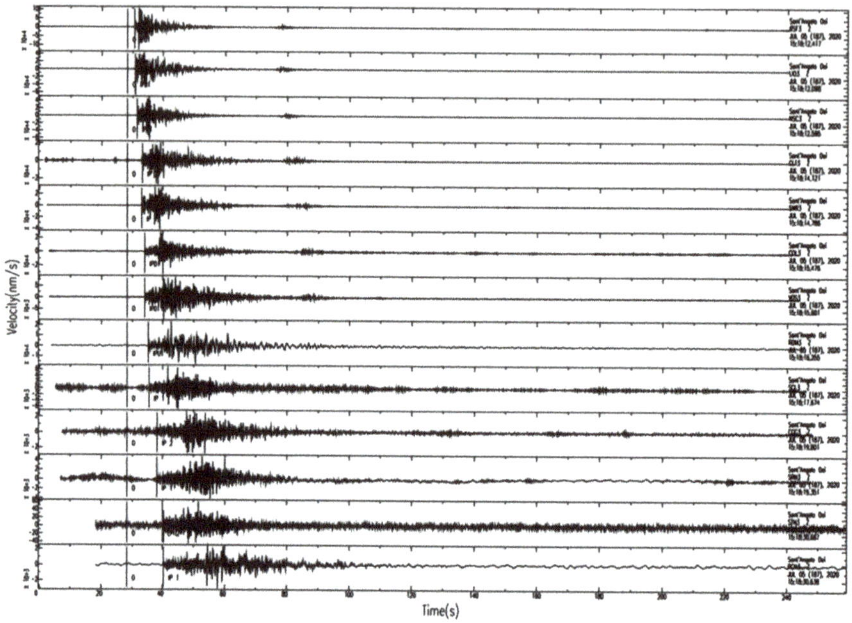

Figure 3. Example of seismic data, represented for the Ml 2.8 event that occurred on 5 July 2020, at 15:18. In the section, stations are ordered from the top to the bottom for increasing epicentral distance. In the coda of the main event, we can recognize smaller magnitude aftershocks.

4. Refined Seismic Catalog

We scanned the continuous helicorder for 4 days starting from 3 July to improve the detection of microearthquakes during the sequence. We used the autocorrelation algorithm FAST (Fingerprint and Similarity Thresholding) [20,58] with the twofold aim of including in the catalog events featuring low signal-to-noise ratio and diminishing the magnitude of completeness for the sequence.

FAST is based on a locality-sensitive hashing algorithm [59] and performs a computationally efficient similarity search, scanning the signal spectrograms, computed in broad frequency bands, with the aim of detecting similar earthquake waveforms. In contrast with template matching, FAST does not require any template for the detection. This feature enables the technique for the detection of microevents during short and spatially constrained sequences.

Since our goal is the search for very-low magnitude events, we applied the technique to the velocimetric records at the five stations closest to the sequence centroid (RSF3, LIO3, NSC3, AND3, MNT3, Figure 4a), limiting our analysis to the vertical component. The distance of the stations from the centroid ranges from 3 to 17 km. We computed the spectrograms, filtering the records between 1 and 20 Hz and separating the continuous waveforms in moving windows having 6.0 s length with a 0.2 s lag. The squared modulus of the Fourier transform of each time window represents a column in the spectrogram, from which we extracted overlapping fingerprints having 32 samples with a lag of 5 samples. To declare a sequence of consecutive and correlated fingerprints as a candidate event, we fixed uncorrelated permitted gaps at a single station to 3 s, this time being roughly related to the maximum difference between S and P arrival times. Detection was declared if uncorrelated gaps at couples of stations had length below 3 s, mimicking maximum P wave arrival time differences across the selected stations. A sensitivity test was performed to determine the best set of the other parameters (number of hash tables, number of hash functions, number of votes, minimum detection threshold; see [58]) minimizing the false negative events. An event was finally declared when detected at least at 2 stations.

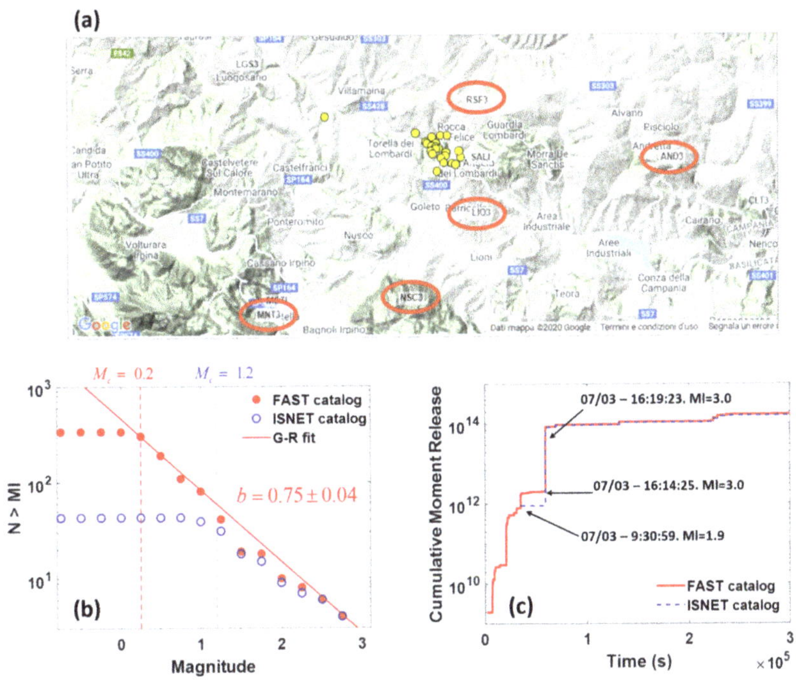

Figure 4. (**a**) Selected stations for the FAST (Fingerprint And Similarity Thresholding) analysis: the dots represent the events automatically detected by ISNet. (**b**) Gutenberg–Richter (GR) analysis for the automatic (blue circles) and FAST (red filled dots) catalogs. For this latter case, the GR is also reported with the b-value estimation. The dashed lines mark the magnitude of completeness for the two catalogs. (**c**) Cumulative moment release according to the two catalogs. The black arrows indicate the first event automatically detected by ISNet (Ml 1.9) and the two main events of the sequence (Ml 3.0).

Scanning the continuous waveforms with FAST, we retrieved 342 events; the new catalog also includes all the 43 events automatically detected by the ISNet procedures. From a visual inspection of the traces, we retrieved only one false positive detection due to a teleseismic event.

For the computation of the magnitude, we imposed that all the events are co-located with the sequence centroid (40.94° N–15.15° E); the local magnitude was then estimated through the ratio of the maximum amplitudes with the events detected and located by the ISNet procedures.

We investigated the magnitude-frequency distribution for the refined catalog, comparing it with the ISNet automatic catalog (Figure 4b). We estimated a magnitude of completeness of Ml 0.2, nearly one unit smaller than that retrieved for the whole seismic network (Ml 1.1, [60]).

We also reported the b-value estimate from the refined catalog as b = 0.75 ± 0.04. When performing this computation on the automatic detections, we retrieved b = 0.70 ± 0.11, which is compatible with the FAST estimate, but shows larger uncertainty. Nevertheless, it is worth to note that some events having magnitude larger than 1.2 have been missed by the automatic procedures, since some of these events fall in the coda of the previous ones. The cumulative seismic moment release (Figure 4c) is not significantly different for the two catalogs yielding an equivalent total magnitude of 3.5. However, we can recognize a foreshock sequence of 23 events preceding the two largest magnitude events in the sequence, that started about 9 h before the first automatic detection (Ml 1.9, 3 July 2020, 09:31).

5. Accurate Earthquake Location

The absolute locations of the sequence events have been obtained using the NLLoc software [1]—http://www.alomax.net/nlloc—that implements a non-linear, global-search probabilistic location in 3D velocity models.

For double-difference (DD) locations, we applied NLDiffLoc [48], a location tool included in the NLLoc software, that allows to perform DD locations by inverting the differential times through a probabilistic, non-linear approach. We also used the tool Loc2ddct, which allows to calculate differential times; initial absolute location of events and corresponding differential travel-times were used as inputs for DD locations. The DD algorithm performs an optimized exploration of the model parameters space using the annealing Metropolis algorithm [1] seeking a solution that maximizes the likelihood function. The latter is based on the misfit between measured and calculated differential phase arrival times. During the exploration, the algorithm computes the posterior probability density function (PDF), which represents the complete solution of the earthquake location problem. The significance and uncertainty of the solution, i.e., the maximum likelihood point, cannot be assessed independently of the complete PDF. Indeed, Gaussian or normal estimators, such as the expectation value and covariance matrix, can be obtained from samples of this function. These estimators can describe location uncertainty in the case of a (non-) linear PDF with a single maximum and an ellipsoidal form [1]. In addition, the software can handle 3D velocity models, with arbitrary complexity and parameterization, required in the case of crustal volumes with strong lateral variations and irregular topography [61].

We analyzed 903 traces at 21 stations from the 43 best recorded events. Within the manual picking, a weight factor inversely proportional to the uncertainty on arrival time picking has been assigned. In this way, we obtained an average of 14 P and S arrival times for each event, ranging from a minimum of 5 phases to a maximum of 31 phases. We processed the first arrival times with NLLoc using the 3D P- and S-wave velocity models optimized for the Irpinia area through an iterative, linearized, tomographic approach in which the P and S arrival times are jointly inverted for earthquake location and velocity determination [62].

For the DD locations, we performed a selection on the events based on the absolute location quality (i.e., at least 8 phases and an azimuthal gap <230°), keeping only 36 events for the analysis. The absolute arrival times of the selected events are combined to obtain about 4000 P and S differential times used as inputs for the DD location. Since the sequence is clustered, we did not impose a maximum distance between event couples.

The absolute locations show events distributed along the NW-SE direction, with an extension of the pattern of about 4–5 km. The depth of the events ranges from 5 to 15 km. The horizontal and depth errors are within 750 m for most of the events. The rms of the resulting absolute locations is within 0.2 s for almost all the events.

From the absolute location, we achieved the final DD locations (Figure 5). After relative location, the events appear more clustered with an extension of the pattern covering a size of 2–3 km and a clear alignment in the NW-SE direction (Figure 5a). The event depth ranges between 7.5 and 12.5 km (Figure 5b). In a section view, these events are distributed according to a direction inclined of about 45–55° with respect to the horizontal plane, with a NE immersion (Figure 5c,d). Horizontal and vertical location errors are within one hundred meters (Figure 5e–g) and the rms is within 0.06 s for most of the events. Since the location errors are estimated analyzing the likelihood function in the vicinity of the best solution, the resulting errors could be underestimated in the case of multiple maxima in the PDF. However, in our case, a DD location rms of 0.06 s indicates that the hypocenters are located within a sphere of about 300 m radius, if considering a P-wave velocity of 5.5 km/s at 8–10 km of depth [62]. Since the rms also accounts for the uncertainty on the origin time, the effective uncertainty on spatial coordinates could be smaller, coherently with estimates from the posterior PDF.

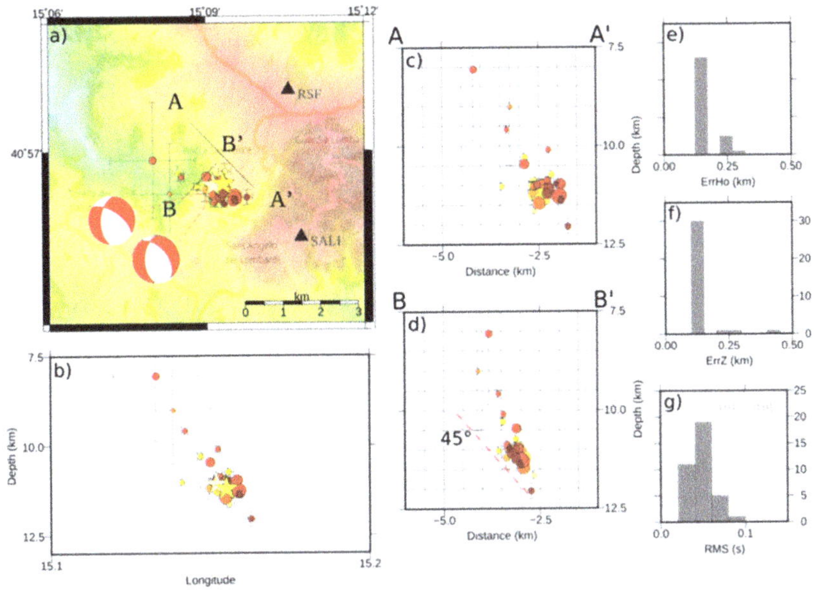

Figure 5. Double-difference (DD) locations. (**a**) DD location of 36 events of the sequence located with NLDiffLoc in a 3D velocity model. The color of the event hypocenters is associated with time of occurrence (from yellow to red) and the size to their magnitude. The stars represent the two Ml 3.0 events. For these events, we also reported the focal mechanism. (**b**) East–west vertical section of the events. (**c**,**d**) Cross-section projection of events along the profiles AA′ and BB′, respectively, indicated in the map of panel (**a**). (**e**) Histogram of horizontal location errors. (**f**) Histogram of vertical location errors. (**g**) Histogram of DD location rms. We reported the mean and the standard deviation of the rms distribution.

6. Source Parameters

We chose the Brune source model [63] to infer source parameters (seismic moment, earthquake size, stress drop, and radiated energy released during the seismic event), which corresponds to an instantaneous pulse applied to the fault. We applied the SPAR (Source PARameters estimator) technique [64], which inverts the observed displacement spectra relying on a probabilistic framework based on the conjunction of states of information between data and model spaces. The inversion strategy allows to jointly retrieve source parameters along with their uncertainties and to investigate the between-parameter correlations. In this approach, theoretical Green's functions are evaluated in a simplified model accounting for both geometrical spreading and anelastic attenuation. We assumed a constant, frequency independent quality factor, which corresponds to the median value Q = 226 retrieved for the area [65]. For the Irpinia region, it has been demonstrated that a frequency independent quality factor is a model preferable to a frequency dependent quality factor following a power law [65]. Beyond the uncertainty estimation, the shape of the posterior PDF is also used to evaluate the quality of the retrieved estimations and eventually to discard unconstrained solutions. This technique has been efficiently tested and validated on the main events (M > 4.0) of the 2016–2017 central Italy sequence [64].

In Figure 6a, the displacement amplitude spectra have been plotted for the Ml 3.0 event occurred on 3 July 2020, at 16:14. The several curves refer to the diverse stations at which the parameters are constrained during the inversion. For this event, we estimated an average moment magnitude of Mw = 2.91 $\pm$ 0.02, a corner frequency of fc = 5.0 $\pm$ 0.2 Hz, and a spectral fall-off γ = 2.42 $\pm$ 0.04. Source parameters have been estimated for 36 events in the sequence, the same events for which we got accurate double-difference locations. In Figure 6b, we represent the retrieved corner frequencies as a function of the moment magnitude (red points). Despite the large scattering in the data, we recognize that larger magnitude events (Mw > 2.0) show larger stress drops (between 1 and 10 MPa). At smaller moment magnitudes (1.0 < Mw < 2.0), the stress drop is on average smaller, with most of the events having a stress drop between 0.1 and 1 MPa. However, about one-third of the events in this magnitude range shows a stress drop larger than 1 MPa, excluding a saturation of the corner frequency at small magnitude. In Figure 6b, we also represent the corner frequencies as a function of the moment magnitude retrieved for ~720 events occurred in the Irpinia, with magnitude between 1.0 and 3.5 (black points) [65]. We found that our results are very consistent with the findings of [65]. Assuming self-similarity, we retrieved an average stress drop of 0.64 MPa. It is worth to note that estimations of corner frequency for the smallest magnitude events (M < 1.5) may be biased due to limited bandwidth when inverting the spectra. For events with Ml > 1.5, we also retrieve consistent estimates of the moment magnitude as compared to the local magnitude.

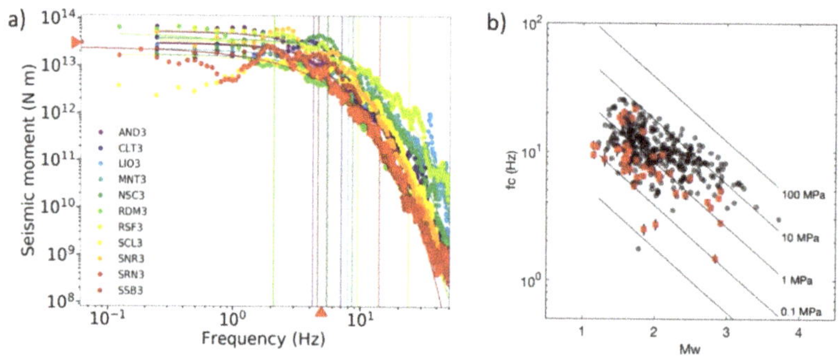

Figure 6. (**a**) Displacement spectra for the main event (Ml 3.0, 3 July 2020, 16:14) of the sequence; for this event, we estimated an average moment magnitude of Mw = 2.91 ± 0.02, a corner frequency of fc = 5.0 ± 0.2 Hz, and a spectral fall-off of γ = 2.42 ± 0.04. The intra-event variability on the seismic moment and corner frequency ranges over one order of magnitude. The red arrows point to the mean values of the parameters. (**b**) Plot of the corner frequency as a function of the moment magnitude (this study—red points; seismicity of the Irpinia region from [65]—black points). Straight lines individuate the curves along which the stress drop is constant.

From the computation of the corner frequencies, we also retrieved the source radius for each of the analyzed events. In Figure 7a, we show the source radius as a function of the moment magnitude: we found that the radius ranges in the interval 150–400 m for the largest magnitude events in the sequence and between 30 and 60 m for an Mw 1.5 event. Looking at the double difference locations, we see that the spatial extension of the events in the sequence is of the same order of magnitude (hundreds of meters) as the size of the largest magnitude events, indicating that the regions fractured by the different events are contiguous and stress transfer is likely to be the main mechanism for event production.

We finally estimated the radiated seismic energy, relying on a time domain estimator [21], which is grounded on the computation of the squared velocity in the time domain integrated over the S-wave time window [66], after its correction for attenuation along the path. The attenuation model has been determined considering ~2300 local earthquakes recorded by INFO over the last ten years [21].

The velocity integral is computed considering band-pass filtered signals between 0.5 and 40 Hz, and a time window starting 0.1 s before the S-wave onset and ending at different percentages of the cumulated energy as a function of the source to site distance R: (i) 90% when R < 25 km; (ii) 80% when 25 km < R < 50 km; (iii) 70% when R > 50 km. In addition, we imposed a minimum time window length of 5 s and a maximum time window length of 20 s. The radiated energy is finally obtained by averaging the estimates over several recording stations (minimum of 3 stations; the sum of signal to noise ratio for the three components ≥200). In Figure 7b, we represent the radiated energy as a function of the seismic moment. The energy-to-moment scaling observed for the sequence (red squares) is in good agreement with values estimated for the Irpinia seismicity (gray points). The radiated energy trend with seismic moment is also consistent with an average stress drop in the sequence smaller than the value assumed by Kanamori as global average (from 2 to 6 MPa) while the steeper slope suggests a possible deviation from self-similar behavior.

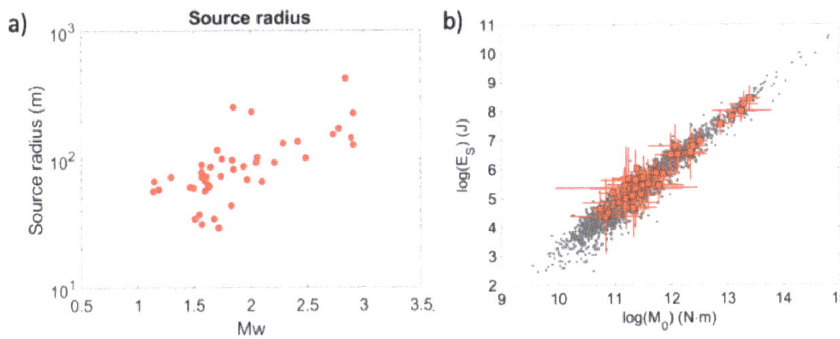

Figure 7. (**a**) Source radius as a function of the moment magnitude for the events in the sequence. The source radius ranges from about 50 m for a Mw 1.5 event to about 250 m for a Mw 3.0 event. (**b**) Scaling of the radiated energy as a function of the seismic moment. The red points correspond to events of this sequence, the gray points represent background events recorded by ISNet.

7. Focal Mechanism Solutions

We computed 21 fault plane solutions from the inversion of P-wave polarities for earthquakes with local magnitude ranging between 1.2 and 3.0. Focal mechanism solutions are calculated from the inversion of P-wave polarities using FPFIT code [67] for all the events showing at least 5 P-wave polarities. The two main Ml 3.0 events show a similar normal-fault kinematics with a minor strike-slip component. The nodal planes have NW-SE trending and a dip of about 50–60°. The remaining solutions show a common normal faulting style with a minor and variable strike-slip component.

We classified the fault plane solutions according to the plunge of P- and T-axes to derive the tectonic regime in which the seismic sequence originated. As shown in Figure 8, most of the solutions (~66%) belong to a pure normal fault regime while few solutions (~34%) belong to a normal strike-slip regime. In agreement with this classification, P-plunges range between 45° and 85° and T-plunges range between 0° and 85°. Moreover, T-axis orientations show an azimuth ranging between 25° and 97°, in accordance with the regional stress tensor calculated for the Irpinia region [47].

We calculated the composite focal mechanism using the data of the two main earthquakes (Ml 3.0), as shown in Figure 8. The normal fault kinematics of the solution with NE-SW trending nodal planes (320, 55, −120; 185, 45, −54) is compatible with the fault plane solutions calculated for single earthquakes and well fits all the available polarity data. Despite the uncertainty of the solutions due to the small number of available data and the small size of the events, a good agreement across the solutions is reached. A large group of seismic stations show the same type of P-wave polarity highlighting a similar rupture kinematics and fault plane geometry during the evolution of the seismic sequence. In the lower panel of Figure 8, all the available polarities data are plotted on the composite focal mechanism solution obtained from the inversion of P-wave polarities for the two Ml 3.0 earthquakes of the sequence. Polarities from six stations (LIO3, MNT3, RDM3, SALI, SSB3, VDS3) are the same during the whole sequence, and 4 stations (AND3, CLT3, COL3, RSF3) show a dominant polarity. Only two stations (NSC3 and SNR3) display a large variability, not always consistent with composite fault plane solutions. The results are summarized in

Table 1. Discrepancies and variability can be explained with the station location on the focal sphere, close to nodal planes: if uncertainties associated with location and focal mechanism are considered, a little rotation of the rupture plane can justify the polarity variation.

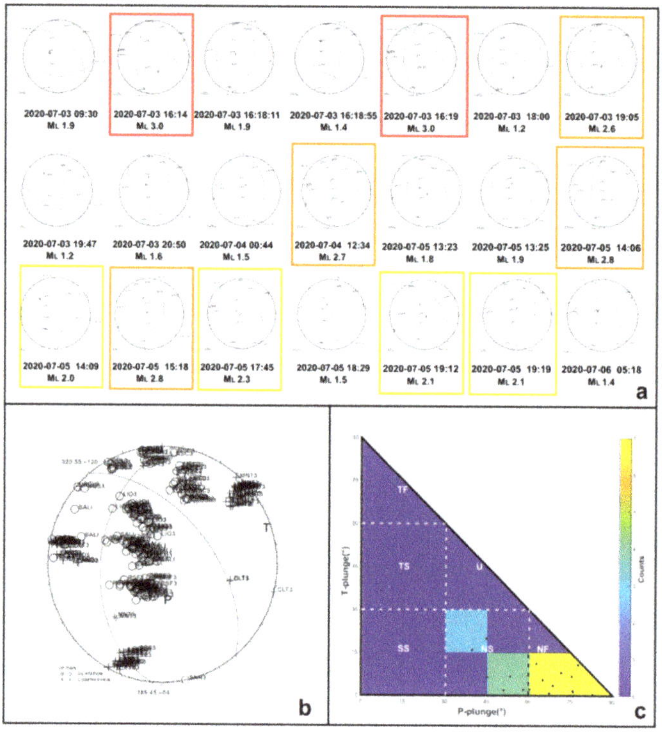

Figure 8. (**a**) Fault plane solutions computed for 21 earthquakes of the sequence. Earthquakes with Ml 3.0, 2.5 ≤ Ml < 3.0, and 2.0 ≤ Ml < 2.5 are highlighted in red, orange, and yellow, respectively. (**b**) Composite focal mechanism solution obtained from the inversion of P-wave polarities for the two main events. (**c**) Fault plane solution classification according to the plunge of P- and T-axes with specific tectonic regimes (Legend: NF, normal fault; NS, normal-oblique; SS, strike-slip; TF, thrust fault; TS, thrust oblique; U, unknown). The number of earthquakes (color bar) is counted in bins of 15° × 15°.

Table 1. 235 P-wave polarity data available for the whole seismic sequence. Data are organized by seismic stations.

Station	Polar.	No.	Polar.	No.
AND3	UP	9	DOWN	1
CLT3	UP	16	DOWN	1
COL3	UP	1	DOWN	6
LIO3	UP	0	DOWN	30
MNT3	UP	27	DOWN	0
NSC3	UP	8	DOWN	18
RDM3	UP	6	DOWN	0
RSF3	UP	1	DOWN	33
SALI	UP	0	DOWN	38
SCL3	UP	1	DOWN	1
SNR3	UP	4	DOWN	11
SSB3	UP	22	DOWN	0
VDS3	UP	0	DOWN	1

Focal mechanism solutions of the main earthquakes show the activation of a NW-SE striking fault structure with 50–60° dip, in agreement with seismogenic sources of the Irpinia region. Despite the low magnitude of the analyzed earthquakes, fault plane solutions reveal a normal fault tectonic regime, consistent with the regional stress field. Considering the spatial distribution of the hypocenters, the NE-dipping nodal plane can be assumed as the preferential one along which the seismic sequence originated, showing the same orientation of the adjacent seismogenic fault segment activated during the 1980 Irpinia earthquake.

8. Ground Motion

The ground motion characteristics were investigated through the main properties of peak values both in terms of acceleration and velocity. A first rough estimation of the frequency content associated with ground motion records can be obtained evaluating the ratio of peak ground acceleration (PGA) to peak ground velocity (PGV) [68,69]. We reported an average ratio value of 4.1 ± 1.6 g/ms^{-1} which allows to classify the records in the class of high acceleration—low velocity. Such a high ratio suggests that the records are mainly characterized by a short duration, high predominant frequencies, and narrow-band spectra [70].

We also compared the peak values to the ground motion prediction equation (GMPE) inferred for the Irpinia region [71]. The comparison for the Ml 3.0 event of the sequence (3 July 2020, 16:19) is shown in Figure 9a. We observed a good agreement between predictions and observations; these latter are almost all included within one standard deviation in the predictions of the GMPEs.

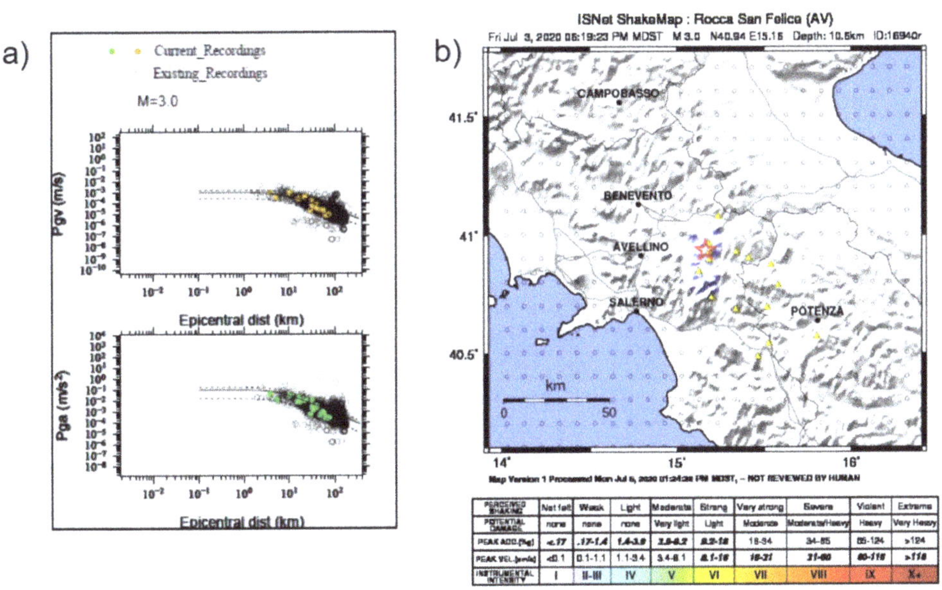

Figure 9. (**a**) Comparison between observed peak ground values and the GMPEs (Ground Motion Prediction Equations) [71] for the Ml 3.0, 3 July 2020, 16:19 earthquake. Lines identify the mean GMPE, and the mean plus/minus one standard deviation. Open circles represent peak observations. (**b**) Shakemap® computed for instrumental intensity, for the same event. The red star corresponds to the event epicenter.

We also present, in Figure 9b, the Shakemap® [72] for the same event, in terms of instrumental intensity, automatically generated at ISNet. The intensity distribution is slightly elongated toward the south direction, with respect to the earthquake epicenter.

The maximum instrumental intensity has been estimated to be III, which is consistent with reports provided by the INGV.

9. Early Warning Analysis

During the sequence, two earthquake early warning (EEW) systems, PRESTo and SAVE, were operating at INFO. PRESTo [23] is an open-source software platform for regional (network-based) EEW which integrates algorithms for real-time data collection, event detection, rapid earthquake location, magnitude estimation, and real-time ground motion prediction in the area of interest.

For the current setting at the INFO, the alert in PRESTo is released if at least five stations have detected the event, independently of the estimated event magnitude or shaking intensity in the area.

SAVE is an on-site, P-wave-based EEW approach [73] which has been conceived to operate either with a single station (i.e., a single sensor located at the target site) or with a set of co-located seismic nodes within a small area around the target to protect. SAVE processes the vertical component of both accelerometers and (broadband) velocimeters, and predicts the expected ground shaking at the recording site issuing a local alert level, together with a qualitative assessment of the earthquake magnitude and source-to-site distance, based on measurements on the early portion of the P-wave.

Unfortunately, during the early days of the sequence, some stations of the network suffered from a temporary failure of the communication system. This resulted in tremendous delays of data transmission and in the partial loss of a few recorded data, which prevented EEW systems from correct operation. For this reason, we could not evaluate the performance of the systems for the whole sequence, but each system automatically detected a smaller number of events, depending on the availability of real-time recorded data.

PRESTo detected a total of 21 events of the sequence, but only 10 of these events (with local magnitude between 1.4 and 2.8) did not suffer for real-time data communication problems and had real-time EEW estimates. The list of the detected events is available in Table 2. For each detected event, the table shows the comparison between the local (Ml) magnitude (as computed by INFO) and the estimated magnitude by PRESTo at the first alert. The difference between the first PRESTo estimate and the local magnitude is also reported in the table (ΔM). From the analysis of the PRESTo outputs, we found that the first magnitude estimate is available on average 3.9 s after the first P wave detection at the network and at the same time, the average difference between the estimated magnitude and the bulletin one is 0.1 unit.

Table 2. List of events detected by PRESTo. The event is identified by its ID, as shown in the ISNet bulletin (http://isnet-bulletin.fisica.unina.it/cgi-bin/isnet-events/isnet.cgi), the local magnitude, the first estimate from PRESTo and the difference between the two magnitudes.

Event ID	Ml	M_PRESTo	ΔM
16975	1.4	1.5	0.1
16973	2.1	2.1	0
16972	2.1	2.5	0.4
16967	2.8	3.4	0.6
16964	2	2	0
16963	2.8	3	0.2
16962	1.9	1.9	0
16961	1.8	1.2	−0.6
16958	1.3	1.2	−0.1
16956	2.7	3.1	0.4

Figure 10a shows the results of PRESTo during the Ml 2.7, event, that occurred on 4 July 2020 at 12:34, while Figure 10b shows the evolution of real-time estimates of magnitude, location, and number of data providing information to the system. As it can be seen from the plot, the first estimates of location and magnitude are available at about 5 s from the origin time, using 5 stations. After 2 s, the estimates of location (epicenter and depth) and magnitude converge and stabilize to the real values. We computed the available lead-time at the main cities of the Campania region, as the difference between the theoretical arrival time of the S-wave and the time of the first alert release. The lead-times are reported in Table 3. The lead-time ranges from 5.8 s at the city of Avellino (~25 km from the event epicenter) to 17.5 s at Naples (~70 km away from the epicenter).

Table 3. Theoretical lead-time. The table shows the epicentral distance and the theoretical lead-time for the main cities of the Campania region.

City	Epicentral Distance (km)	Lead-Time (s)
Avellino	25.4	5.8
Benevento	33.4	7.7
Caserta	65.4	16.6
Napoli	69.1	17.5
Salerno	39.6	9.1

Finally, for this event, we theoretically estimated the radius of the blind zone (i.e., the area where the S-wave arrives before the alert release) which turned out to be around 15 km from the epicenter.

As for SAVE, at the time of the sequence, the system was running at 3 stations of the network: RSF3, COL3, and AVG3. The station RSF3, which is the closest to the sequence, detected 24 events, COL3 detected 11 events, while no events were detected at the farthest station AVG3.

Figure 11a shows a screenshot of the same Ml 2.7 event analyzed by PRESTo at RSF3. For this event, using 3 s of P-wave, the on-site EEW system estimated a local intensity of III, and no warning was declared at that station. For all the detected events, the magnitude and distance estimates were not available, due to the low signal-to-noise ratio that prevented from the use of the average period of the P-wave as a proxy for the earthquake magnitude. Instead, for all the events, SAVE predicted the expected ground motion intensity at each site. Figure 11b,c shows the cumulative performance in terms of lead-time and predicted vs. observed intensity for RFS3 (Panel b) and COL3 (Panel c).

Intensity estimates differ from the real ones of 0 or 1 unit. For the detected events, the available lead-time at the considered stations ranges between 1 to 6 s.

Finally, at the time of the sequence, the ISNet EWapp [74] was under testing by a limited number of users. The smartphones were distributed over the area, but the number was too small to perform a reliable statistical analysis. However, we can report here that the ISNet EWapp received the alerts provided by PRESTo and correctly predicted the expected intensity at each site. For all the events, the predicted intensity never exceeded the threshold of intensity IV at the location of the smartphones.

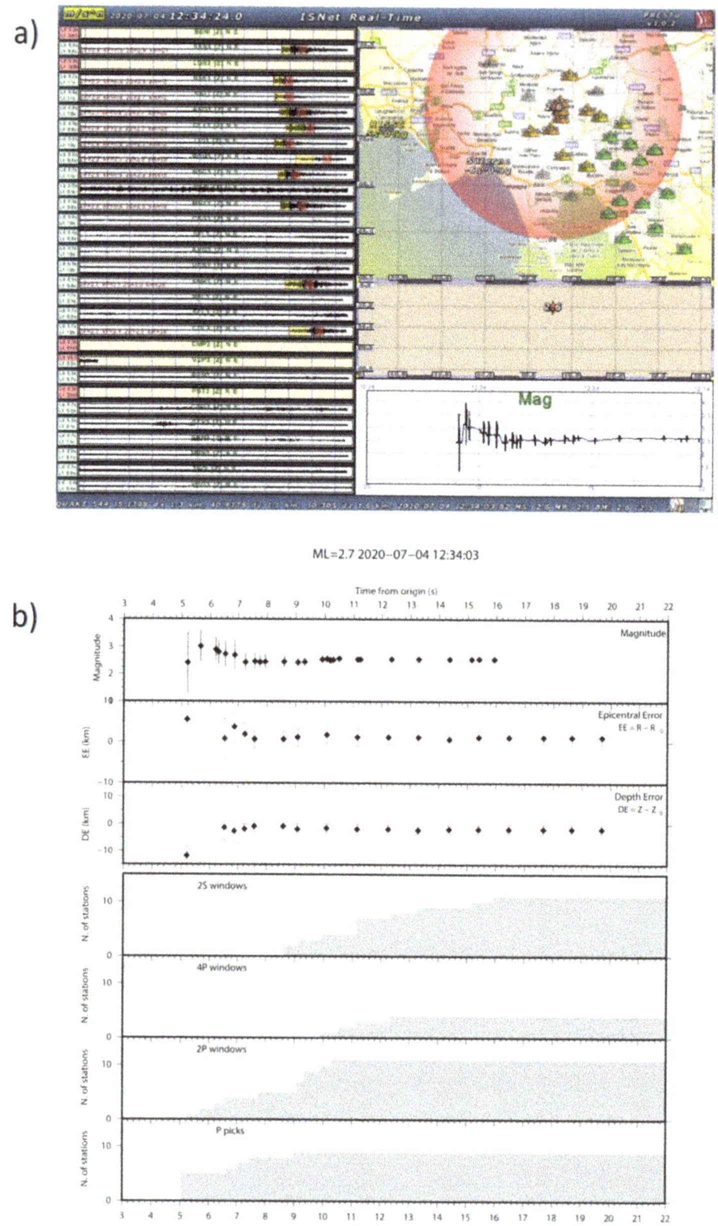

Figure 10. Performance of PRESTo during the Ml 2.7 event that occurred on 4 July 2020 at 12:34. (**a**) Screenshot of PRESTo during the event. (**b**) Real-time estimates of magnitude and location, and the number of available data as a function of time. From top to bottom, the figure shows: magnitude, epicentral error, depth error, number of stations with available 2 s of S wave (2S), 4 s of P wave (4P), 2 s of P wave (2P), and number of available P-wave picks.

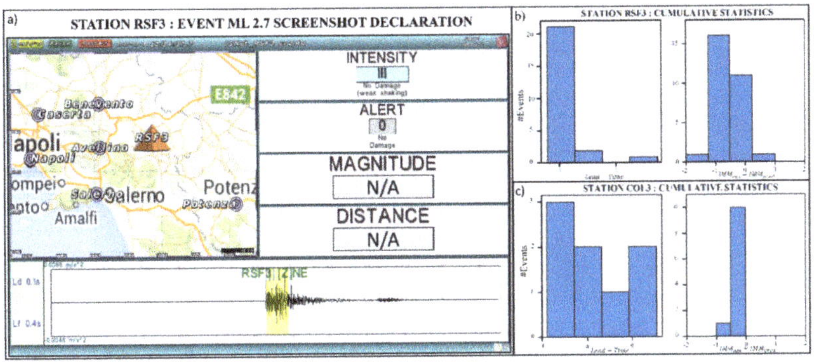

Figure 11. Performance of SAVE. (**a**) Screenshot of SAVE during the Ml 2.7 event that occurred on 4 July 2020, at 12:34, recorded at RSF3. (**b**) Performance of SAVE at RSF3 and (**c**) COL3. Performances are shown in terms of lead-time (left side) and difference between predicted and observed intensities.

10. Discussion and Conclusions

In this study, we have analyzed a seismic sequence, detecting more than 340 events with magnitude in the range $-0.5 < \text{Ml} < 3$. The sequence occurred at the northern tip of the main segment of the 1980, M 6.9, Irpinia earthquake, at the boundary of the Monte Cervialto portion of the fault. Double difference locations and focal mechanisms agree in recognizing that the sequence ruptured an asperity along NE-dipping plane, with a dip angle ranging between 50° and 60°. Despite large uncertainties in the smallest magnitude event locations, the hypocenter distribution delineates a structure dipping at 50–55° at depths of 10 to 12 km and becoming steeper at shallower depths (between 7.5 and 9 km), with a dip angle of 60–65°, eventually being a portion of a listric fault. Strike and dip directions are consistent with the ones of the first fault segment that ruptured during the 1980 earthquake [42], indicating that this asperity is just ahead of the northern endpoint of the 1980 event, on the continuation of that segment. Most of the events within the sequence occurred at depths between 10.5 and 11.5 km, indicating that most of the slip was released at a depth comparable with that of the 1980 hypocenter [40]; according to tomographic models of the area [15,34], most of the events either developed in the basement or at its top with few events rupturing the upper Apulian carbonates. This is a common feature of the sequences that occurred in this region [4].

The size of the asperity, inferred from earthquake location is of the order of 800–900 m, corresponding to an approximate source radius of 400–450 m (Figure 5). On the other hand, the cumulative seismic moment results in an equivalent seismic event with moment magnitude of 3.5. If we estimate the source radius for this latter event, using the average stress drop retrieved from the analysis of source parameters, we get a value fully consistent with the extension inferred from earthquake location. When representing on the fault plane the seismic events within their own size, most of the events look contiguous, indicating that the sequence ruptured a single patch along the fault plane. Looking at the inter-event distance (median value of 370 m and standard deviation of 340 m), as compared to the size of main events, we argue that the dominant triggering mechanism within the sequence is the dynamic and static stress transfer, that allows the nucleation of individual events in the sequence.

From the stress drop analysis, we also recognize a strong heterogeneity in the stress release. Despite the large uncertainty in the evaluation of the stress drop, we reported a dual behavior for the earthquakes in the sequence: largest magnitude events featured a stress drop of 1–10 MPa, while stress drop of most of the small magnitude events ranges in the interval 0.1–1 MPa. Those values are fully consistent with the estimates retrieved for the seismicity of the whole region [65]. The largest stress drop values in the sequence are comparable with the stress drop estimated for the 1980 earthquake [75], while smaller values of stress drop in the area are ascribed to high pore-pressure that locally decreases the normal stress [65]. This variability has been observed also in the computation of the apparent stress over a much larger time interval and at the scale of the whole network [21].

It is worth to note that the retrieved stress drop is based on the estimation of the corner frequency and seismic moment from the inversion of the observed displacement spectra, fixing the path correction due to anelastic attenuation to the value found in [65]. To check the robustness of the stress drop estimates, we performed synthetic tests changing the value of the quality factor. Specifically, we have investigated the hypothesis that small events have the same stress drop as the largest magnitude events in the sequence, but their stress drop appears smaller because of an incorrect correction for path effects. Retrieving a stress drop one order of magnitude smaller than the input value at a source-receiver travel time of 5 s, requires an average quality factor as low as 130. This average value is very unlikely in the quality factor distribution represented in [65], when five or more estimates of the source parameters are available. Alternatively, an apparent stress drop at small magnitudes could be due to a complex, frequency dependent quality factor. This hypothesis cannot be rejected a-priori and requires further modeling and testing with respect to the analysis performed in [65], which is beyond the scope of this study.

The sequence started with a series of cracklings with M < 1 events and small stress drops, possibly indicating a fluid induced sequence initiation promoted by low normal stress and local fault lubrication. After the occurrence of many of those small magnitude events, the stress accumulated in the main asperity has reached the yield strength, releasing a large amount of stress in the two mainshocks of the sequence. In the later stage of the sequence, we reported again small stress drop events, some of which also occurred at few kilometers away from the sequence centroid, in the Apulian carbonates layer; these events could be related to local fluid diffusion, activating small pre-existent fractures prone to rupture.

Since the sequence occurred close to the main segment that ruptured during the 1980 earthquake, we may question why this sequence did not nucleate a large earthquake similar to the 0 s event of the Irpinia earthquake. This may be related to the fact that the sequence does not occur on the prolongation of the Monte Cervialto segment, but on a subparallel fault, or the Monte Cervialto fault has a stepover, with a geometrical discontinuity that prevented the jump on the main segment activated during the 1980 earthquake. Additionally, it is possible that the endpoint of the Monte Cervialto fault is a mechanical barrier, with large yield strength not overcome after stress release during the sequence. Investigating the geometrical details of these faults at such a small scale requires the development of new tools that will allow location and characterization of those events of small magnitude (M < 0.5) buried in the noise, and here detected with autocorrelation techniques.

Despite missing some information at the decametric scale, detection of events with Ml < 1, featuring low signal to noise ratio, is very useful in characterizing the time evolution of the sequence and in identifying potential foreshock activity. This sequence has indeed started 16 h before the main events, with initial seismic activity mainly characterized by small cracklings with Ml < 0.5. Thus, only such high-resolution catalogs, extracted with automatic autocorrelation techniques, enable to catch the foreshock initial phase of the sequence, as also pointed out by other studies [17]. It is worth to note that we performed an offline analysis of the events, with autocorrelation techniques running after the occurrence of the main events of the sequence. The near real-time analysis looking at the continuous data before the occurrence of the mainshocks merits further investigation, which is beyond the scope of this study. Within the improved catalog, we also reported a b value of b = 0.70, which is significantly different from the one related to the background seismicity of the area (b = 0.93, [60]). This decrease in the b-value can be related to an increase of the differential stress [76], associated with fluid pressure.

Since the sequence occurred in a sector where the network density is higher and the EEW systems worked at least for a part of the sequence, analysis of the system performance can highlight advantages and limits of actual systems. Specifically, we can individuate a sort of "minimum size" of the blind zone for the regional system, which has been estimated to be 15 km. This size represents the actual limit of regional EEW systems, within which no actions can be activated at the occurrence of an earthquake. This poses also a lower limit in magnitude (Ml < 6), below which EEW is not a viable risk reduction tool, because the blind zone almost superimposes with the area damaged by the earthquake. Onsite systems can provide a still positive, albeit very small lead-time for targets within the blind zone of the regional system, as also shown for the 2016–2017 central Italy sequence [77]. A contraction of the blind zone can be also obtained by estimating the earthquake size on an expanding P-wave time window, using for instance, the shape of the logarithm of the peak displacement in the time domain (LPDT curves [78]). When applied to this sequence, the average discrepancy between the predicted moment magnitude and the value provided by SPAR is 0.36, which is slightly larger than the average uncertainty on the magnitude estimate. This technique also provides a first rough estimate of the stress drop of the event, modeling the LPDT curve with a displacement function characterized by symmetric triangular function. For the sequence, the predicted average stress drop is 0.9 MPa, which is close to the estimate retrieved from spectral analysis.

Author Contributions: Conceptualization, G.F.; data curation, L.E., F.C. and R.R.; formal analysis, G.F., G.M.A., A.C., S.C., G.D.L., L.E., A.E., M.P. and A.S.; methodology, G.F.; software, L.E., M.P. and F.C.; supervision, A.Z.; validation, S.G., A.G.I., S.N., G.R. and S.T.; writing—original draft, G.F. and G.M.A.; writing—review & editing, A.C., S.C., G.D.L., A.E., M.P., A.S., S.G., A.G.I., S.N., G.R. and A.Z. All authors have read and agreed to the published version of the manuscript.

Funding: Data: data products and service provision of this research are funded by MUR, Ministero Università e Ricerca, through the project EPOS-Italia, and by DPC, Dipartimento di Protezione Civile, through a collaboration agreement with the University of Naples Federico II. Part of the research has been founded by the national project PRIN FLUIDS, grant number 20174 × 3P29.

Institutional Review Board Statement: Not Applicable.

Informed Consent Statement: Not Applicable.

Data Availability Statement: Publicly available datasets were analyzed in this study. This data can be found here: http://isnet-bulletin.fisica.unina.it/cgi-bin/isnet-events/isnet.cgi.

Acknowledgments: We are grateful to three anonymous reviewers who contributed to improve the manuscript. The geological map of Figure 2 has been redrawn from the map of [79].

Conflicts of Interest: The authors declare no conflict of interest. The funders had no role in the design of the study; in the collection, analyses, or interpretation of data; in the writing of the manuscript, or in the decision to publish the results.

References

1. Lomax, A.; Virieux, J.; Volant, P.; Berge-Thierry, C. Probabilistic Earthquake Location in 3D and Layered Models. In *Advances in Seismic Event Location*; Modern Approaches in Geophysics, Volume 18; Thurber, C.H., Rabinowitz, N., Eds.; Springer: Dordrecht, The Netherlands, 2000. [CrossRef]
2. Satriano, C.; Lomax, A.; Zollo, A. Real-Time Evolutionary Earthquake Location for Seismic Early Warning. *Bull. Seism. Soc. Am.* **2008**, *98*, 1482–1494. [CrossRef]
3. Waldhauser, F.; Ellsworth, W.L. A double-difference earthquake location algorithm: Method and application to the northern Hayward Fault, California. *Bull. Seismol. Soc. Am.* **2000**, *90*, 1353–1368. [CrossRef]
4. Stabile, T.A.; Satriano, C.; Orefice, A.; Festa, G.; Zollo, A. Anatomy of a microearthquake sequence on an active normal fault. *Sci. Rep.* **2012**, *2*, 410. [CrossRef] [PubMed]
5. Rowe, C.A.; Thurber, C.H.; White, R.A. Dome growth behavior at Soufriere Hills Volcano, Montserrat, revealed by relocation of volcanic event swarms, 1995–1996. *J. Volcanol. Geotherm. Res.* **2004**, *134*, 199–221. [CrossRef]
6. Valoroso, L.; Chiaraluce, L.; Piccinini, D.; Di Stefano, R.; Schaff, D.; Waldhauser, F. Radiography of a normal fault system by 64,000 high-precision earthquake locations: The 2009 L'Aquila (central Italy) case study. *J. Geophys. Res. Solid Earth* **2013**, *118*, 1156–1176. [CrossRef]
7. Chen, X.; Shearer, P.M.; Abercrombie, R.E. Spatial migration of earthquakes within seismic clusters in Southern California: Evidence for fluid diffusion. *J. Geophys. Res. Space Phys.* **2012**, *117*. [CrossRef]

8. Miller, S.A. Aftershocks are fluid-driven and decay rates controlled by permeability dynamics. *Nat. Commun.* **2020**, *11*, 5787. [CrossRef]
9. Abercrombie, R.E. Earthquake source scaling relationships from − 1 to 5 ML using seismograms recorded at 2.5-km depth. *J. Geophys. Res.* **1995**, *100*, 24015–24036. [CrossRef]
10. Prieto, G.A.; Shearer, P.M.; Vernon, F.L.; Kilb, D. Earthquake source scaling and self-similarity estimation from stacking P and S spectra. *J. Geophys. Res.* **2004**, *109*. [CrossRef]
11. Abercrombie, R.E.; Rice, J.R. Can observations of earthquake scaling constrain slip weakening? *Geophys. J. Int.* **2005**, *162*, 406–424. [CrossRef]
12. Oth, A.; Parolai, S.; Bindi, D.; Wenzel, F. Source Spectra and Site Response from S Waves of Intermediate-Depth Vrancea, Romania, Earthquakes. *Bull. Seism. Soc. Am.* **2009**, *99*, 235–254. [CrossRef]
13. Miller, S.A. Properties of large ruptures and the dynamical influence of fluids on earthquakes and faulting. *J. Geophys. Res. Solid Earth* **2002**, *107*, ESE-3. [CrossRef]
14. Dublanchet, P.; Godano, M.; Bernard, P. Inferring fault mechanical conditions from the source parameters of a complex microseismic multiplet in the Corinth rift, Greece. *J. Geophys. Res. Solid Earth* **2015**, *120*, 7655–7682. [CrossRef]
15. Amoroso, O.; Russo, G.; De Landro, G.; Zollo, A.; Garambois, S.; Mazzoli, S.; Parente, M.; Virieux, J. From velocity and attenuation tomography to rock physical modeling: Inferences on fluid-driven earthquake processes at the Irpinia fault system in southern Italy. *Geophys. Res. Lett.* **2017**, *44*, 6752–6760. [CrossRef]
16. Bouchon, M.; Durand, V.; Marsan, D.; Karabulut, H.; Schmittbuhl, J. The long precursory phase of most large interplate earthquakes. *Nat. Geosci.* **2013**, *6*, 299–302. [CrossRef]
17. Trugman, D.T.; Ross, Z.E. Pervasive Foreshock Activity Across Southern California. *Geophys. Res. Lett.* **2019**, *46*, 8772–8781. [CrossRef]
18. Ende, M.P.A.V.D.; Ampuero, J.-P. On the Statistical Significance of Foreshock Sequences in Southern California. *Geophys. Res. Lett.* **2020**, *47*. [CrossRef]
19. Gibbons, S.J.; Ringdal, F. The detection of low magnitude seismic events using array-based waveform correlation. *Geophys. J. Int.* **2006**, *165*, 149–166. [CrossRef]
20. Yoon, C.E.; O'Reilly, O.; Bergen, K.J.; Beroza, G.C. Earthquake detection through computationally efficient similarity search. *Sci. Adv.* **2015**, *1*, e1501057. [CrossRef]
21. Picozzi, M.; Bindi, D.; Zollo, A.; Festa, G.; Spallarossa, D. Detecting long-lasting transients of earthquake activity on a fault system by monitoring apparent stress, ground motion and clustering. *Sci. Rep.* **2019**, *9*, 16268. [CrossRef]
22. Iannaccone, G.; Zollo, A.; Elia, L.; Convertito, V.; Satriano, C.; Martino, C.; Festa, G.; Lancieri, M.; Bobbio, A.; Stabile, T.A.; et al. A prototype system for earthquake early-warning and alert management in southern Italy. *Bull. Earthq. Eng.* **2010**, *8*, 1105–1129. [CrossRef]
23. Satriano, C.; Elia, L.; Martino, C.; Lancieri, M.; Zollo, A.; Iannaccone, G. PRESTo, the earthquake early warning system for Southern Italy: Concepts, capabilities and future perspectives. *Soil Dyn. Earthq. Eng.* **2011**, *31*, 137–153. [CrossRef]
24. Doglioni, C. Geological remarks on the relationships between extension and convergent geodynamic settings. *Tectonophysics* **1995**, *252*, 253–267. [CrossRef]
25. Patacca, E.; Scandone, P. Geology of the southern Apennines. *Boll. Soc. Geol. It.* **2007**, *7*, 75–119.
26. Cavazza, W.; Roure, F.M.; Spakman, W.; Stampfli, G.M.; Ziegler, P.A. *The TRANSMED Atlas. The Mediterranean Region from Crust to Mantle: The Mediterranean Region from Crust to Mantle: Geological and Geophysical Framework of the Mediterranean and the Surrounding Areas*; Springer: Berlin/Heidelberg, Germany, 2004.
27. Casero, P.; Roure, F.; Vially, R. Tectonic framework and petroleum potential of the southern Apennines. *Gener. Accumul. Prod. Eur. Hydrocarb.* **1991**, *1*, 381–387.
28. Bally, A.; Burbi, L.; Cooper, C.; Ghelardoni, R. Balanced cross sections and seismic reflection profiles across the central Apennines. *Memorie della Societa Geologica Italiana* **1986**, *35*, 275–310.
29. Patacca, E.; Sartori, R.; Scandone, P. Tyrrhenian basin and Apenninic arcs: Kinematic relations since late Tortonian times. *Memorie della Societa Geologica Italiana* **1990**, *45*, 425–451.
30. Frepoli, A.; Amato, A. Contemporaneous extension and compression in the Northern Apennines from earthquake fault-plane solutions. *Geophys. J. Int.* **1997**, *129*, 368–388. [CrossRef]
31. D'Agostino, N.; Avallone, A.; Cheloni, D.; D'Anastasio, E.; Mantenuto, S.; Selvaggi, G. Active tectonics of the Adriatic region from GPS and earthquake slip vectors. *J. Geophys. Res. Solid Earth* **2008**, *113*. [CrossRef]
32. Ascione, A.; Mazzoli, S.; Petrosino, P.; Valente, E. A decoupled kinematic model for active normal faults: Insights from the 1980, MS = 6.9 Irpinia earthquake, southern Italy. *GSA Bull.* **2013**, *125*, 1239–1259. [CrossRef]
33. Chiarabba, C.; Jovane, L.; Distefano, R. A new view of Italian seismicity using 20 years of instrumental recordings. *Tectonophysics* **2005**, *395*, 251–268. [CrossRef]
34. Improta, L.; De Gori, P.; Chiarabba, C. New insights into crustal structure, Cenozoic magmatism, CO2degassing, and seismogenesis in the southern Apennines and Irpinia region from local earthquake tomography. *J. Geophys. Res. Solid Earth* **2014**, *119*, 8283–8311. [CrossRef]
35. Pantosti, D.; Schwartz, D.P.; Valensise, G. Paleoseismology along the 1980 surface rupture of the Irpinia Fault: Implications for earthquake recurrence in the southern Apennines, Italy. *J. Geophys. Res. Solid Earth* **1993**, *98*, 6561–6577. [CrossRef]

36. Buck, W.R. flexural rotation of normal faults. *Tectonics* **1988**, *7*, 959–973. [CrossRef]
37. King, G.; Ellis, M. The origin of large local uplift in extensional regions. *Nature* **1990**, *348*, 689–693. [CrossRef]
38. Ghisetti, F.; Vezzani, L. Normal faulting, transcrustal permeability and seismogenesis in the Apennines (Italy). *Tectonophysics* **2002**, *348*, 155–168. [CrossRef]
39. Westaway, R.; Jackson, J. Surface faulting in the southern Italian Campania-Basilicata earthquake of 23 November 1980. *Nat. Cell Biol.* **1984**, *312*, 436–438. [CrossRef]
40. Westaway, R.; Jackson, J. The earthquake of 1980 November 23 in Campania–Basilicata (southern Italy). *Geophys. J. Int.* **1987**, *90*, 375–443. [CrossRef]
41. Bernard, P.; Zollo, A. The Irpinia (Italy) 1980 earthquake: Detailed analysis of a complex normal faulting. *J. Geophys. Res. Solid Earth* **1989**, *94*, 1631–1647. [CrossRef]
42. Pantosti, D.; Valensise, G. Faulting mechanism and complexity of the November 23, 1980, Campania-Lucania Earthquake, inferred from surface observations. *J. Geophys. Res. Solid Earth* **1990**, *95*, 15319–15341. [CrossRef]
43. Amoruso, A.; Crescentini, L.; Scarpa, R. Faulting geometry for the complex 1980 Campania-Lucania earthquake from levelling data. *Geophys. J. Int.* **2005**, *162*, 156–168. [CrossRef]
44. Amato, A.; Chiarabba, C.; Malagnini, L.; Selvaggi, G. Three-dimensional P-velocity structure in the region of the MS = 6.9 Irpinia, Italy, normal faulting earthquake. *Phys. Earth Planet. Inter.* **1992**, *75*, 111–119. [CrossRef]
45. Amato, A.; Selvaggi, G. Aftershock location and P-velocity structure in the epicentral region, of the 1980 Irpinia earthquake. *Annali di Geophysica* **1993**, *36*, 237–243.
46. Patella, D.; Petrillo, Z.; Siniscalchi, A.; Improta, L.; Di Fiore, B. Magnetotelluric profiling along the CROP-04 section in the Southern Apennines. In *CROP PROJECT: Deep Seismic Exploration of the CENTRAL Mediterranean and Italy*; Finetti, I.R., Ed.; Elsevier: Amsterdam, The Netherlands, 2005; pp. 263–280.
47. De Matteis, R.; Matrullo, E.; Rivera, L.; Stabile, T.A.; Pasquale, G.; Zollo, A. Fault Delineation and Regional Stress Direction from the Analysis of Background Microseismicity in the southern Apennines, Italy. *Bull. Seism. Soc. Am.* **2012**, *102*, 1899–1907. [CrossRef]
48. De Landro, G.; Amoroso, O.; Stabile, T.A.; Matrullo, E.; Lomax, A.; Zollo, A. High-precision differential earthquake location in 3-D models: Evidence for a rheological barrier controlling the microseismicity at the Irpinia fault zone in southern Apennines. *Geophys. J. Int.* **2015**, *203*, 1821–1831. [CrossRef]
49. Adinolfi, G.M.; Cesca, S.; Picozzi, M.; Heimann, S.; Zollo, A. Detection of weak seismic sequences based on arrival time coherence and empiric network detectability: An application at a near fault observatory. *Geophys. J. Int.* **2019**, *218*, 2054–2065. [CrossRef]
50. Vassallo, M.; Festa, G.; Bobbio, A.; Serra, M. Low shear velocity in a normal fault system imaged by ambient noise cross correlation: The case of the Irpinia fault zone, Southern Italy. *J. Geophys. Res. Solid Earth* **2016**, *121*, 4290–4305. [CrossRef]
51. Chiarabba, C.; Chiodini, G. Continental delamination and mantle dynamics drive topography, extension and fluid discharge in the Apennines. *Geology* **2013**, *41*, 715–718. [CrossRef]
52. Italiano, F.; Martelli, M.; Martinelli, G.; Nuccio, P.M. Geochemical evidence of melt intrusions along lithospheric faults of the Southern Apennines, Italy: Geodynamic and seismogenic implications. *J. Geophys. Res. Solid Earth* **2000**, *105*, 13569–13578. [CrossRef]
53. Chiodini, G.; Granieri, D.; Avino, R.; Caliro, S.; Costa, A.; Minopoli, C.; Vilardo, G. Non-volcanic CO2Earth degassing: Case of Mefite d'Ansanto (southern Apennines), Italy. *Geophys. Res. Lett.* **2010**, *37*. [CrossRef]
54. Miller, S.A.; Collettini, C.; Chiaraluce, L.; Cocco, M.; Barchi, M.; Kaus, B.J.P. Aftershocks driven by a high-pressure CO2 source at depth. *Nature* **2004**, *427*, 724–727. [CrossRef] [PubMed]
55. Lucente, F.P.; De Gori, P.; Margheriti, L.; Piccinini, D.; Di Bona, M.; Chiarabba, C.; Agostinetti, N.P. Temporal variation of seismic velocity and anisotropy before the 2009 MW6.3 L'Aquila earthquake, Italy. *Geology* **2010**, *38*, 1015–1018. [CrossRef]
56. Johnson, C.E.; Bittenbinder, A.; Bogaert, B.; Dietz, L.; Kohler, W. Earthworm: A flexible approach to seismic network processing. *Iris Newsl.* **1995**, *14*, 1–4.
57. Cocco, M.; Chiarabba, C.; Di Bona, M.; Selvaggi, G.; Margheriti, L.; Frepoli, A.; Lucente, F.P.; Basili, A.; Jongmans, D.; Campillo, M. The April 1996 Irpinia seismic sequence: Evidence for fault interaction. *J. Seism.* **1999**, *3*, 105–117. [CrossRef]
58. Bergen, K.J.; Beroza, G.C. Detecting earthquakes over a seismic network using single-station similarity measures. *Geophys. J. Int.* **2018**, *213*, 1984–1998. [CrossRef]
59. Andoni, A.; Indyk, P. Near-Optimal Hashing Algorithms for Approximate Nearest Neighbor in High Dimensions. In Proceedings of the 47th Annual IEEE Symposium on Foundations of Computer Science (FOCS'06), Berkeley, CA, USA, 21–24 October 2006; pp. 459–468.
60. Vassallo, M.; Festa, G.; Bobbio, A. Seismic Ambient Noise Analysis in Southern Italy. *Bull. Seism. Soc. Am.* **2012**, *102*, 574–586. [CrossRef]
61. Michelini, A.; Lomax, A. The effect of velocity structure errors on double-difference earthquake location. *Geophys. Res. Lett.* **2004**, *31*. [CrossRef]
62. Amoroso, O.; Ascione, A.; Mazzoli, S.; Virieux, J.; Zollo, A. Seismic imaging of a fluid storage in the actively extending Apennine mountain belt, southern Italy. *Geophys. Res. Lett.* **2014**, *41*, 3802–3809. [CrossRef]
63. Brune, J.N. Tectonic stress and the spectra of seismic shear waves from earthquakes. *J. Geophys. Res.* **1970**, *75*, 4997–5009. [CrossRef]

64. Supino, M.; Festa, G.; Zollo, A. A probabilistic method for the estimation of earthquake source parameters from spectral inversion: Application to the 2016–2017 Central Italy seismic sequence. *Geophys. J. Int.* **2019**, *218*, 988–1007. [CrossRef]
65. Zollo, A.; Orefice, A.; Convertito, V. Source parameter scaling and radiation efficiency of microearthquakes along the Irpinia fault zone in southern Apennines, Italy. *J. Geophys. Res. Solid Earth* **2014**, *119*, 3256–3275. [CrossRef]
66. Festa, G.; Zollo, A.; Lancieri, M. Earthquake magnitude estimation from early radiated energy. *Geophys. Res. Lett.* **2008**, *35*. [CrossRef]
67. Reasenberg, P.; Oppenheimer, D. *FPFIT, FPPLOT and FPPAGE: Fortran Computer Programs for Calculating and Displaying Earthquake Fault-Plane Solutions*; Open File Report 85–739; US Geological Survey: Menlo Park, CA, USA, 1985; pp. 85–739.
68. Zhu, T.J.; Heidebrecht, A.C.; Tso, W.K. Effect of peak ground acceleration to velocity ratio on ductility demand of inelastic systems. *Earthq. Eng. Struct. Dyn.* **1988**, *16*, 63–79. [CrossRef]
69. Ameri, G.; Emolo, A.; Pacor, F.; Gallovič, F. Ground-motion simulations for the 1980 M 6.9 Irpinia earthquake (Southern Italy) and scenario events. *Bull. Seismol. Soc. Am.* **2011**, *101*, 1136–1151. [CrossRef]
70. Kwon, O.S.; Elnashai, A. The effect of material and ground motion uncertainty on the seismic vulnerability curves of RC structure. *Eng. Struct.* **2006**, *28*, 289–303. [CrossRef]
71. Emolo, A.; Convertito, V.; Cantore, L. Ground-motion predictive equations for low-magnitude earthquakes in the Campania–Lucania area, Southern Italy. *J. Geophys. Eng.* **2011**, *8*, 46–60. [CrossRef]
72. Wald, D.J.; Quitoriano, V.; Heaton, T.H.; Kanamori, H.; Scrivner, C.W.; Worden, C.B. TriNet "ShakeMaps": Rapid Generation of Peak Ground Motion and Intensity Maps for Earthquakes in Southern California. *Earthq. Spectra* **1999**, *15*, 537–555. [CrossRef]
73. Caruso, A.; Colombelli, S.; Elia, L.; Picozzi, M.; Zollo, A. An on-site alert level early warning system for Italy. *J. Geophys. Res. Solid Earth* **2017**, *122*, 2106–2118. [CrossRef]
74. Colombelli, S.; Carotenuto, F.; Elia, L.; Zollo, A. Design and implementation of a mobile device app for network-based earthquake early warning systems (EEWSs): Application to the PRESTo EEWS in southern Italy. *Nat. Hazards Earth Syst. Sci.* **2020**, *20*. [CrossRef]
75. Deschamps, A.; King, G.C.P. The Campania-Lucania (southern Italy) earthquake of 23 November 1980. *Earth Planet. Sci. Lett.* **1983**, *62*, 296–304. [CrossRef]
76. Scholz, C.H. The frequency-magnitude relation of microfracturing in rock and its relation to earthquakes. *Bull. Seismol. Soc. Am.* **1968**, *58*, 399–415.
77. Festa, G.; Picozzi, M.; Caruso, A.; Colombelli, S.; Cattaneo, M.; Chiaraluce, L.; Elia, L.; Martino, C.; Marzorati, S.; Supino, M.; et al. Performance of Earthquake Early Warning Systems during the 2016–2017 Mw 5–6.5 Central Italy Sequence. *Seism. Res. Lett.* **2018**, *89*, 1–12. [CrossRef]
78. Nazeri, S.; Colombelli, S.; Zollo, A. Fast and accurate determination of earthquake moment, rupture length and stress release for the 2016–2017 Central Italy seismic sequence. *Geophys. J. Int.* **2019**, *217*, 1425–1432. [CrossRef]
79. Bigi, G.; Coli, M.; Cosentino, D.; Parotto, M.; Praturlon, A.; Sartori, R.; Scandone, P.; Turco, E. Structural Model of Italy scale 1:500.000, sheet 4. In *C.N.R., Progetto Finalizzato Geodinamica*; SELCA: Firenze, Italy, 1992.

geosciences

Article

Assessing Current Seismic Hazards in Irpinia Forty Years after the 1980 Earthquake: Merging Historical Seismicity and Satellite Data about Recent Ground Movements

Aldo Piombino [1,*], Filippo Bernardini [2] and Gregorio Farolfi [1]

1 Department of Earth Sciences, University of Florence, 50125 Florence, Italy; gregorio.farolfi@unifi.it
2 Istituto Nazionale di Geofisica e Vulcanologia (INGV), Sezione di Bologna, 40128 Bologna, Italy; filippo.bernardini@ingv.it
* Correspondence: aldo.piombino@unifi.it

Abstract: Recently, a new strain rate map of Italy and the surrounding areas has been obtained by processing data acquired by the persistent scatterers (PS) of the synthetic aperture radar interferometry (InSAR) satellites—ERS and ENVISAT—between 1990 and 2012. This map clearly shows that there is a link between the strain rate and all the shallow earthquakes (less than 15 km deep) that occurred from 1990 to today, with their epicenters being placed only in high strain rate areas (e.g., Emilia plain, NW Tuscany, Central Apennines). However, the map also presents various regions with high strain rates but in which no damaging earthquakes have occurred since 1990. One of these regions is the Apennine sector, formed by Sannio and Irpinia. This area represents one of the most important seismic districts with a well-known and recorded seismicity from Roman times up to the present day. In our study, we merged historical records with new satellite techniques that allow for the precise determination of ground movements, and then derived physical dimensions, such as strain rate. In this way, we verified that in Irpinia, the occurrence of new strong shocks—forty years after one of the strongest known seismic events in the district that occurred on the 23 November 1980, measuring Mw 6.8—is still a realistic possibility. The reason for this is that, from 1990, only areas characterized by high strain rates have hosted significant earthquakes. This picture has been also confirmed by analyzing the historical catalog of events with seismic completeness for magnitude $M \geq 6$ over the last four centuries. It is easy to see that strong seismic events with magnitude $M \geq 6$ generally occurred at a relatively short time distance between one another, with a period of 200 years without strong earthquakes between the years 1732 and 1930. This aspect must be considered as very important from various points of view, particularly for civil protection plans, as well as civil engineering and urban planning development.

Keywords: Irpinia; seismic hazard; earthquake; strain rate; GNSS; InSAR

Citation: Piombino, A.; Bernardini, F.; Farolfi, G. Assessing Current Seismic Hazards in Irpinia Forty Years after the 1980 Earthquake: Merging Historical Seismicity and Satellite Data about Recent Ground Movements. *Geosciences* **2021**, *11*, 168. https://doi.org/10.3390/geosciences11040168

Academic Editors: Sabina Porfido, Giuliana Alessio, Germana Gaudiosi, Rosa Nappi, Alessandro Maria Michetti and Jesus Martinez-Frias

Received: 26 October 2020
Accepted: 30 March 2021
Published: 7 April 2021

Publisher's Note: MDPI stays neutral with regard to jurisdictional claims in published maps and institutional affiliations.

1. Introduction

This study is based on the analysis of a fine-scale ground velocity map of Italy determined by the fusion of Global Navigation Satellite Systems (GNSS) with synthetic aperture radar interferometry (InSAR) data derived from satellites [1]. The dataset derives from a period of observation between 1990 and 2012. The InSAR dataset is part of the "Piano Straordinario di Telerilevamento" (Special program for Remote Sensing, promoted by the Italian Ministry of Environment). Due to the quasi-polar orbit of the satellites, space-borne InSAR observations can only determine the East–West (E–W) and Up–Down (U–D) components of the movement of persistent scatterers. However, there are millions of scatterers that are unreachable, due to the fact that only a few hundred GNSS stations exist. The North–South (N–S) component is provided by a C^2 continuous bi-cubic interpolation function that is well suited to interpolate sparse GNSS stations displaced inside the study site and surrounding areas. To do this it uses a hierarchical structure at different refinement levels.

The fusion of GNSS with InSAR is a method based on the calibration of InSAR with GNSS measures derived from permanent stations and survey campaigns [2,3]. The results are a coherent fine-scale ground velocity map with a spatial resolution that is unreachable using the previous velocity field maps determined with the GNSS technique alone. By using this technique, Farolfi, Piombino, and Catani [1] provided new information about the complex geodynamics of the Italian peninsula and thanks to the high spatial resolution of the ground movements map, identified interesting patterns of small areas with respect to the surrounding ones. Moreover, their work confirmed the division of peninsular Italy into two sectors, with opposed E–W components of movement in the Stable Europe Frame (Figure 1). This has been depicted by older studies based only on GNSS station movements ([4] and references therein): the western block (Tyrrhenian) is moving westward, while the eastern one (Adriatic) shows an eastward movement. The relative motion of these blocks implies their divergence; the effect of which is represented by the numerous currently active normal fault systems along the central and southern Apennines—which are close to the border between these two sectors—and the associated seismicity.

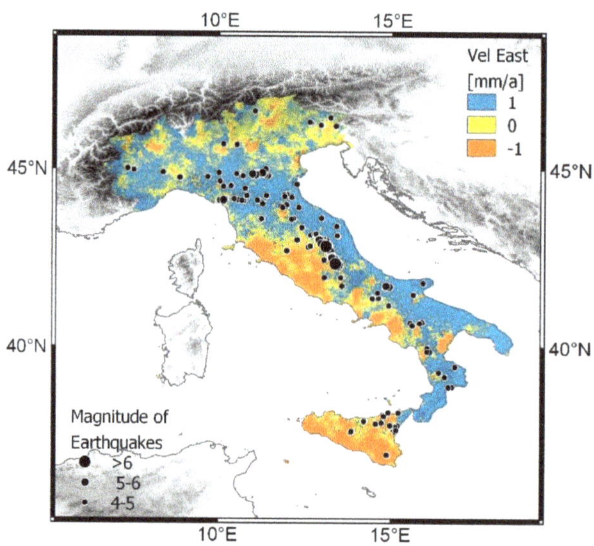

Figure 1. Map of the East component of the ground velocity field of the Italian Peninsula, derived from Global Navigation Satellite Systems (GNSS) and synthetic aperture radar interferometry (InSAR) during more than two decades of observation (1990–2017). The area of the Central Apennines presents major earthquakes from 1990 to present day. From the figure above, it is clear that the main seismogenic areas are linked to the boundary that divides the two blocks with opposite E–W components of velocity.

The Apennine chain, an approximately linear belt hosting the most rapidly slipping normal faults, and the most damaging earthquakes, are coincident with the areas in which the morphological surface height, when averaged on a horizontal scale of tens of kilometers, is greatest [5]. In this area, the first studies based on the relative movements of the GNSS stations have already determined a medium value of a ca. 3 mm/a extension, linked to the differential movements between the two blocks. This also allows the emplacement of melt intrusions along deep-rooted faults [6]—the last occurrence of this kind probably triggered the 2013/2014 Matese seismic swarm [7]—and the widespread emission of deep-originated CO_2 [8]. This regime is dissecting the former Cenozoic east-verging thrust belt related to the west-dipping subduction of the Apulian lithosphere [9]. This compressive regime ended at 650 ka in the middle Pleistocene [10].

The E–W component of InSAR movements [1] has also confirmed the frame depicted by [11], in which the Ortona–Roccamonfina is not a single lineament, but a 30 km wide deformation channel: this channel is characterized by prevalent west-directed velocities in the stable Europe frame, nested in the Adriatic eastward-moving block.

The vertical component of the InSAR data highlights the current general uplift occurring in most of Southern Italy, even if this uplift is lower than in the Central Apennines (especially in the "Abruzzo Dome" [1]), confirming a wealth of the geological literature. Conversely, few areas show subsidence, mainly because of human groundwater exploitation. In this frame, the highest uplift values of the whole Southern Apennines—exceeding 1.8 mm/a—are present in the chain segment between Benevento and Potenza. This area of higher-than-surroundings uplift roughly corresponds to the Irpinia sector, in a belt just west of the Campania–Puglia border. Thus, it is possible to call this area the "Irpinian Dome" (Figure 2). The Ufita and Marzano faults represent the surface traces of the two different patterns of the East–West ground velocity component (Figure 3 (top)).

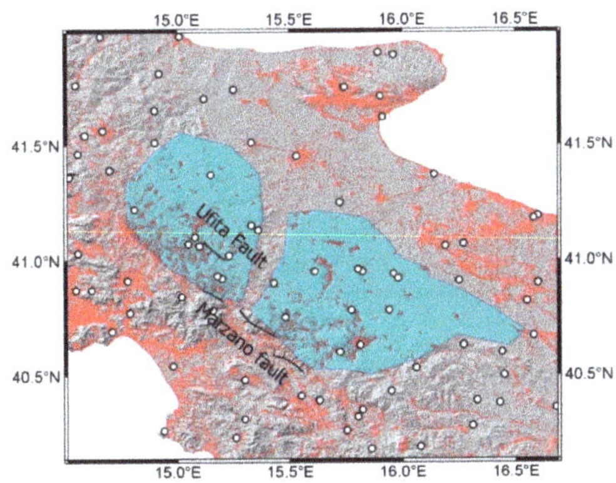

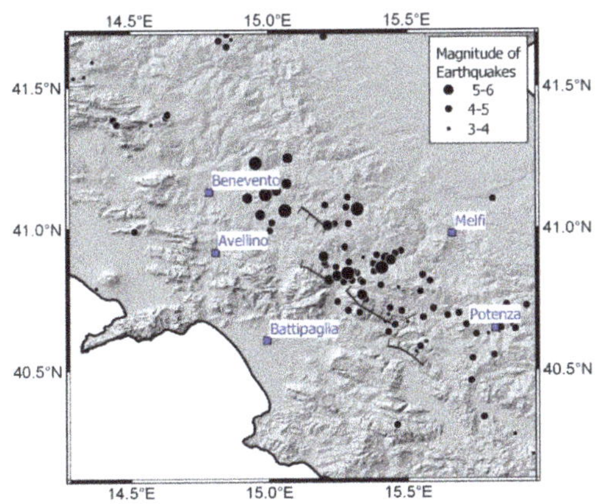

Figure 2. (**Top**) Map of the distribution of persistent scatterers (PS) (red points) and the GNSS permanent stations (white points) involved in the detection of ground surface movements of the study area. The main geodynamic features are represented in the background of the map: the Irpinian Dome is the cyan area and the Ufita and Marzano faults are represented with black hashed lines. (**Bottom**) Map of the main seismic events (black circles) that occurred from 1466 to 2017 with the main towns highlighted (dark blue squares).

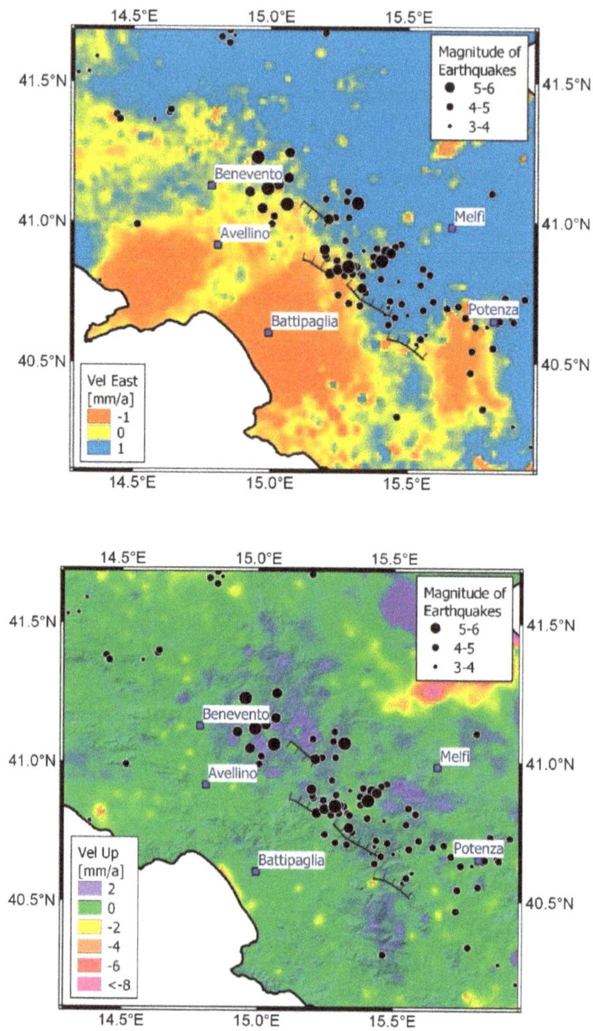

Figure 3. (**Top**) The East component of the ground velocity field of the Irpinia–Sannio area with the main earthquakes of $M \geq 3$ occurred since 1900. For the earthquakes of $M \geq 4$, the label represents the year of the occurrence. (**Bottom**) Map of the vertical component of the ground velocity in Irpinia with the main earthquakes that occurred from 1466 to 2017. The main towns are drawn and labeled with dark blue squares, and the main faults (see Figure 2 (top)) are represented with black hashed lines.

The uplifting area is divided into two different parts and, between them, exists a narrow corridor of lower uplift <1 mm/a, (Figure 3 (bottom)). It is interesting to note that this corridor is placed near the epicentral areas of the 1930 and 1980 earthquakes.

The geographical axis of the Irpinian dome is placed east of the main NE-dipping faults, on the surface projection of the hanging wall. Any useful information of the N–S component of the ground movement can be detected by InSAR satellites because their quasi-polar orbits only make the detection of vertical and E–W velocity components possible.

2. Relationship between Strain Rate and Earthquakes

The strain rate provides a measure of the superficial deformation, and for this reason, is useful information for studying and analyzing geodynamics. Many authors have produced strain rate maps of the Italian territory using GPS station data. In the last decade, for example, Riguzzi et al., (2012) [12] estimated the strain rate, using the GPS velocity solution, of the Italian area—provided by Devoti et al., (2011) [13].

Palano (2015) [14] carried out an analysis of the stress and strain-rate fields of Italy. He performed a comparison of GPS inferred strain-rate data and 308 stress datasets interpolated at each node of a regular grid.

Montone and Mariucci (2016) [15] provided an updated present day stress map for the Italian territory combining seismicity, data retrieved from a breakout analysis in deep wells, and fault data. Starting from this base Mastrolembo and Caporali (2017) [16] presented a direct comparison of the principal horizontal directions of stress and strain-rate directions of extension, estimated at the position of each stress measurement in their data set. For this, they used GPS data coming from over 500 stations distributed on the Italian peninsula, however, they did not provide a general map.

This work instead benefits from a new fine-scale strain rate field of the whole continental Italy and Sicily (Figure 4) [17], determined from the surface ground movements map obtained by the satellite InSAR observations between 1990 and 2012 [1]. The two-dimensional velocity gradient tensor is calculated by applying the infinitesimal strain approach [18,19] with a grid of 20 km × 20 km. The known horizontal incremental velocity vector V_i of the i-vertex polygon is defined as:

$$V_i = A_i + \frac{\partial V_i}{\partial x_j} x_j = A_i + t_{ij} x_j \tag{1}$$

where A_i is the unknown velocity at the origin of the coordinate system, x_j is the position of the station, and t_{ij} is the displacement gradient tensor. Following the tensor theory, we separated the second-rank tensor into a symmetric and an anti-symmetric tensor. Then, t_{ij} can be additively decomposed as follows:

$$t_{ij} = \frac{(t_{ij} + t_{ji})}{2} + \frac{(t_{ij} - t_{ji})}{2} = e_{ij} + \omega_{ij} \tag{2}$$

The symmetric and anti-symmetric parts of the infinitesimal strain rates can be associated with the infinitesimal strain e_{ij} and rotation ω_{ij} tensors. Principal strains $e_{1,}e_2$ were computed as:

$$e_1, e_2 = \frac{1}{2}(e_{ii} + e_{jj}) \pm \frac{1}{2}\sqrt{(e_{ii} - e_{jj})^2 + 4e_{ij}^2} \tag{3}$$

and the horizontal second invariant of the strain rate (SR) tensor was also evaluated as the scalars and is presented in Figure 4:

$$SR = \sqrt[2]{{e_1}^2 + {e_2}^2} \tag{4}$$

The determination of the second invariant of the strain rate provides important additional information to support the analysis of the geodynamics and the earthquake distribution of the study area. A recent study [17] based on the analysis of the seismic events that have occurred since 1990 in the Italian peninsula, shows that the probability of earthquakes occurring is linked to SR by a linear correlation. More specifically, the probability that a strong seismic event will occur doubles with the doubling of SR. Then, the SR is used as an independent and quantitative tool to spatially forecast seismicity.

The results of this study agree with these former studies, especially for the detection of the high strain rate along the Central and Southern Apennines axis and in Northern Sicily [19].

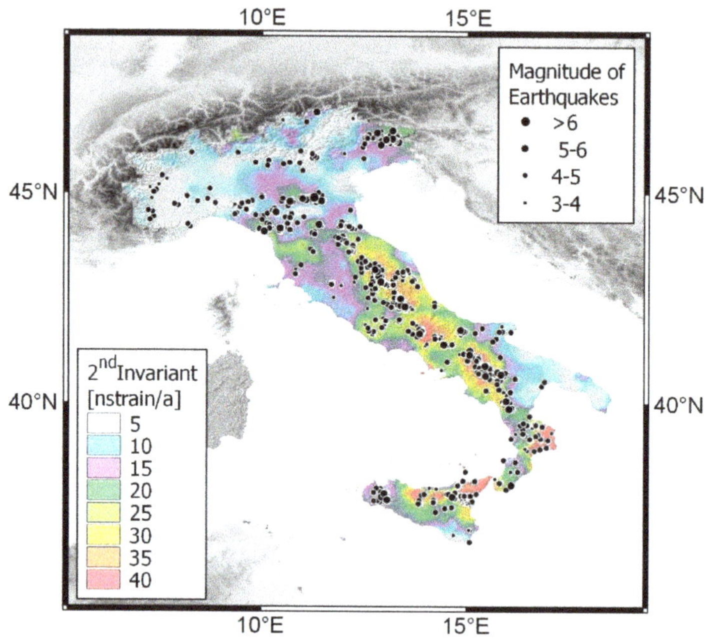

Figure 4. Map of the horizontal strain rate field of the Italian peninsula, determined by an infinitesimal approach from the horizontal velocity field derived from GNSS and InSAR during more than two decades of observation (1990–2017). Main earthquakes that occurred from 1990 to 2017 are represented on the map.

This new theory, based on observables, identifies significant earthquake (M > 5.5) prone areas with high strain rate areas. It gives a new perspective for the interpretation of recent earthquakes and this theory also predicts the events that occurred after the observation period of the study. For example, in the outer side of the Alps, the strongest earthquakes recorded in the past decades occurred in just two of the four areas of this sector showing high strain rate value: the M 5.3 20 December 1991 Graubunden [20] and the M 5.3 22 March 2020 Zagabria earthquake [21]. The others selected areas located around the cities of Marseille and Innsbruck. Furthermore, after 1990 in Italy, all earthquakes with M > 5.1 and a hypocenter less than 15 km deep only occurred in areas showing a high strain rate: 1997, 2009 and 2016 Central Apennines seismic sequences, the 2012 Emilia earthquakes, and the 2013 Lunigiana earthquake (on the surface of the Po Plain the *SR* is lower than in the buried and seismic Apenninic units because of its attenuation in the plastic Neogene sedimentary cover).

Outside these areas no shallow significant earthquakes occurred until 1990, even though strong events occurred after 1940, such as Friuli (the M 6.5 6 May 1976), Western Sicily (the Mw 6.4 15 January 1968), and Valais (the M 6.1 25 January 1946). In addition to these high strain rate areas that have been hit by strong earthquakes, there are others that, while showing high values of this value, have not been hit by relevant earthquakes since 1990. In recent years, only areas characterized by high strain rates have been affected by significant earthquakes, therefore, it is not unreasonable to empirically hypothesize that significant seismic events of the next decades have a greater chance of occurring only in the areas characterized by high strain rates. The year 1990 is taken as a milestone because, after beginning the survey in 1991, the former earthquakes do not influence the data.

3. Stain Rate in Irpinia

Irpinia is one of the main areas of the core of the Central and Southern Apennines chain. The differential movements between the two blocks, in which the Italian peninsula is divided imply a medium strain rate of 50 nstrain/a [5]. Here, the main fault systems are the Ufita, Monte Marzano, and Caggiano faults [22] (Figure 2 (top)). However, deformation is also linked to faults with a highly different orientation, well constrained in the historical record. For this issue, it is interesting to note that the focal mechanism solutions of the 1930 and 1962 earthquakes are significantly different from the kinematics of the typical large earthquakes that occurred along the crest of the Southern Apennines. Instead, these are well-fitted by the Mw 6.9 23 November 1980 earthquake, caused by predominant normal faulting along NW–SE-striking planes. The fault linked to the Mw 6.7 23 July 1930 earthquake is blind and its magnitude and focal mechanism are debated ([23] and references therein). Many focal mechanisms have been proposed, from a "classical" NW–SE to an ESE–WNW striking plane. These belong to an array of oblique dextral slips on the EW-trending planes crossing the whole Southern Apennines which is dissecting the orogen in various contiguous sectors. The level of the transcurrent component is debated as well. However, the effects of the earthquake presented in [24] fit better with a NW–SE striking fault.

The 1962 sequence is composed of three different shocks at 18:09, 18:19, and 18:44 UTC, the second being the most destructive (Io IX MCS, Mw 6.1, [25]). Additionally, identification of the faults responsible for these earthquakes is difficult because of the lack of reported surface faulting. Only in 2016 was a reliable focal mechanism produced [25] with two solutions: dominant strike-slip rupture along a north-dipping, E–W striking plane, or along a west-dipping, N–S striking plane. Its depth is still controversial, varying between 7 and 35 km. Therefore, the focal mechanism solutions of the 1962 earthquakes are significantly different from the kinematics of the typical large earthquakes occurring along the crest of the Southern Apennines, well-fitted instead by the Mw 6.9 23 November 1980 earthquake, caused by predominant normal faulting along NW–SE-striking planes.

Irpinia is one of the areas in Italy showing a higher strain rate (Figures 4 and 5) during the 1991–2011 InSAR survey: currently north of it, in the Sannio sector, the strain rate is at a low level with a value of 20 nstrain/a 10 km north of Benevento. However, to the SE of Benevento, the value increases to 35 nstrain/a in less than 30 km at Grottaminarda, reaching the highest levels (48 nstrain/a) 15 km south of the epicenter of the 23 November 1980 earthquake. Therefore, Irpinia is still currently one of the areas with a higher strain rate in Italy, with values always >32 nstrain/a, and showing a maxima over the hanging wall of the Monte Marzano fault system. The southward strain rate dramatically drops to 35 nstrain/a near Polla. However, while north of Irpinia along the chain axis the value drops rapidly under 30 nstrain/a, the southward values remain above this value for much longer, up to the Pollino line (the border between Central Apennines and Calabria–Peloritani arc). In the picture of the EW-trending lithospheric faults dissecting the Apenninic orogen, these sudden strain rate drops north and south of Irpinia can be related to different strain rate conditions occurring in the adjacent sectors.

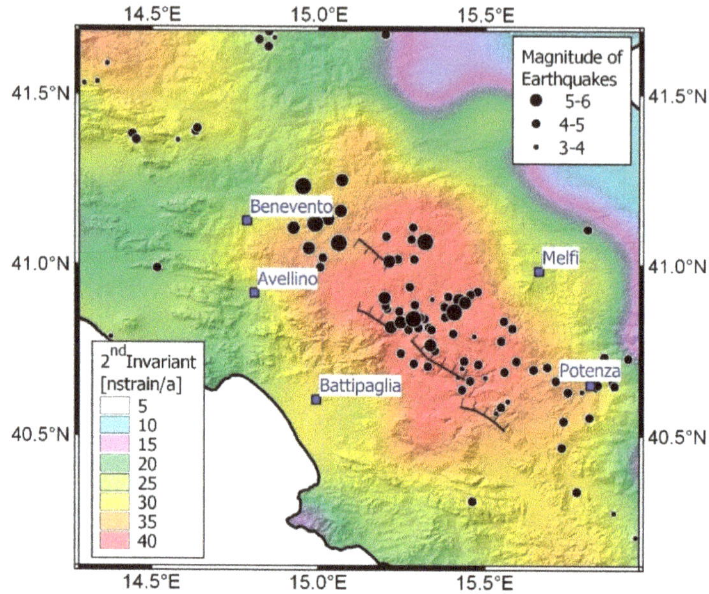

Figure 5. Map of the horizontal strain rate field of Irpinia derived from GNSS and InSAR during more than two decades of observation (1990–2017). The main earthquakes that occurred from 1466 to 2017 are represented on the map. The main towns of the area are drawn and labeled with dark blue squares, and the main faults (see Figure 2 (top)) are represented with black hashed lines. Since the strongest shallow events that occurred inland in Italy from 1990 to today are placed only in areas characterized by high strain rates [17], the high strain rate detected in Irpinia implies—from a theoretical point of view—a scenario where a new strong earthquake seems more likely. This can be somehow counterintuitive, because this area hosted most of the strongest earthquakes in southern Italy after 1908, in 1930, 1962, and 1980: only the Mw 6.4 1968 Belice and the Mw 6.0 1978 Patti gulf events (both in Sicily) reached similar magnitudes [26]. Only in the NE Sicily 1978 earthquake was the strain rate as high as in Irpinia. Therefore, from this point of view, we can hypothesize that in Irpinia, the probability of a new strong event is still very high.

4. The Historical Record of Earthquakes in Irpinia

Additionally, historical seismicity can allow this—somehow unexpected—statement, given the time intervals between Irpinian earthquakes. "Irpinia" is a historical–geographical area of southern Italy, located in the Campania region, approximately corresponding to the territory of the current province of Avellino, which in turn, largely recalls the historic province of Principato Ultra of the Kingdom of Naples.

The Irpinia area is one of the most seismically active sectors of the entire Italian territory. The seismogenic belt that runs along the Apennine chain, in fact, crosses the northern and eastern part of the province of Avellino, where strong earthquakes have frequently occurred over centuries.

The most important historical seismic events are placed in the hanging wall of the Monte Marzano fault system [22]. If we take a polygon with vertices at the coordinate points 41.314° N, 14.971° E; 41.105° N, 14.874° E; 40.739° N, 15.352° E; 41.056° N, 15.574° E (depicted in dark red in Figure 6 (bottom)), corresponding to the Apennine seismic belt site of the major historical and instrumental seismicity, the parametric catalog of Italian earthquakes CPTI15 [26] reports about twenty earthquakes with magnitude Mw $\geq$ 5.0, starting from the year 1000 (see Table 1). Of these, seven have a Mw between 6.0 and 6.8. It must be said that the catalog can be considered complete, for the strongest events (Mw $\geq$ 6.0) only for the last 400 years, namely from 1620 up to today [27]. From the diagram in Figure 6 (top), it can be seen that until the end of the 17th century, the seismic history of the Irpinia sector is largely incomplete and poorly documented. This, obviously, is not because there were no earthquakes at all, but because only little and partial historical information about that area for those ancient periods exists today. Only a couple of earthquakes are known (in 1466 and 1517) to have occurred in this period, plus two events before the year 1000, which occurred in the year 989 and 62 CE [28]; thus, outside the reference window of the historical catalog. Both these events originated from the monte Marzano Fault [29]. The earthquake of 5 December 1456 [30] was deliberately not taken into consideration in the present study, because it is a complex event that affected a very large area of southern Italy, causing damage from Puglia to Abruzzo, and whose epicenter is not well located nor defined. Probably, that earthquake was made up of several shocks that occurred in different sectors of the central–southern Apennines a few days apart, and Irpinia was only one of the several areas that were struck [28].

Table 1. List of the main Irpinia earthquakes (Mw > 5.0) extracted from the CPTI15 catalog [26]. For the description of the various parameters see this catalog at https://emidius.mi.ingv.it/CPTI15-DBMI15/index_en.htm (accessed on 30 January 2021) As attested by Rovida et al. [27], this historical record can be considered complete since 1620 for M 6.0+ earthquakes.

Year Mo Da	Epicentral Area	Lat	Lon	Io (MCS)	Mw
1466 01 15	Irpinia–Basilicata	40.765	15.334	8–9	6.0
1517 03 29	Irpinia	41.011	15.210	7–8	5.3
1692 03 04	Irpinia	40.903	15.196	8	5.9
1694 09 08	Irpinia–Basilicata	40.862	15.406	10	6.7
1702 03 14	Sannio–Irpinia	41.120	14.989	10	6.6
1732 11 29	Irpinia	41.064	15.059	10–11	6.8
1741 08 06	Irpinia	41.049	14.970	7–8	5.4
1794 06 12	Irpinia	41.108	14.924	7	5.3
1853 04 09	Irpinia	40.818	15.215	8	5.6
1905 11 26	Irpinia	41.134	15.028	7–8	5.2
1910 06 07	Irpinia–Basilicata	40.898	15.421	8	5.8
1930 07 23	Irpinia	41.068	15.318	10	6.7
1962 08 21	Irpinia	41.248	15.069		5.7
1962 08 21	Irpinia	41.230	14.953	9	6.2
1962 08 21	Irpinia	41.158	15.065		5.3
1980 11 23	Irpinia–Basilicata	40.842	15.283	10	6.8
1980 11 24	Irpinia–Basilicata	40.811	15.268		5.0
1981 01 16	Irpinia–Basilicata	40.890	15.439		5.2
1982 08 15	Irpinia	40.832	15.244	6	5.3

A lack of seismic events in the historical record for a given area can be due to the following reasons:

(a) an area of genuinely low long-term seismicity;
(b) either the incompleteness or a too-short time-span of the earthquake catalog;
(c) a quiescent period in an area characterized by temporal clustering, followed by a long recurrence interval [31].

The seismic history of Irpinia is better documented, starting from the end of 1600, and as minor events ($4.0 \leq Mw < 5.0$) can be considered well documented only starting from the end of the 19th century (Figure 6)

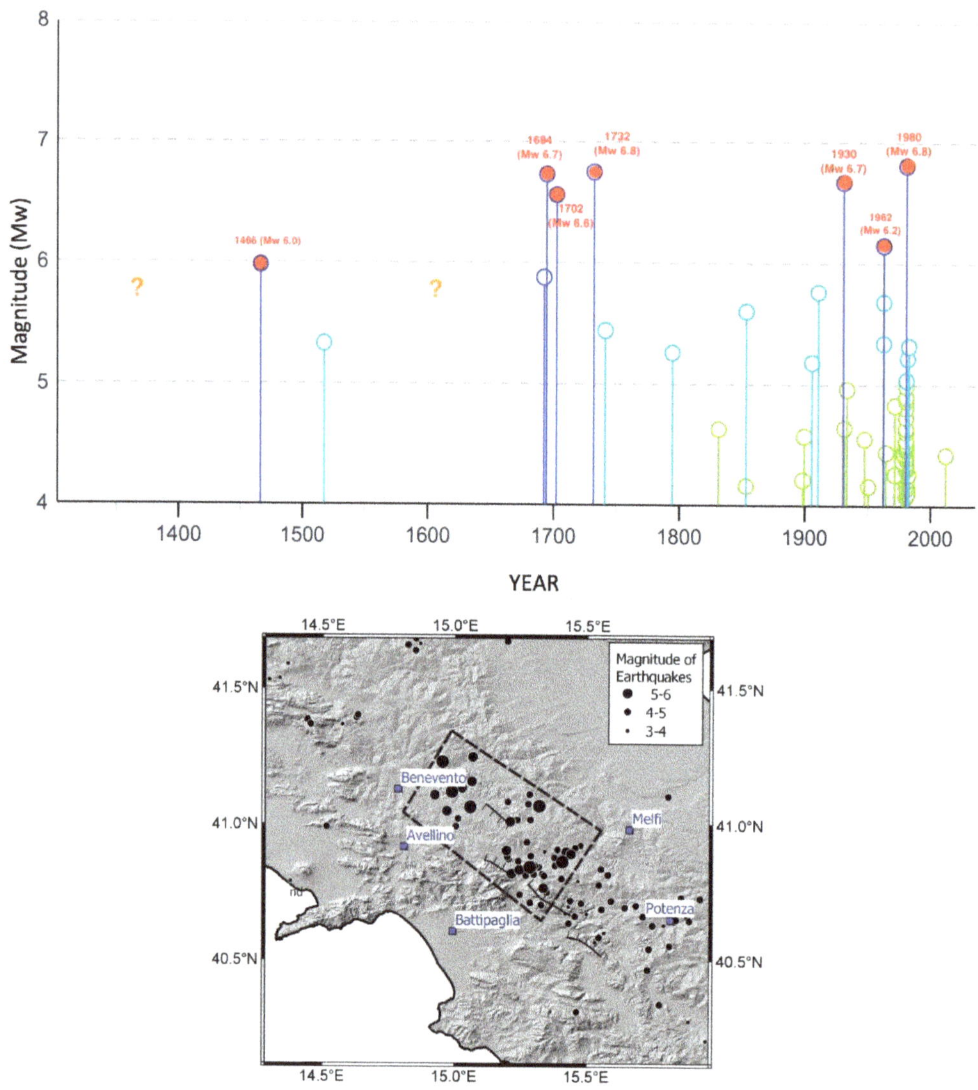

Figure 6. (**Top**) Seismic history of the Irpinia area as from the CPTI15 catalog [26]. The dark red polygonal area shows the Irpinian main seismic belt, the coordinates of its vertices are: 41.314° N, 14.971° E; 41.105° N, 14.874° E; 40.739° N, 15.352° E; 41.056° N, 15.574° E. Such a historical record is well documented only for the last 400 years (namely since 1620 [27]), whereas for the previous centuries it is poorly known, with long timespans lacking information. The figure also shows that inside the catalog completeness span-time of 400 years (since 1620), there are a couple of evident clusters of strong earthquakes (Mw $\geq$ 6.0) which are 200-years apart. The first, between 1694 and 1732, when three M 6.5+ events occurred over a period of 38 years. The second, between 1930 and 1980, when three M 6.0+ events occurred over a period of 50 years. (**Bottom**). Map of the main historical seismic events of Irpinia from 1466 until 1982, reported in Table 1. The main towns of the area are drawn and labeled with dark blue squares, and the main faults (see Figure 2 (top)) are represented with black hashed lines.

From its seismic history it can also be seen that, over the 400 year time-span of seismic catalog completeness for M $\geq$ 6.0 events, in Irpinia, the strongest earthquakes (Mw $\geq$ 6.0) tend to group over time, spaced from long phases characterized by lower and less frequent seismicity (Figure 6 (top and bottom)). At the turn of the seventeenth and eighteenth centuries, over a period of 40 years, Irpinia was affected by four damaging earthquakes, three of which occurred in just 10 years (1692, 1694, 1702, and 1732). Of these, the ones that occurred in 1694 (considered as a sort of twin of the 1980 earthquake), in 1702 and in 1732 were large events of Mw > 6.5. Each of these caused extensive destruction over large areas and many casualties. Another cluster of strong earthquakes is the one that hit the sector in the twentieth century, between 1930 and 1980 (three events with Mw $\geq$ 6.0 over a period of 50 years). So, Irpinia belongs to the belt of very high seismic hazard running along the Central and Southern Apennines (Figure 7).

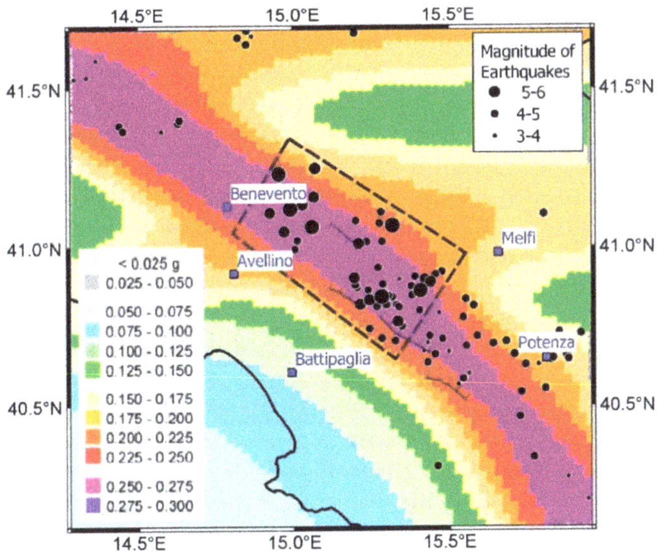

Figure 7. Map of seismic hazards in Irpinia (see [32]) and adjoining areas (colors in the background), derived mostly from the historical seismic records, as shown by the overlapping of strong seismic events within the map itself. The main towns of the area are drawn and labeled with dark blue squares, and the main faults (see Figure 2 (top)) are represented with black hashed lines.

It is unlikely that in the 200-year time-span between 1732 and 1930 there were large (M 6.0+) earthquakes in the Irpinia area, since these are not present in the historical record. In the same time interval, not just "minor" earthquakes are well documented in the very same area (i.e., the 1741 Mw 5.4, 1794 Mw 5.3, and 1853 Mw 5.6 Irpinian events; see Table 1, Figure 5), but strong events are also well known to have struck other adjacent Apennine areas (the 1805 Mw 6.7 Matese earthquake; those of 1851 Mw 6.5 and 1857 Mw 7.1 in Basilicata [26]). Therefore, it can be assumed that the historical seismicity of Irpinia has been characterized by periods of intense activity, with strong earthquakes over a few years or decades, interspersed with long periods of minor-to-moderate activity, with earthquakes of magnitude lower than 6.0.

The spatio-temporal clustering of earthquakes in the Southern Apennines is well documented in the scientific literature. By comparing the number of earthquakes on record in the last five to seven centuries, with the number implied by slip-rates on active normal faults averaged over 18 kyrs in the Southern Apennines, Papanikolaou and Roberts [33] demonstrated that the long history of earthquakes in the Italian Apennines may indeed contain evidence for earthquake clustering. In particular, according to Papanikolaou and Roberts [33], Irpinia and northern Basilicata show a very high number of earthquakes and this indicates that this area may be in a temporal earthquake cluster phase. Meanwhile, the sector located slightly further south, up to the Pollino massif, could be in a temporary anti-clustering process.

The strain rate map in Farolfi et al. [17], in which the Irpinia–Basilicata sector is characterized by a much higher strain rate than the Pollino sector (and the intermediate sector, Vallo di Diano, shows intermediate values), fits well with these results.

In the Central and Northern Apennines, earthquake clustering is known to exist. For example, Tondi and Cello [34] observed a time interval of ca. 350 years among the beginning of seismic clusters in the Central Apennines Fault System. The current sequence, which started in 1997 and continued with the events of 2009 and 2016, has arrived on time if we consider that two main seismic clusters in the past began in the years 1349 and 1688 [35]. In the Northern Apennines, a major seismic crisis occurred between 1915 and 1921 [36], while in this area, the historical record before 1915 is composed of a few destructive events [26]. Additionally, the two-year period 2012–2013 showed a high level of activity, not only in the area of the Emilia seismic sequence, but also in the Garfagnana sector, accompanied by a high strain rate.

Currently, the most likely explanation for seismic clustering is the "stress transfer" between faults [36] and references therein, due to coseismic movement rearranging the Coulomb failure stress on other nearby faults [37]. However, this explanation falls short when there is the occurrence of an isolated, single event (such as the Mw 6.1 6 November 1599 Valnerina, and the Mw 6.4 13 January 1832 Valle Umbra earthquakes) that do not trigger a level of Coulomb stress transfer, resulting in strong earthquakes on other neighboring faults.

For other researchers, there is a sort of "domino effect" between the crustal blocks that make up the Apennines [38].

In conclusion, we suppose that the deformation rate value, as described in Farolfi et al. [18], represents a *conditio sine qua non* for the occurrence of strong earthquakes (M > 5.5). This hypothesis was corroborated by the observation that all the strongest shocks of the last three decades in the Italian territory are located in areas characterized by a high rate of deformation [17].

5. Conclusions

In the last twenty years, the main shallow earthquakes (depth $\leq$ 15 km) in Italy and the Alps have occurred only in some of the horizontal strain rate zones, as depicted by Montone and Mariucci [15]. Meanwhile, the strain rate is currently low in other areas affected by recent earthquakes that occurred before the 1990–2012 survey, such as Belice (1968) and Friuli (1976). These areas are also where the higher seismic events from 1915 to now have occurred. The area of the 1915 Marsica earthquake also shows lower-than-surroundings strain rate values, such as in the Central and Eastern sections of the Northern Apennines (in the Western sector, higher seismicity barely corresponds to a slightly higher strain rate). In this picture, the high strain rate level indicates that the scenario of a new strong shake in Irpinia is not unlikely. Additionally, the historical record is in agreement with this, given the short temporal distance between strong (M6+) seismic events in Irpinia during the 400 years of catalog completeness (i.e., from 1620 to present), and a long 200-years period without M6+ seismic events occurring between 1732 and 1930.

Moreover, by merging historical seismicity and InSAR satellite data, we think that, in the future, a hypothesis related to the following scenarios should be explored:

- the short time gap between strong events in Irpinia during the 1694–1732 and 1930–1980 periods is linked to periods of continuous high strain rates;
- instead, the long seismic gap (a lack of strong seismicity) between 1732 and 1930 could have originated from a strain rate drop after the 1732 earthquake.

Author Contributions: Conceptualization, A.P., F.B.; historical and geographical investigation, F.B., A.P.; methodology, software, validation, formal analysis, data analysis, G.F.; investigation, resources, writing—original draft preparation, writing—review and editing, visualization, F.B., A.P. and G.F.; supervision, project administration, A.P.; funding acquisition, F.B. All authors have read and agreed to the published version of the manuscript.

Funding: This research received no external funding.

Data Availability Statement: The Parametric Catalogue of Italian Earthquake CPTI15 (version 3.0) is free and available at: https://emidius.mi.ingv.it/CPTI15-DBMI15/description_CPTI15_en.htm (accessed on 30 January 2021). As to any other dataset or maps here provided in this study, they can be available upon request.

Conflicts of Interest: The authors declare no conflict of interest.

References

1. Farolfi, G.; Piombino, A.; Catani, F. Fusion of GNSS and Satellite Radar Interferometry: Determination of 3D Fine-Scale Map of Present-Day Surface Displacements in Italy as Expressions of Geodynamic Processes. *Remote. Sens.* **2019**, *11*, 394. [CrossRef]
2. Farolfi, G.; Bianchini, S.; Casagli, N. Integration of GNSS and satellite InSAR data: Derivation of fine-scale vertical surface motion maps of Po Plain, Northern Apennines and Southern Alps, Italy. *IEEE Trans. Geosci. Remote. Sens.* **2019**, *57*, 319–328. [CrossRef]
3. Farolfi, G.; Del Soldato, M.; Bianchini, S.; Casagli, N. A procedure to use GNSS data to calibrate satellite PSI data for the study of subsidence: An example from the north-western Adriatic coast (Italy). *Eur. J. Remote. Sens.* **2019**, *52*, 54–63. [CrossRef]
4. Farolfi, G.; Del Ventisette, C. Contemporary crustal velocity field in Alpine Mediterranean area of Italy from new geodetic data. *GPS Solut.* **2015**, *20*, 715–722. [CrossRef]
5. D'Agostino, N.; Jackson, J.A.; Dramis, F.; Funiciello, R. Interactions between mantle upwelling, drainage evolution and active normal faulting: An example from the central Apennines (Italy). *Geophys. J. Int.* **2001**, *147*, 475–497. [CrossRef]
6. Italiano, F.; Martelli, M.; Martinelli, G.; Nucci, M. Geochemical evidence of melt intrusions along lithospheric faults of the Southern Apennines, Italy: Geodynamic and seismogenic implications. *J. Geophys. Res.* **2000**, *105*, 13569–13578. [CrossRef]
7. Di Luccio, F.; Chiodini, G.; Caliro, S.; Cardellini, C.; Convertito, V.; Pino, N.A.; Tolomei, C.; Ventura, G. Seismic signature of active intrusions in mountain chains. *Sci. Adv.* **2018**, *4*, e1701825. [CrossRef]
8. Chiodini, G.; Cardellini, C.; Di Luccio, F.; Selva, J.; Frondini, F.; Caliro, S.; Rosiello, A.G.; Beddini, G.; Ventura, G. Correlation between tectonic CO_2 Earth degassing and seismicity is revealed by a 10-year record in the Apennines, Italy. *Sci. Adv.* **2020**, *6*, 35. [CrossRef] [PubMed]
9. Doglioni, C.; Harabaglia, P.; Martinelli, G.; Mongelli, F.; Zito, G. A geodynamic model of the Southern Apennines accretionary prism. *Terra Nova* **1996**, *8*, 540–547. [CrossRef]
10. Rosenbaum, G.; Lister, G.S. Neogene and Quaternary rollback evolution of the Tyrrhenian Sea, the Apennines, and the Sicilian Maghrebides. *Tectonics* **2004**, *23*, TC1013. [CrossRef]
11. Di Bucci, D.; Corrado, S.; Naso, G. Active faults at the boundary between Central and Southern Apennines (Isernia, Italy). *Tectonophysics* **2002**, *359*, 47–63. [CrossRef]
12. Riguzzi, F.; Crespi, M.; Devoti, R.; Doglioni, C.; Pietrantonio, G.; Pisani, A.R. Geodetic strain rate and earthquake size: New clues for seismic hazard studies. *Phys. Earth Planet. Inter.* **2012**, *206*, 67–75. [CrossRef]
13. Devoti, R.; Esposito, A.; Pietrantonio, G.; Pisani, A.R.; Riguzzi, F. Evidence of large scale deformation patterns from GPS data in the Italian subduction boundary. *Earth Planet. Sci. Lett.* **2011**, *311*, 230–241. [CrossRef]
14. Palano, M. On the present-day crustal stress, strain-rate fields and mantle anisotropy pattern of Italy. *Geophys. J. Int.* **2015**, *200*, 969–985. [CrossRef]
15. Montone, P.; Mariucci, M.T. P-wave velocity, density, and vertical stress magnitude along the crustal Po Plain (Northern Italy) from sonic log drilling data. *Pure Appl. Geophys.* **2015**, *172*, 1547–1561. [CrossRef]
16. Mastrolembo, B.; Caporali, A. Stress and strain-rate fields: A comparative analysis for the Italian territory. *Boll. Geofis. Teor. Appl.* **2017**, *58*, 265–284.
17. Farolfi, G.; Keir, D.; Corti, G.; Casagli, N. Spatial forecasting of seismicity provided from Earth observation by space satellite technology. *Sci. Rep.* **2020**, *10*, 1–7. [CrossRef] [PubMed]
18. Dermanis, A.; Livieratos, E. Applications of deformation analysis in geodesy and geodynamics. *Rev. Geophys.* **1983**, *21*, 41–50. [CrossRef]
19. Farolfi, G.; Del Ventisette, C. Strain rates in the Alpine Mediterranean region: Insights from advanced techniques of data processing. *GPS Solut.* **2017**, *21*, 1027–1036. [CrossRef]
20. INGV—Italian Seismic Bulletin. Available online: http://terremoti.ingv.it/en/bsi (accessed on 30 January 2021).
21. Markušic, S.; Stanko, D.; Korbar, T.; Belic, N.; Penava, D.; Kordic, B. The Zagreb (Croatia) M5.5 Earthquake on 22 March 2020. *Geosciences* **2020**, *10*, 252. [CrossRef]
22. Galli, P.; Peronace, E. New paleoseismic data from the Irpinia Fault. A different seismogenic perspective for southern Apennines (Italy). *Earth Sci. Rev.* **2014**, *136*, 175–201. [CrossRef]

23. Pino, N.A.; Palombo, B.; Ventura, G.; Perniola, B.; Ferrari, G. Waveform modeling of historical seismograms of the 1930 Irpinia earthquake provides insight on "blind" faulting in Southern Apennines (Italy). *J. Geophys. Res.* **2008**, *113*, B05303. [CrossRef]
24. Serva, L.; Esposito, E.; Guerrieri, L.; Porfido, S.; Vittori, E.; Comercia, V. Environmental effects from five his-torical earthquakes in southern Apennines (Italy) and macroseismic intensity assessment: Contribution to IN-QUA EEE Scale Project. *Quat. Int.* **2007**, *173*, 30–44. [CrossRef]
25. Vannoli, P.; Bernardi, F.; Palombo, B.; Vannucci, G.; Console, R.; Ferrari, G. New constraints shed light on strike-slip faulting beneath the southern Apennines (Italy): The 21 August 1962 Irpinia multiple earthquakes. *Tectonophysics* **2016**, *691*, 375–384. [CrossRef]
26. Rovida, A.; Locati, M.; Camassi, R.; Lolli, B.; Gasperini, P.; Antonucci, A. *Italian Parametric Earthquake Catalogue (CPTI15)*; Version 3.0; Istituto Nazionale di Geofisica e Vulcanologia (INGV): Rome, Italy, 2021. [CrossRef]
27. Rovida, A.; Locati, M.; Camassi, R.; Lolli, B.; Gasperini, P. The Italian earthquake catalogue CPTI15. *Bull. Earthq. Eng.* **2020**, *18*, 2953–2984. [CrossRef]
28. Rovida, A.; Locati, M.; Antonucci, A.; Camassi, R. (Eds.) *Italian Archive of Historical Earthquake Data (ASMI)*; Istituto Nazionale di Geofisica e Vulcanologia (INGV): Rome, Italy, 2017. [CrossRef]
29. Galli, P. Roman to Middle Age Earthquakes Sourced by the 1980 Irpinia Fault: Historical, Archaeoseismological, and Paleoseismological Hints. *Geosciences* **2020**, *10*, 286. [CrossRef]
30. Fracassi, U.; Valensise, G. Unveiling the Sources of the Catastrophic 1456 Multiple Earthquake: Hints to an Unexplored Tectonic Mechanism in Southern Italy. *Bull. Seismol. Soc. Am.* **2007**, *97*, 725–748. [CrossRef]
31. Marco, S.; Stein, M.; Agnon, A.; Ron, H. Long-term earthquake clustering: A 50,000 year paleoseimic record in the Dead Sea Graben. *J. Geophys. Res.* **1996**, *101*, 6179–6191. [CrossRef]
32. Stucchi, M.; Meletti, C.; Montaldo, V.; Akinci, A.; Faccioli, E.; Gasperini, P.; Malagnini, L.; Valensise, G. *Pericolosità Sismica di Riferimento per Il Territorio Nazionale MPS04*; Data Set; Istituto Nazionale di Geofisica e Vulcanologia (INGV): Rome, Italy, 2004. [CrossRef]
33. Papanikolaou, I.D.; Roberts, G. Clustering and anticlustering in the Southern Apennines as evidenced from geological fault slip-rate seismic hazard maps and the historical record. In Proceedings of the 2nd INQUA-IGCP-567 International Workshop on Active Tectonics, Earthquake Geology, Archaeology and Engineering, Corinth, Greece, 19–24 September 2011; pp. 174–177.
34. Tondi, E.; Cello, G. Spatiotemporal evolution of the Central Apennines fault system (Italy). *J. Geodyn.* **2003**, *36*, 113–128. [CrossRef]
35. Viti, M.; Mantovani, E.; Cenni, N.; Vannucchi, A. Interaction of seismic sources in the Apennine belt. *Phys. Chem. Earth* **2013**, *63*, 25–35. [CrossRef]
36. Bonini, M.; Corti, G.; Delle Donne, D.; Sani, F.; Piccardi, L.; Vannucci, G.; Genco, R.; Martelli, L.; Ripepe, M. Seismic sources and stress transfer interaction among axial normal faults and external thrust fronts in the Northern Apennines (Italy): A working hypothesis based on the 1916–1920 time–space cluster of earthquakes. *Tectonophysics* **2016**, *680*, 67–89. [CrossRef]
37. Stein, R.S.; King, G.C.P.; Lin, J. Change in failure stress on the southern San Andreas fault system caused by the 1992 magnitude 5 7.4 Landers earthquake. *Science* **1992**, *258*, 1328–1332. [CrossRef] [PubMed]
38. Mantovani, E.; Viti, M.; Cenni, N.; Babbuccia, D.; Tamburelli, C.; Baglione, M.; D'Intinosante, V. Seismotectonics and present seismic hazard in the Tuscany–Romagna–Marche–Umbria Apennines (Italy). *J. Geodyn.* **2015**, *89*, 1–14. [CrossRef]

geosciences

MDPI

Article

Ground Response and Historical Buildings in Avellino (Campania, Southern Italy): Clues from a Retrospective View Concerning the 1980 Irpinia-Basilicata Earthquake

Lucia Nardone [1,*], Fabrizio Terenzio Gizzi [2] and Rosalba Maresca [3]

1 National Institute of Geophysics and Volcanology, Osservatorio Vesuviano, 80124 Naples, Italy
2 Institute of Heritage Science, (Italian) National Research Council (ISPC-CNR), 85100 Potenza, Italy; fabrizioterenzio.gizzi@cnr.it
3 Department of Sciences and Technologies, University of Sannio, 82100 Benevento, Italy; maresca@unisannio.it
* Correspondence: lucia.nardone@ingv.it

Received: 30 October 2020; Accepted: 14 December 2020; Published: 18 December 2020

Abstract: Cultural heritage represents our legacy with the past and our identity. However, to assure heritage can be passed on to future generations, it is required to put into the field knowledge as well as preventive and safeguard actions, especially for heritage located in seismic hazard-prone areas. With this in mind, the article deals with the analysis of ground response in the Avellino town (Campania, Southern Italy) and its correlation with the effects caused by the 23rd November 1980 Irpinia earthquake on the historical buildings. The aim is to get some clues about the earthquake damage cause-effect relationship. To estimate the ground motion response for Avellino, where strong-motion recordings are not available, we made use of the seismic hazard disaggregation. Then, we made extensive use of borehole data to build the lithological model so being able to assess the seismic ground response. Overall, results indicate that the complex subsoil layers influence the ground motion, particularly in the lowest period (0.1–0.5 s). The comparison with the observed damage of the selected historical buildings and the maximum acceleration expected indicates that the damage distribution cannot be explained by the surface geology effects alone.

Keywords: ground response; 2D numerical analysis; soil-structure effects; cultural heritage

---

1. Introduction

Avellino is a town of historical interest, which is located in a structural depression of the Campanian Apennines, in the Irpinia region (Southern Italy) (Figure 1). This is an area with a high seismic hazard where the peak ground acceleration (PGA) with a 10% probability of being exceeded in 50 years has been estimated in the range 0.20–0.27 g (http://esse1.mi.ingv.it/d2.html). The hazard reflects the seismic history of the locality, which was affected by several strong earthquakes [1] (Figure 2). The 23rd November 1980 Irpinia-Basilicata earthquake (Me = 6.7; Mw = 6.9 [2]) was the last significant event that strongly hit Avellino (Is = VIII of the Mercalli-Cancani-Sieberg Macroseismic Scale, (MCS) [3]). The earthquake, which caused about 3000 casualties, hit a wide area of the Campania and Basilicata regions that recorded serious damage, especially the provinces of Avellino, Potenza, and Salerno.

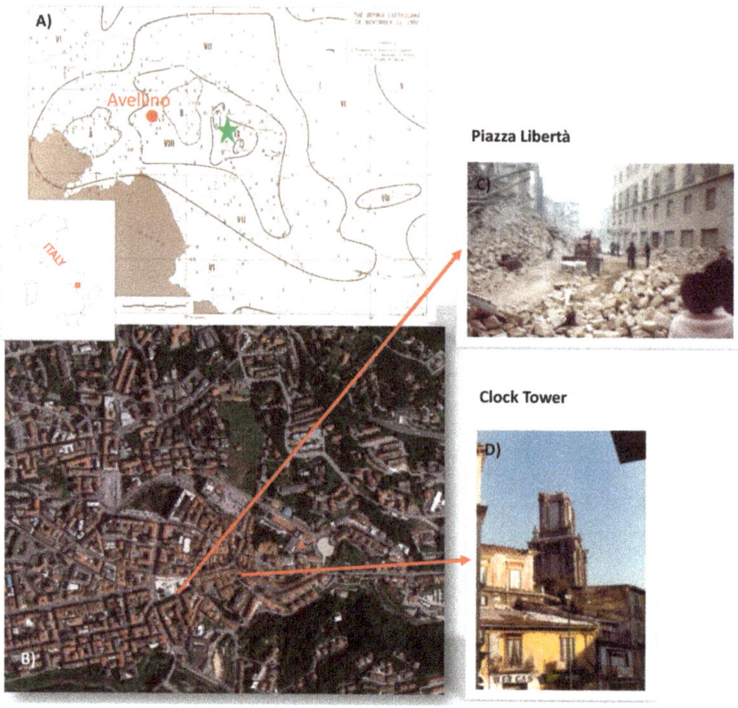

Figure 1. (**A**) Isoseismal lines of the 23rd November 1980 Irpinia-Basilicata earthquake (modified after [3]). The green star identifies the epicenter of the earthquake. (**B**) Geographical localization of the historical center of Avellino. Examples of the damage of the earthquake are shown in the two pictures: (**C**) Piazza Libertà (Data Source: sisma80.it/avellino.html accessed on 30 November 2020) and (**D**) Torre dell'Orologio (Data Source: www.avellinesi.it, accessed on 30 November2020).

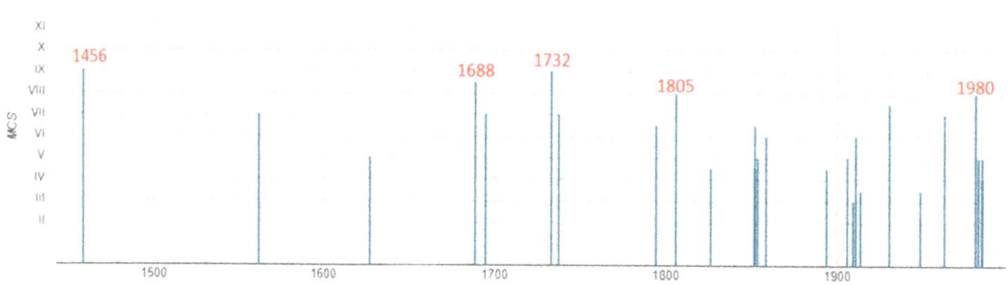

Figure 2. Seismic history of Avellino. The locality was hit by about ten earthquakes with site intensity greater than or equal to VII MCS (Data Source: [1], intensities in MCS scale). The dates of the earthquakes with Is ≥ VIII are reported in red (5 December 1456, 5 June 1688, 29 November 1732, 26 July 1805, 23 November 1980).

The town of Avellino was founded in Roman times (Abellinum), in the same site where Atripalda town is located today. In the Middle Ages, Avellino developed around the Cathedral, built in the XII century, later restored and transformed several times due to earthquake consequences. Near the end of century XVIII, the new districts extended mainly towards the west. Numerous religious and civil architectures remain in Avellino downtown at present, testifying with their restoration history and scars the seismic hazard of the territory as well as the vulnerability of the heritage.

To get some clues about the cause-effect relationship, in this work we investigated the ground response and the damage levels in some historical buildings of Avellino, taking advantage of a detailed knowledge of the surface geology of the Avellino urban area. Previous works highlighted resonance effects in a wide period range (0.08–0.62 s) due to the complexity of the surface geology [4], and also soil/building resonance effects [5].

Starting from the geological model based on previous investigations, we performed a 2D ground response analysis for the sites where the heritage lies, using an equivalent linear finite element approach. However, considering that strong-motion recordings in Avellino are not available, the definition of the seismic input in the assessment of the ground response was performed making use of a probabilistic approach, using the seismic hazard disaggregation. This method is aimed at defining seismic events by classes of magnitude and distance from the site, which provides the largest contribution to ground motion exceedance for defined hazard levels. Finally, we correlated retrospectively the estimated seismic response to the damage produced by the 1980 earthquake for the historical buildings.

2. Previous Studies

Earthquake damage distribution depends on multiple aspects: energy of the earthquake and its focal mechanism, attenuation in the crust, site effects, and vulnerability of buildings. At the municipality scale, the impact of an earthquake on the territory is essentially due to local geological effects and the vulnerability of buildings. This aspect relates to the seismic risk, which is of a higher level for historic cities with architectural and religious heritage as several studies performed also on more recent earthquakes highlighted [6–11]. The main problem associated with the use of macroseismic data lies in the difficulty of separating the different contributions. To overcome this difficulty an integrated approach, which combines macroseismic, geological, and geophysical data, is advantageous, mostly for seismic hazard assessment. The vulnerability, which is involved in seismic risk by affecting the damage distribution in the urban environment, concerns the condition of the buildings before the earthquake, which may not be easy to come by. An example of the multi-disciplinary approach is offered by [11] who studied the effects caused by the 1930 Irpinia earthquake and other two historical ones in a small historical town of the Basilicata region. These authors correlated the uneven damage to the soil conditions, combining geological and geophysical surveys with the analysis of archive sources and vulnerability data. Guidoboni et al. [12] depicted the damage scenario for the city of Palermo (Sicily, Southern Italy) for three damaging historical earthquakes. They correlated the damage distribution as inferred by historical sources with the near-surface geology, operating with a geological database in the GIS framework. They found some correlation between damage level and sediment thickness.

Many authors have correlated macroseismic data with Horizontal to Vertical Spectral Ratio (HVSR) seismic noise measurements. This has been seen as an effective tool for verifying the contribution of the local geology on earthquake damage. Noise HVSR measurements provide the resonant frequencies that are a function of the surface geology [13–15]. Free-field noise HVSR measurements are very useful in the absence of earthquake recordings, but they do not allow correlating the amount of damage with seismic site amplification. Maresca et al. [5] found that the measured noise HVSR amplifications were not effective in correlating with the damage produced by the 1980 Irpinia earthquake in Avellino. Additionally, they found that amplitude peaks were mostly associated with high-velocity contrast zones. To study the effects of earthquakes in the urban environment some authors carried out noise measurements in buildings coupled with noise free-field measurements (review paper by [5,16–18]). Soil/building resonance effects produce amplification of ground motion, resulting in more damage in case of strong earthquakes.

In regions of moderate seismicity, it is important to be able to anticipate the effects of a large event by simulating the ground motion it may generate [19]. Several studies make use of geological and geophysical prospecting data for modeling of the local seismic response, even for microzoning planning [20–23]. Nunziata and Costanzo [24] computed the ground motion at the historical center of Napoli for two historical earthquakes, based on a detailed physical model of subsoil. These authors used a hybrid technique based on the mode summation and finite difference methods, fitting synthetics with a recording of a moderate earthquake, for a preventive definition of the expected shaking. If strong motion data are available, the modeled seismic response can be effectively validated by data and then compared to the damage produced [25,26]. Reference [27] correlated earthquake damage to the computed site amplification, focusing the attention on two historic centers of the Abruzzi region, which were damaged by the 2016–2017 Central Italy seismic sequence. These authors correlated amplification phenomena, which they evaluated through detailed seismic microzonation studies [28,29], with damage.

Earthquake damage in Italy often concerns villages built on top of a relief. More easily, studies on topographical effects have focused on seismic propagation in a homogeneous medium [30,31]. In more realistic cases the methodological approach for modeling the local seismic response takes into account both the stratigraphic and the topographical effect [32–35].

One of the aspects connected to the seismic response modeling is that of having one or more reference earthquakes for using as input motion in the computing. However, strong motion recordings may be unavailable. An alternative approach involves Probabilistic Seismic Hazard Analysis (PSHA). The seismic hazard model for Italy (MPS04) was drawn up by the Istituto Nazionale di Geofisica e Vulcanologia (http://esse1.mi.ingv.it) [36,37], and its main use concerns the definition of the design spectra of the Italian Building Code [38,39]. The MPS04 model provides PGA and spectral accelerations computed for 10 periods (from 0.1 to 2 s), for 9 probabilities of exceedance in 50 years (from 2% to 81%, corresponding to return periods from 2475 to 44 years), considering rocky soil conditions and flat topography. It was estimated on a grid of points, located every 0.05 degrees. Through disaggregation of seismic hazard, we gain the magnitudes and distances, which contribute the most to the hazard at a specific site [40–42]. The main advantage of this approach is to guarantee the definition of a group of earthquakes scenario; in fact, the selection of a unique design earthquake does not allow taking into account all the features of ground motion expected at that site [43]. In this study, to estimate the ground response in the historical center of Avellino, where strong-motion recordings are not available, we operated seismic hazard disaggregation. So, we drew on the seismic hazard database, and on the strong motion ITACA database [44] produced for the Italian territory in the framework of the agreements between the Italian Department of Civil Protection (Dipartimento della Protezione Civile, DPC) and the Istituto Nazionale di Geofisica e Vulcanologia (INGV).

3. Cultural Heritage and 1980 Earthquake Damage

The current location of Avellino does not correspond to the place where the ancient one stood. The ancient Abellinum was in fact destroyed following the wars between the Byzantines and the Longobard and the consequent Longobard conquest of large continental portions of Italy after they invaded the peninsula (568 AD). The new inhabited center, which was built at a distance of about 4 km from the old one, quickly assumed considerable importance linked to its strategic position in the geopolitical framework of the time.

From an urbanistic point of view, the town rapidly extended around the Cathedral and even today the ancient and limited medieval layout of the town, of the mainly modern matrix, is recognizable in the area between the ruins of the medieval Castle and the Cathedral.

The building of the Cathedral took place starting from 1132 on the remains of the Longobard one. The façade is in neoclassical style and the interior is a Latin cross with three naves. However, the building has undergone many restorations as a result of the damage caused by the manifold earthquakes that hit the site historically. Among these, we remember the Irpinia earthquake of 29 November 1732 (Me = 6.6, Is = IX MCS [1]) as a result of which the Cathedral partially collapsed [1] subsequently restored and reopened for cult in 1736 [45]. Before the earthquake of 1980, the 23 July 1930 Irpinia earthquake (Me = 6.7, Is = VII–VIII MCS [1]) caused significant damage to the building to make it unsafe. With the earthquake of 1980, the Cathedral was made totally unfit for use: the tympanum of the façade and part of the dome collapsed, serious injuries were opened in the walls and a chapel on the left collapsed [1].

Among the other religious buildings that suffered significant damage as a consequence of the 1980 earthquake, we remember the Church of Sant'Antonio Abate, Santa Maria di Costantinopoli and Santa Maria del Rifugio. The Church of Sant'Antonio Abate is located in a district that was severely damaged by the earthquake of 1980. Located in the ancient center of the town, the church was built in the sixteenth century and underwent subsequent and significant alterations. The interior has a single nave, with decorative and stylistic elements of a later age. During the 1980 earthquake, there was very significant damage to all the structures of the church, with partial collapses and cutting lesions on all the walls, especially in the corner walls [46]. The Church of Santa Maria di Costantinopoli was built in the 16th century and has a plan in the shape of a Latin cross. The building was restored several times following various earthquakes such as that of 1688, 1732, and 1930. As a result of the 1980 earthquake there was very significant damage, with the roof collapsing on the right side of the apse area and collapses that had affected the higher parts of the building, the masonry had undergone various injuries both in the longitudinal and transverse walls of the transept, towards the east, while on the west side and in the apse there were considerable lesions due to the rotation to the east of the post wall south. The square-plan bell tower had completely collapsed [46,47]. The Church of Santa Maria del Rifugio was built in the seventeenth century and has a single nave with a triumphal arch. The 1980 earthquake caused serious damage with a crack pattern that caused significant water infiltrations, with damage to the works of art. The Church of Santa Maria del Carmine, which was designed as a private chapel, stood in the historic center until the earthquake of 23 November 1980. It had a single nave plan with two small side chapels, with a valuable altar, the most valuable artistic element of the Church [46,47].

Regarding non-religious buildings, the effects were different. The Balestrieri palace, dating back to the nineteenth century, is developed on four floors and defines, together with other buildings and the Torre dell'Orologio (Clock Tower), the early medieval urban core. Important damage occurred to the palace, while very serious damage occurred to the Clock Tower, with the collapse of the terminal part bearing the clock (see the picture in Figure 1) and internal injuries to the two orders below the collapsed part with widespread detachment of ashlars constituting the covering of the tower and different portions of the masonry in unsafe summits. The palace Festa, dating back to the seventeenth century, is on two floors above the ground floor, is located in the upper part of the historic center of Avellino and the main façade faces the palace Greco founded between the end of the sixteenth and the beginning of the seventeenth century, in the near the Cathedral [46]. Both suffered significant damage.

4. Damage Levels and Classification

To classify the damage suffered by the architectonic heritage after the 1980 earthquake, we made use of the inclusion/exclusion criteria indicated in Table 1. The classification takes advantage of the European Macroseismic Scale (EMS-98, [48]), simplifying and grouping the damage levels that it indicates. The SLD damage level reflects the Grade 1 fixed in the Scale, the other two damage levels were established grouping Grade 2 and Grade 3 (MHD) as well as merging Grade 4 and Grade 5 (VHDC). The first level is the slight damage, where no structural damage occurs; the second, intermediate level of damage, includes structural damage from slight to moderate, the heaviest level is the third in which are included buildings that suffered heavy and very heavy structural damage, including partial and total collapse.

Table 1. Damage level classification: outline of the main inclusion/exclusion criteria used for the architectonic heritage of Avellino.

Slight Damage (SLD)	Moderate to Heavy Damage (MHD)	Very Heavy Damage to Collapse (VHDC)
No structural damage Slight non-structural damage Capillary cracks	Slight to moderate structural damage Cracks with breadth and frequency in walls increasing from moderate to heavy level Isolate collapse of non-structural elements	Heavy and very heavy structural damage Partial and total collapse

According to these pre-fixed levels, Table 2 reports the synthesis of the damage level for each building.

Table 2. The historical monuments of Avellino damaged by the 1980 earthquake with the main information and the level of damage assessed based on bibliographic sources (SLD, Slight Damage; MHD, Moderate to Heavy Damage; VHDC, Very Heavy Damage to Collapse).

Site Code	Local Name	Heritage Typology	Latitude (N)	Longitude (E)	Century of First Building	Damage Level
M01	Duomo	Cathedral	40.915295	14.797219	XII	**VHDC**
M02	Chiesa e Convento di S. Maria delle Grazie	Church and Convent	40.920927	14.795772	XVI	**MHD**
M03	Chiesa del S.S. Rosario	Church	40.913279	14.787485	XX	**MHD**
M04	Chiesa di S. Francesco Saverio	Church	40.916301	14.795762	XVIII	**MHD**
M05	Torre dell'Orologio	Clock Tower	40.915034	14.79597	XVII	**VHDC**
M06	Palazzo de Conciliis	Palace	40.915392	14.798211	XVIII	**MHD**
M07	Chiesa di S. Maria di Costantinopoli	Church	40.914128	14.797541	XVI	**VHDC**
M08	Chiesa di S. Antonio Abate	Church	40.913221	14.795846	XVI	**VHDC**
M09	Palazzo Balestrieri	Palace	40.915345	14.795645	XIX	**MHD**
M10	Palazzo Festa	Palace	40.915497	14.796319	XVII	**MHD**
M11	Palazzo Greco	Palace	40.915301	14.796331	XVI–XVII	**MHD**
M12	Chiesa di Santa Maria del Carmine	Church	40.913997	14.794816	XV, no longer existing	**VHDC**
M13	Chiesa del Gesù Sacramentato	Church	40.915075	14.793691	XVIII	**MHD**
M14	Chiesa di Santa Maria del Rifugio	Church	40.914127	14.7942	XVII	**MHD**

5. Ground Motion Modelling

5.1. Disaggregation of Seismic Hazard and Input Motion Selection

The study of the seismic response requires accelerometric recordings to design the seismic input motion representative of the seismic hazard at the site. In the study area, strong-motion recordings are lacking. Indeed, the AVL accelerometric station, belonging to the Italian strong motion network (ISMD), operates in Avellino since 2015 (http://terremoti.ingv.it/instruments/station/AVL), but no strong earthquakes occurred in the region since that time. For this reason, we made use of a probabilistic approach, as suggested for microzoning and engineering planning.

Our goal was to select suitable records, which reflect: (i) the seismogenic features of the sources; (ii) the ground motion intensity measures; (iii) and the soil conditions appropriate to the site. We referred to the seismic hazard database of the national territory (http://esse1.mi.ingv.it). This database provides the ground shaking on a hard rock on 16,852 sites (grid nodes) spaced with 0.05 degrees step, for nine return periods (RP). Following the New Italian Building Code (NIBC), [38,39], the ground shaking is expressed in terms of maximum horizontal acceleration (a_g), and its spectral parameters (maximum value of the amplification factor of the horizontal acceleration spectrum [F_0]; period corresponding to the beginning of the constant velocity section in the horizontal acceleration spectrum [T_C]).

Table 3 shows the values of these parameters for the historical center of Avellino, with geographic coordinates 40.915325N and 14.795909E, included between nodes 32,764, 32,765, 32,986, and 32,987 of the grid, for a return period of 475 years.

Table 3. Ground shaking parameters for the historical center of Avellino for a return period of 475 years.

a_g [g]	F_0	T_c [s]
0.197	2.371	0.368

Through disaggregation of seismic hazards, we separated contributions from different seismogenic sources. The disaggregation process allows us to define the contribution of seismogenic sources located at distance *R* from the historical center of Avellino, able to generate earthquakes of magnitude *M*, thus identifying earthquakes that dominate the hazard scenario. Disaggregation results are expressed in the *M*-*R*-ε space, where ε expresses the degree of scattering of empirical parameters, and it is equal to the number of (logarithmic) standard deviations by which the (logarithmic) ground motion deviates from the median value predicted by the attenuation equation used in the analysis [49,50]. Based on the disaggregation of the peak horizontal acceleration value (a_g) for rock condition with a probability of exceeding 10% in 50 years, we derived the percentage contribution to hazard, as a function of mean magnitude *M*, mean site-source distance *R*, and ε (Figure 3), (http://esse1.mi.ingv.it/), [51,52].

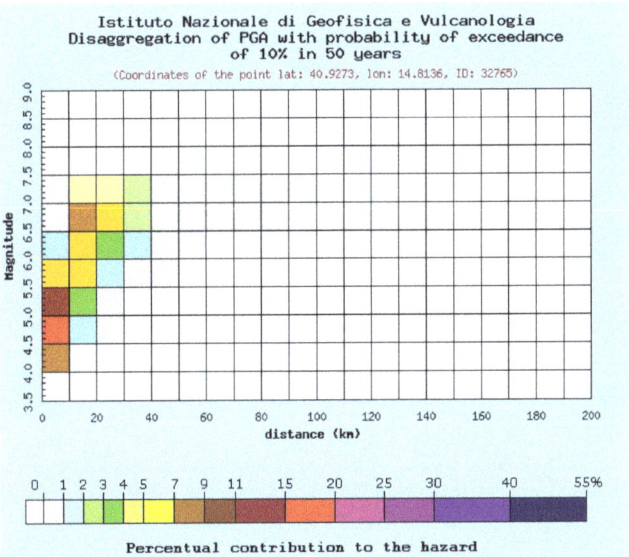

Figure 3. Percentage of contribution to seismic hazard of the 32,765 node which is the nearest to the historical center of Avellino (from http://esse1-gis.mi.ingv.it), as a function of magnitude (M) and distance (R). Mean values: M = 5.8; R = 14.0; ε = 1.13.

The M-R disaggregation grid (Figure 3) highlights that the maximum contribution to seismic hazard in the study area is due to earthquakes with a magnitude between 4.5 and 7.0, located within a 30 km distance. Events with greater magnitude are expected progressively at greater distances, with lower contribution to hazard.

The choice of seismic records must be made in terms of magnitude and distance bins, site classification, and supporting a good match with the target acceleration response spectrum. This last is defined based on the site features (level of seismic hazard, soil, and topographic conditions, buildings design). In this study, we used the tool REXELite [53] to select the set of time histories compatible with the target acceleration response spectrum. This tool generated the target response spectrum, according to Eurocode 8 (EC8; [54]), and to NIBC. It was built taking into account the site (A,B,C,D, or E) and topographic (T1,T2,T3,T4) codifications, based on the building features (Building

Nominal Life (vn), Building functional type (cu), and Damage Limit State). To build the target response spectrum, we chose the Damage Limit State SLV (life-saving limit state) for a structure located in the historical center of Avellino, on soil type A, flat topography (T1), with a Nominal Life of 50 years, which corresponds to the design for a 50-year return period according to the code (Figure 4).

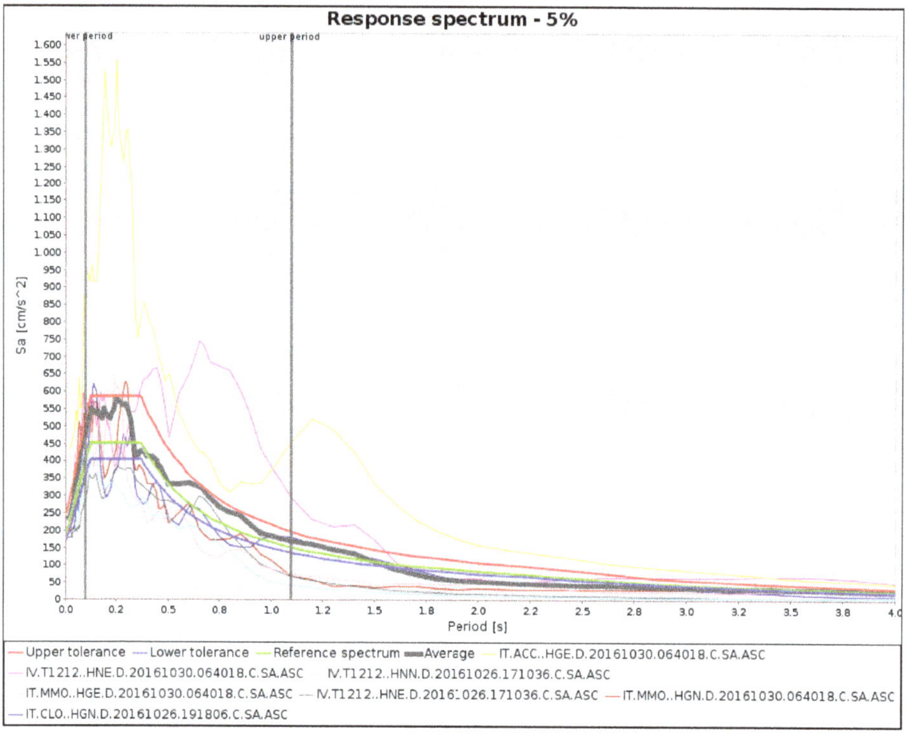

Figure 4. Horizontal reference spectrum (green line) within upper (30%, red line) and lower (10%, blue line) tolerance limits, and not-scaled acceleration spectra, compatible with the reference response spectrum. The thick grey line is the average of the seven selected acceleration spectra.

Time-histories were extracted from the Italian Strong-Motion Database (ISMD), [44,55], whose average was compatible with the target spectrum. Records were selected through seismic hazard disaggregation, by expressing the mean annual rate of exceedance as a function of magnitude and source-to-site distance (Figure 3). Moreover, only normal fault type events were selected, consistently with predominant focal mechanisms in the study area, as reported by the Seismogenic Zonation ZS9 [37,56]. Disaggregation has provided a scenario event for the 475-year return period with magnitude M = 5.8 and distance R = 14 km. A set of seven accelerograms that satisfied the selection criteria was identified (Figures 4 and 5), assigning as tolerance for the average spectra 10% lower and 30% upper in the period range 0.1–1.1 s. The list of seismic events related to these recordings is shown in Table 4. All the selected seismograms belong to the 1996 Amatrice-Visso-Norcia seismic sequence. Accelerograms produced by different seismogenic zones that meet the required features were not available in the database.

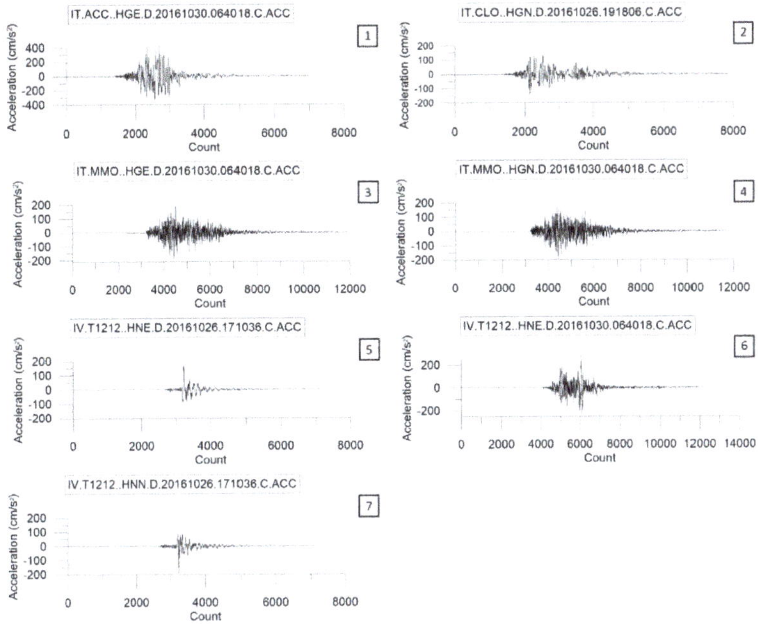

Figure 5. Not scaled accelerometric time series selected with REXELite tool, as representative of earthquakes scenario for the historical center of Avellino. Numbers in the small boxes indicate the event-numbers listed in the Table 4.

Table 4. List of seismic events, which produced the accelerometric waveforms chosen for this study.

N	Epicentral Area	Origin-Time (mm/dd/yyyy; hh:mm)	Latitude	Longitude	Depth (km)	Mw	Station Code	Epicentral Distance (km)
1	Norcia (PG)	10/30/16 6:40	42.8322	13.1107	9.2	6.5	ACC.HGE	18.6
2	Visso (MC)	10/26/16 19:18	42.9087	13.1288	7.5	5.9	CLO.HGN	10.8
3	Norcia (PG)	10/30/16 6:40	42.8322	13.1107	9.2	6.5	MMO.HGE	19.2
4	Norcia (PG)	10/30/16 6:40	42.8322	13.1107	9.2	6.5	MMO.HGN	19.2
5	Castelsantangelo sul Nera (MC)	10/26/16 17:10	42.8747	13.1243	8.1	5.4	T1212.HNE	15.2
6	Norcia (PG)	10/30/16 6:40	42.8322	13.1107	9.2	6.5	T1212.HNE	10.5
7	Castelsantangelo sul Nera (MC)	10/26/16 17:10	42.8747	13.1243	8.1	5.4	T1212.HNN	15.2

5.2. Geological Model

The historical center of Avellino is located on a complex geology raised area (Figures 6 and 7). To have a detailed picture of a complex geomorphological high, we made an accurate selection of the data available for the historical center, especially the drilled boreholes. We used valuable information acquired both from the PUC (Piano Urbanistico Comunale) and from the geotechnical investigation, available on the web, carried out on sites where constructions were done.

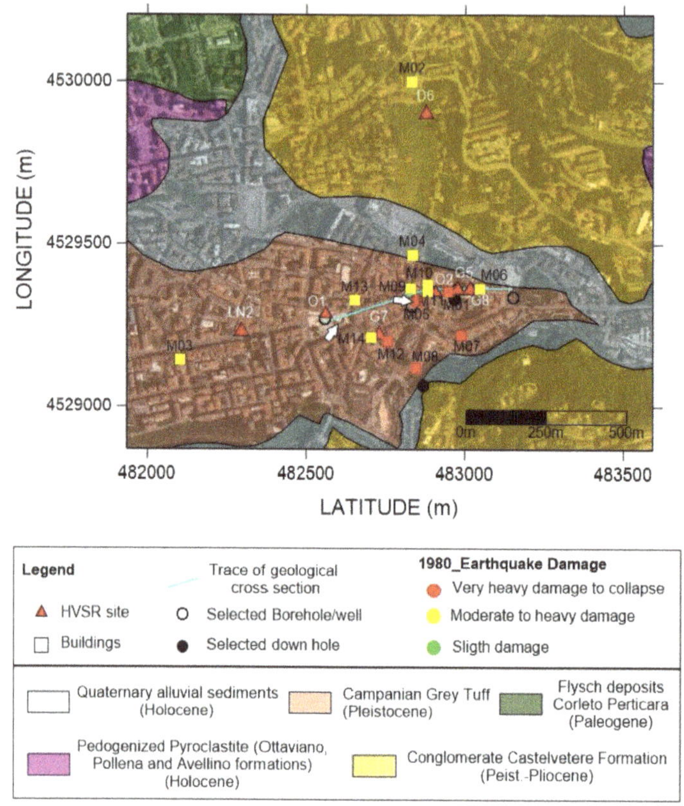

Figure 6. Geological map of the historical center of Avellino (modified from [5]). The white arrows indicate the position of the Piazza Libertà and the Torre dell'Orologio (Clock Tower, M05). Coordinates are in UTM (meters).

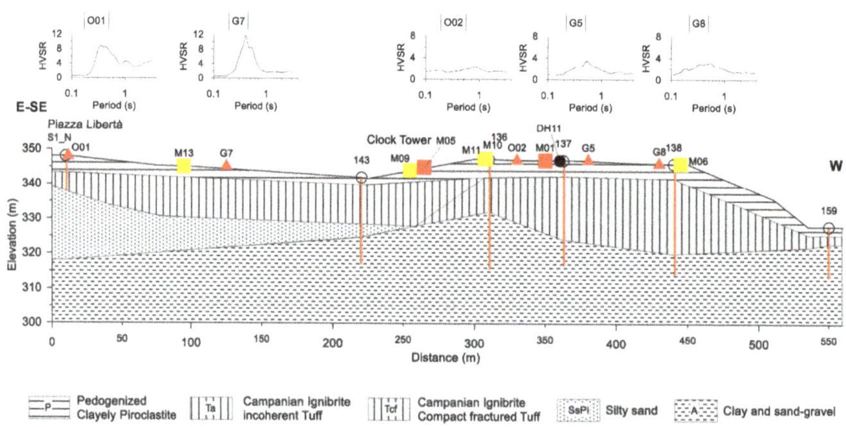

Figure 7. Lithostratigraphic cross-section along the trace indicated in Figure 6. The noise Horizontal to Vertical Spectral Ratio (HVSR) measurement sites with the HVSR curves are also reported. The orange vertical lines indicate the depth of each well.

We revised the geological cross-section reported in [4], using the stratigraphic wells located in the historical center, thus providing the new lithostratigraphic cross-section (Figure 7). The trace of the section was chosen to highlight the differences in both lithological and morphological settings, following the boreholes and seismic surveys. Moreover, most of the architectonic heritage listed in Table 2 stands along this section.

The lithostratigraphic cross-section intercepts the stratigraphic wells signed as S1_N, 143, 136, 137, 138, 159, and the downhole D11. The seismic bedrock in the study area is represented by the Miocenic clay and sand-gravel (A), which outcrops to the north and east of the studied area. Tuffs belonging to the Campanian Ignimbrite formation (39,000 years) follow in the succession. These tuffs consist of lapideous, weakly cemented, or fractured (Tcf), and incoherent lithofacies (Ta). A layer of silty sand (SsPl in Figure 7 Early Pleistocene) overlaying the clayey Miocene basement, is intercepted in the area between S1_N and 143. Pyroclastic deposits (P) cover all the previous lithological units.

All the wells intercept the top of the seismic bedrock, except S1_N well, which is located in Piazza Libertá, where the depth of the seismic bedrock is higher than 30 m. The lithostratigraphic cross-section points out that the subsoil of the historical center is particularly complex, mostly for the presence of different lithofacies of the same tuff formation, and for the irregular geometries of discontinuity surfaces.

5.3. Ground Response Analysis

We carried out a 2D numerical modeling of the subsoil along the section shown in Figure 7, using the LSR2D software provided by Stacec s.r.l. (http://stacec.it/Prodotto/92/lsr-2d). The algorithm implements the Finite Element Method and performs total stress analyses in the time domain. By applying a standard linear-equivalent approach it computes the horizontal response spectrum at the surface. The analysis domain is subdivided into triangular elements, or cells, which belong to five kinds of lithotypes (as reported in the legend of Figure 7). For each lithotype, the software requires the following parameters:

- volume weight, shear modulus, damping at low strain, Poisson's ratio;
- normalized shear modulus (G/G_0) and damping ratio (D) curves versus shear strain (γ);
- constant α for the calculation of the characteristic value of the shear deformation starting from the maximum value of γ (α) (typically equal to 0.65).

As output, the code provides:

- maximum accelerations in all nodes of the mesh;
- maximum tangential stresses and strains in each element;
- time history of the acceleration in the selected nodes (vertical and horizontal components).

In this way, we got effective modeling of a complex subsoil model. The S-wave velocity models were reconstructed from the DH11 and DH8 downholes as well as from the HVSR inversion procedures reported in [4]. In light of this, we can assign the ranges of velocity for all the lithotypes. The volume weights were averaged using the geotechnical data collected from PUC.

The dynamic soil properties (Table 5), namely the G/G_0 and D/D_0 curves, were derived from the literature or suggested by recent microzonation measures obtained in other Campanian areas where the same lithotypes are present. In detail, for the volcanic lithotypes (P, Tcf, and Ta) we used the modulus and the damping curves obtained from microzonation studies at Ischia Island. For the lithotype SsPl, which behaves like clay sands, we applied the curves proposed by [57]. Rocky lithotype, characterized by high values of stiffness, was considered as linear elastic lithotype with 5% damping (D).

Table 5. Mechanical and dynamic lithotype parameters assigned to each lithological unit used for the computational analysis. The curves marked by an asterisk are from the Level 3 Seismic Microzonation of Casamicciola Terme, Lacco Ameno, and Forio (Ischia Island, Naples) (http://www.commissarioricostruzioneischia.it).

Lithotype	S wave Velocity (m/s)	Density (kg/m^3)	Poisson Ratio	G/G$_0$ and D/D$_0$ Curves
P	180	1500	0.4	Piroclastic_CMS_Ischia*
Tcf	635	2000	0.35	Tuff_CMS_Ischia*
Ta	475	1500	0.4	Tuff_CMS_Ischia*
SsPl	300	1800	0.35	[57]
A	825	1900	0.35	Average Rock

The seismic input motion (Table 3) is applied simultaneously to all the nodes at the base boundary of the analysis domain and the equation system is solved in the time domain using the constant average acceleration (CAA) method. The results of the 2D numerical simulations were expressed in terms of the Amplification Factor (AF), calculated as the ratio between the integral of the output pseudo-acceleration spectrum and the integral of the corresponding spectrum of the input signal, over three different period intervals (0.1–0.5 s, 0.4–0.8 s, and 0.7–1.1 s)

$$AF = \frac{\int_{T_1}^{T_2} PSA_{out}(T)dT}{\int_{T_1}^{T_2} PSA_{in}(T)dT}, \tag{1}$$

where T_1 and T_2 are the minimum and maximum period in the interval. Using Equation (1), for all the output points and each period interval, the logarithmic mean of the amplification factor values relating to the seven input accelerograms was calculated, as reported in the following equation:

$$\begin{gathered} m_{\ln} = \tfrac{1}{7}\textstyle\sum_{i=1}^{7} \ln(FA_i) \\ FA_{rif} = e^{m_{\ln}} \end{gathered} \tag{2}$$

We inserted 14 control points along the topographic surface (red points in Figure 8), choosing these as the points with the most relevant stratigraphic variations, or corresponding to the position of the buildings. The side boundary nodes were constrained along the Z direction and damped along the X direction (red box in Figure 8), to avoid the presence of reflected waves over the edges.

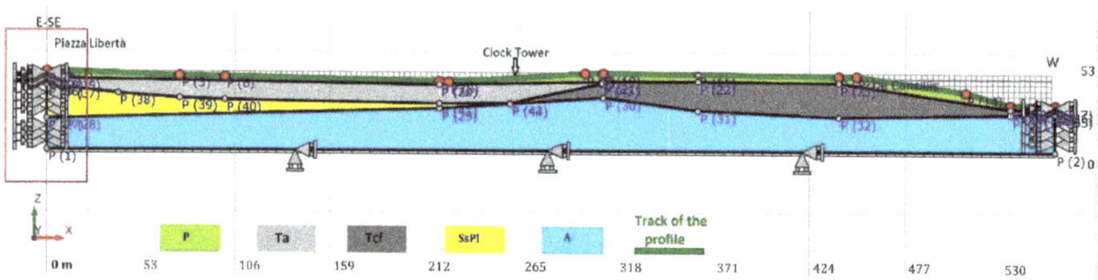

Figure 8. Subsoil model adopted for the 2D ground response analysis. The red box highlights the damped conditions imposed at the lateral boundaries. Mechanical and dynamic lithotypes parameters are reported in Table 5.

The output of the program, after 10 iterations, returned the acceleration model shown in Figure 9. The distribution of the acceleration values (along the X direction) shows that the maximum occurs where the stratigraphic variations along the vertical are more significant. The low acceleration and amplification factor values at the edges of the section are the effect of damping conditions imposed at boundaries.

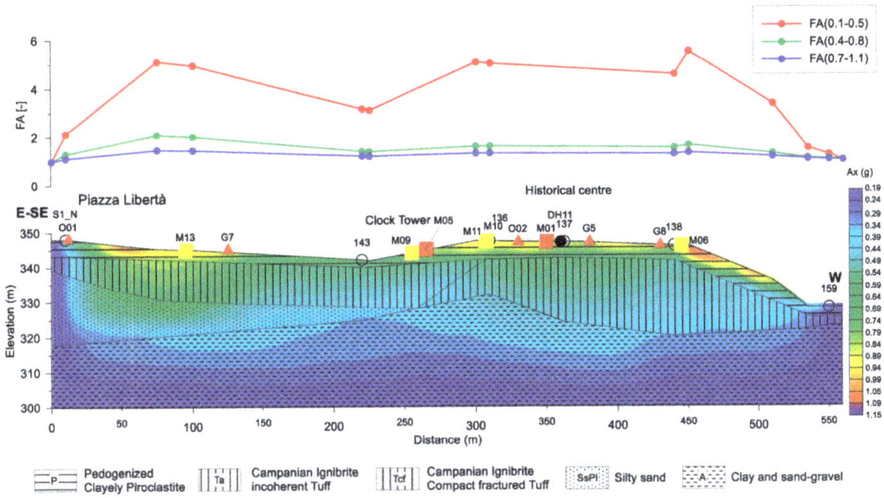

Figure 9. **Top**: 2D amplification factors obtained with Equation (1) in three different period ranges. **Bottom**: contour map of the maximum acceleration (peak ground acceleration—PGA) over imposed to the geological section.

The higher acceleration values obtained in the area between Piazza Libertà and the historical center are due to the presence of loose and incoherent deposits (Loose Tuff (Ta), Pyroclastic deposits (P)) at the top of the sequence. This effect also influences the HVSR curves of O01 and G7 sites (Figure 7), where the higher amplitude has been related to a high impedance contrast [5]. The scheme is different in the historical center, where the Tuff is compact or fractured, with Vs equal to 635 m/s; here the higher acceleration values are probably linked to coupled stratigraphic and topographic effects. This is particularly evident in the area between M06 (Palazzo de Conciliis) and the well 159, where the amplification is related to edge effects.

The highest values of the amplification factor, along the geological cross-section, are in the low period range 0.1–0.5 s (red curve in Figure 9). This means that the ground motion is significantly amplified in the ranges of the period of engineering interest. The HVSR curves of O01 and G7 sites, that highlight amplifications as high as 8 in the range 0.3–0.5 s (Figure 7), experimentally confirm these results.

Along the section, the amplification decreases by almost 60 percent for the intermediate (0.4–0.8 s, the green curve in Figure 9) and high (0.7–1.1 s, the blue curve in Figure 9) periods. The results obtained are substantially compatible with HVSR observations: the HVSR curves show almost flat spectral ratios for periods higher than 1 s, whereas amplification peaks appear in the period range 0.1–0.3 s.

6. Discussion and Conclusions

The historic center of Avellino is characterized by a complex morpho-lithological structure. It is located on a topographic relief, formed by the raising of the Miocene basement, on which the tuffaceous Campanian Ignimbrite formation was emplaced. We chose a west-east trending section for a twofold reason: (i) to highlight the variability of both morphological and geological features, (ii) to correlate the lithological pattern of the subsoil to the sites hosting the most important architectonical heritage of Avellino.

We calculated the 2D ground response in a linear-equivalent regime using realistic and detailed modeling along the representative section. We used a probabilistic approach to define the input motion and used the software REXELite to automatically select real ground acceleration records from the Italian Strong-Motion Database, compatible with a target acceleration response spectrum.

The modeled ground acceleration shows the highest values in the period range of engineering interest (0.1–0.4 s) and is controlled by the seismic impedance contrast in the subsoil within the first tens of meters, as well as by the topography. This result is in agreement with the experimental data, consisting of HVSR ambient noise measurements, previously acquired. At higher periods, the accelerations decrease by almost 60 percent.

The town of Avellino experienced a different level of damage to heritage during the 23 November 1980 earthquake. In the historical center, the damage was from the MHD to the VHDC level (see Table 2). Our study was focused on identifying clues about possible relationships between the ground motion amplification and the damage level observed. In a previous work [5], some of the authors of this article performed noise HVSR measurements in a wider area of the town, both free-field and inside civil buildings. We verified the existence of soil/building resonance effects in the 0.22–0.83 s period band, that have influenced the distribution of the damage produced by this earthquake.

The historical buildings taken into consideration here are variable in typology, shape, and structure, and, at present, we do not have their estimates of the vibration modes in the dynamic regime. By comparing the level of damage suffered by these monuments with the ground response computed in the historical center of Avellino, we suggest that the variability in the damage distribution couldn't be solely explained with effects due to surface geology. Even topographical effects, particularly evident at the edge of the ignimbritic high (M06 location, in the sections shown in Figures 7 and 9), did not have a decisive impact, as the MHD damage level of Palazzo de Conciiliis demonstrated. Nearby monuments, Balestrieri Palace (M09) and the Clock Tower (M05) have suffered different damage levels. The tower, the tallest of the selected buildings likely to be placed in the highest examined period band (0.4–0.8 s), suffered extensive damage, contrary to what the modeling foresees. The Cathedral (M01) the Greco Palace (M11) and the Festa Place (M10), which are placed where the same local geological features are present, show the different extent of the damage. This diversity could be due to the different dynamic responses of the structures and hence to their degree of vulnerability.

From all the results we can infer that the state of conservation of the buildings has played a key role in the occurrence of damage to the historical heritage in response to the 1980 Irpinia-Basilicata earthquake. It is worth noting that [58] also have already proved that the vulnerability of buildings due to their poor quality and poor maintenance caused the most damage in response to this strong earthquake. To assess whether soil/building resonance effects could have contributed to the damage experienced by the historical buildings, most of them of irregular shape and structure, dynamic regime measures would be required, which are currently not available. Therefore, this research is to be considered a preliminary study that could provide useful clues for the preservation of the artistic and architectural heritage of the town of Avellino. Further analyses will aim to investigate the dynamic behavior of the historical buildings as well as their vulnerability, thus making it possible to unequivocally discern among the different factors that threaten the heritage in the event of an earthquake occurrence.

Finally, the research of factors that correlate damage distribution to surface geology plays a crucial role in the appropriate design of seismic risk mitigation interventions aimed at the conservation of our architectural heritage.

Author Contributions: Conceptualization, R.M.; methodology, L.N. and F.T.G.; software, L.N.; validation, L.N., F.T.G., and R.M.; formal analysis, L.N., F.T.G., and R.M.; investigation, F.T.G.; writing—original draft preparation, R.M.; writing—review and editing, L.N., F.T.G., and R.M. All authors have read and agreed to the published version of the manuscript.

Funding: This research received no external funding.

Acknowledgments: We wish to thanks three anonymous reviewers for their useful comments, which contributed to improving the final version of the manuscript.

Conflicts of Interest: The authors declare no conflict of interest.

References

1. Guidoboni, E.; Ferrari, G.; Mariotti, D.; Comastri, A.; Tarabusi, G.; Sgattoni, G.; Valensise, G. *CFTI5Med, Catalogo dei Forti Terremoti in Italia (461 a.C.-1997) e Nell'area Mediterranea (760 a.C.-1500)*; Istituto Nazionale di Geofisica e Vulcanologia (INGV): Rome, Italy, 2018. [CrossRef]
2. Bernard, P.; Zollo, A. The Irpinia (Italy) 1980 Earthquake: Detailed Analysis of a Complex Normal Faulting. *J. Geophys. Res.* **1989**, *94*, 1631–1647. [CrossRef]
3. Postpischl, D.; Branno, A.; Esposito, E.; Ferrari, G.; Marturano, A.; Porfido, S.; Rinaldis, V.; Stucchi, M. The Irpinia earthquake of November 23, 1980. In *Atlas of Isoseismal Maps of Italian Earthquakes, CNR-PFG*; CNR: Irpinia, Italy, 1985; pp. 152–157.
4. Maresca, R.; Nardone, L.; Pasquale, G.; Pinto, F.; Bianco, F. Effects of surface geology on seismic ground motion deduced from ambient-noise measurements in the town of Avellino, Irpinia region (Italy). *Pure Appl. Geophys.* **2012**, *169*, 1173–1188. [CrossRef]
5. Maresca, R.; Nardone, L.; Gizzi, F.T.; Potenza, M.R. Ambient noise HVSR measurements in the Avellino historical centre and surrounding area (southern Italy). Correlation with surface geology and damage caused by the 1980 Irpinia-Basilicata earthquake. *Measurement* **2018**, *130*, 211–222. [CrossRef]
6. Gaudiosi, G.; Alessio, G.; Nappi, R.; Noviello, V.; Spiga, E.; Porfido, S. Evaluation of Damages to the Architectural Heritage of Naples as a Result of the Strongest Earthquakes of the Southern Apennines. *Appl. Sci.* **2020**, *10*, 6880. [CrossRef]
7. Lagomarsino, S.; Podestà, S. Damage and vulnerability assessment of churches after the Molise earthquake (2002). *Earthq. Spectr.* **2004**, *20*, S271–S283. [CrossRef]
8. Lagomarsino, S. Damage assessment of churches after L'Aquila earthquake (2009). *Bull. Earthq. Eng.* **2012**, *10*, 73–92. [CrossRef]
9. Sorrentino, L.; Liberatore, L.; Decanini, L.D. The performance of churches in the 2012 Emilia earthquakes. *Bull. Earthq. Eng.* **2014**, *12*, 2299–2331. [CrossRef]
10. Hofer, L.; Zampieri, P.; Zanini, M.A.; Faleschini, F.; Pellegrino, C. Seismic damage survey and empirical fragility curves for churches after the August 24, 2016 Central Italy earthquake. *Soil Dyn. Earthq. Eng.* **2018**, *111*, 98–109.
11. Gallipoli, M.R.; Gizzi, F.T.; Rizzo, E.; Masini, N.; Potenza, M.R.; Albarello, D.; Lapenna, V. Site features responsible for uneven seismic effects in historical centre of Melfi (Basilicata, Southern Italy). *Disaster Adv.* **2012**, *5*, 125–137.
12. Guidoboni, E.; Mariotti, D.; Giammarinaro, M.S.; Rovelli, A. Identification of amplified damage zones in Palermo, Sicily (Italy), during the earthquakes of the last three centuries. *Bull. Seismol. Soc. Am.* **2003**, *93*, 1649–1669. [CrossRef]
13. Gosar, A. Microtremor HVSR study for assessing site effects in the Bovec basin (NW Slovenia) related to 1998 Mw5. 6 and 2004 Mw5. 2 earthquakes. *Eng. Geol.* **2007**, *91*, 178–193. [CrossRef]
14. Del Monaco, F.; Tallini, M.; De Rose, C.; Durante, F. HVNSR survey in historical downtown L'Aquila (central Italy): Site resonance properties vs. subsoil model. *Eng. Geol.* **2013**, *158*, 34–47. [CrossRef]

15. Leyton, F.; Ruiz, S.; Sepúlveda, S.A.; Contreras, J.P.; Rebolledo, S.; Astroza, M. Microtremors' HVSR and its correlation with surface geology and damage observed after the 2010 Maule earthquake (Mw 8.8) at Talca and Curicó, Central Chile. *Eng. Geol.* **2013**, *161*, 26–33. [CrossRef]
16. Mucciarelli, M. Ambient noise measurements on soil and buildings. *Bull. Earthq. Eng.* **2010**, *8*, 481–482. [CrossRef]
17. Castellaro, S.; Padrón, L.A.; & Mulargia, F. The different response of apparently identical structures: A far-field lesson from the Mirandola 20th May 2012 earthquake. *Bull. Earthq. Eng.* **2014**, *12*, 2481–2493. [CrossRef]
18. Moisidi, M. Site-building resonance response in a complex geological setting: Ground motions recorded in the centre of Paleohora Basin and at a rock fractured outcrop site close to the basin edge (SW Crete, Greece). *Phys. Chem. Earth Parts A/B/C* **2017**, *98*, 73–89. [CrossRef]
19. Honoré, L.; Courboulex, F.; Souriau, A. Ground motion simulations of a major historical earthquake (1660) in the French Pyrenees using recent moderate size earthquakes. *Geophys. J. Int.* **2011**, *187*, 1001–1018. [CrossRef]
20. Albarello, D.; Cesi, C.; Eulilli, V.; Guerrini, F.; Lunedei, E.; Paolucci, E.; Pileggi, D.; Puzzilli, L.M. The contribution of the ambient vibration prospecting in seismic microzoning: An example from the area damaged by the April 6, 2009 L'Aquila (Italy) earthquake. *Boll. Geof. Teor. Appl.* **2011**, *52*, 513–538.
21. Khanbabazadeh, H.; Hasal, M.E.; Iyisan, R. 2D seismic response of the Duzce Basin, Turkey. *Soil Dyn. Earthq. Eng.* **2019**, *125*, 105754. [CrossRef]
22. Mouzakiotis, E.; Karastathis, V.; Voulgaris, N.; Papadimitriou, P. Site Amplification Assessment in the East Corinth Gulf Using 3D Finite-Difference Modeling and Local Geophysical Data. *Pure Appl. Geophys.* **2020**, 1–19. [CrossRef]
23. Senatore, M.R.; Boscaino, M.; Pinto, F. The Quaternary geology of the Benevento urban area (Southern Italy) for seismic microzonation purposes. *Ital. J. Geosci.* **2019**, *138*, 66–87. [CrossRef]
24. Nunziata, C.; Costanzo, M.R. Ground Shaking Scenario at the Historical Center of Napoli (Southern Italy) for the 1456 and 1688 Earthquakes. *Pure Appl. Geophys* **2020**, 1–16. [CrossRef]
25. Macerola, L.; Tallini, M.; Di Giulio, G.; Nocentini, M.; Milana, G. The 1-D and 2-D Seismic Modeling of Deep Quaternary Basin (Downtown L'Aquila, Central Italy). *Earthq. Spectra* **2019**, *35*, 1689–1710. [CrossRef]
26. Giallini, S.; Pizzi, A.; Pagliaroli, A.; Moscatelli, M.; Vignaroli, G.; Sirianni, P.; Mancini, M.; Laurenzano, G. Evaluation of complex site effects through experimental methods and numerical modelling: The case history of Arquata del Tronto, central Italy. *Eng. Geol.* **2020**, *2020*, 105646. [CrossRef]
27. Brando, G.; Pagliaroli, A.; Cocco, G.; Di Buccio, F. Site effects and damage scenarios: The case study of two historic centers following the 2016 Central Italy earthquake. *Eng. Geol.* **2020**, *2020*, 105647. [CrossRef]
28. Pergalani, F.; Pagliaroli, A.; Bourdeau, C.; Compagnoni, M.; Lenti, L.; Lualdi, M.; Madiai, C.; Martino, S.; Razzano, R.; Varone, C.; et al. Seismic microzoning map: Approaches, results and applications after the 2016–2017 Central Italy seismic sequence. *Bull. Earthq. Eng.* **2019**, *18*. [CrossRef]
29. Pagliaroli, A.; Gaudiosi, I.; Razzano, R.; Giallini, S.; De Siva, F.; Chiaradonna, A.; Ciancimino, A.; Foti, S. Site response analyses for seismic microzonation: Case-histories, results and applications in Central Italy. In Proceedings of the 7th International Conference on Earthquake Geotechnical Engineering (VII ICEGE), Roma, Italy, 17–20 June 2019.
30. Hudson, J.A.; Boore, D.M. Comments on 'Scattered Surface Waves from a Surface Obstacle'. *Geophys. J. R. Astr. Soc.* **1980**, *60*, 123–127. [CrossRef]
31. Lee, S.J.; Chan, Y.C.; Komatitsch, D.; Huang, B.S.; Tromp, J. Effects of realistic surface topography on seismic ground motion in the Yangminshan region of Taiwan based upon the spectral-element method and LiDAR DTM. *Bull. Seismol. Soc. Am.* **2009**, *99*, 681–693. [CrossRef]
32. Geli, L.; Bard, P.Y.; Jullien, B. The effect of topography on earthquake ground motion: A review and new results. *Bull. Seismol. Soc. Am.* **1988**, *78*, 42–63.
33. Cavallaro, A.; Ferraro, A.; Grasso, S.; Maugeri, M. Topographic effects on the Monte Po hill in Catania (Italy). *Soil Dyn. Earthq. Eng.* **2012**, *43*, 97–113. [CrossRef]
34. Sohrabi-Bidar, A.; Kamalian, M. Effects of three-dimensionality on seismic response of Gaussian-shaped hills for simple incident pulses. *Soil Dyn. Earthq. Eng.* **2013**, *52*, 1–12. [CrossRef]
35. Grelle, G.; Gargini, E.; Facciorusso, J.; Maresca, R.; Madiai, C. Seismic site effects in the Red Zone of Amatrice hill detected via the mutual sustainment of experimental and computational approaches. *Bull. Earthq. Eng.* **2020**, *18*, 1–30.

36. Meletti, C.; Montaldo, V. *Stime Di Pericolosità Sismica Per Diverse Probabilità Di Superamento In 50 Anni: Valori Di Ag. Progetto INGV-DPC S1, Deliverable D2*; Istituto Nazionale di Geofisica e Vulcanologia (INGV): Rome, Italy, 2007. (In Italian)
37. Meletti, C.; Galadini, F.; Valensise, G.; Stucchi, M.; Basili, R.; Barba, S.; Vannucci, G.; Boschi, E. A seismic source zone model for the seismic hazard assessment of the Italian territory. *Tectonophysics* **2008**, *450*, 85–108. [CrossRef]
38. C.S.LL.PP (Consiglio Superiore del Lavori Pubblici). Norme tecniche per le costruzioni. *Gazzetta Ufficiale della Repubblica Italiana*, 4 February 2008. (In Italian)
39. C.S.LL.PP (Consiglio Superiore del Lavori Pubblici). Aggiornamento delle Norme tecniche per le costruzioni. *Gazzetta Ufficiale della Repubblica Italiana*, 20 February 2018. (In Italian)
40. Harmsen, S.; Frankel, A. Geographic deaggregation of seismic hazard in the United States. *Bull. Seismol. Soc. Am.* **2001**, *91*, 13–26. [CrossRef]
41. Akinci, A.; Galadini, F.; Pantosti, D.; Petersen, M.; Malagnini, L.; Perkins, D. Effect of time dependence on probabilistic seismic-hazard maps and deaggregation for the Central Apennines, Italy. *Bull. Seismol. Soc. Am.* **2009**, *99*, 585–610. [CrossRef]
42. Wang, J.P.; Huang, D.; Cheng, C.T.; Shao, K.S.; Wu, Y.C.; Chang, C.W. Seismic hazard analyses for Taipei city including deaggregation, design spectra, and time history with excel applications. *Comput. Geosci.* **2013**, *52*, 146–154. [CrossRef]
43. Jarahi, H. Probabilistic seismic hazard deaggregation for Karaj City (Iran). *Am. J. Eng. Appl. Sci.* **2016**, *9*, 520–529. [CrossRef]
44. Pacor, F.; Paolucci, R.; Luzi, L.; Sabetta, F.; Spinelli, A.; Gorini, A.; Nicoletti, M.; Marcucci, S.; Filippi, L.; Dolce, M. Overview of the Italian strong motion database ITACA 1.0. *Bull. Earthq. Eng.* **2011**, *9*, 1723–1739. [CrossRef]
45. Gambino, N. *Cronistoria dei Restauri Nelle Iscrizioni Commemorative*; Gambino, N., Ed.; La Cattedrale di Avellino: Cava dei Tirreni, Italy, 1985; pp. 85–86. (In Italian)
46. Proietti, G. *Ministero Per I Beni Culturali E Ambientali—Soprintendenza Generale Agli Interventi Post-Sismici In Campania E Basilicata, Dopo La Polvere. Rilevazione Degli Interventi Di Recupero Post-Sismico Del Patrimonio Archeologico, Architettonico Ed Artistico Delle Regioni Campania E Basilicata Danneggiato Dal Terremoto Del 23 Novembre 1980 e del 14 Febbraio 1981 (Anni 1985–1989)*; Tomo II, Avellino, Istituto Poligrafico e Zecca dello Stato: Roma, Italy, 1994. (In Italian)
47. Frattani, P. *Sisma 1980. Effetti Sul Patrimonio Artistico Della Campania E Basilicata. Campania. Supplemento Del Bollettino d'Arte*; Istituto Poligrafico e Zecca dello Stato: Rome, Italy, 1982. (In Italian)
48. Grünthal, G. *European Macroseismic Scale 1998 (EMS-98). Cahiers du Centre Européen de Géodynamique et de Séismologie 15*; Centre Européen de Géodynamique et de Séismologie: Luxembourg, 1998; p. 99.
49. Bazzurro, P.; Allin Cornell, C. Deaggregation of seismic hazard. *Bull. Seismol. Soc. Am.* **1999**, *89*, 501–520.
50. Bommer, J.J.; Acevedo, A.B. The use of real earthquake accelerograms as input to dynamic analysis. *J. Earthq. Eng.* **2004**, *8*, 43–91. [CrossRef]
51. Spallarossa, D.; Barani, S. Disaggregazione della pericolosità sismica in termini di M-R-ε. Progetto DPC-INGV S1, Deliverable D14. 2007. Available online: http://esse1.mi.ingv.it/d14.html (accessed on 10 September 2020).
52. Barani, S.; Spallarossa, D.; Bazzurro, P. Disaggregation of Probabilistic Ground-Motion Hazard in Italy. *Bull. Seismol. Soc. Am.* **2009**, *99*, 2638–2661. [CrossRef]
53. Iervolino, I.; Galasso, C.; Paolucci, R.; Pacor, F. Engineering ground motion record selection in the ITalian ACcelerometric Archive. *Bull. Earthq. Eng.* **2011**, *9*, 1761–1778. [CrossRef]
54. C.E.N. *Eurocode 8: Design Provisions for Earthquake Resistance of Structures, Part 1.1: General Rules, Seismic Actions and Rules for Buildings*; European Committee for Standardization: Brussels, Belgium, 2003.
55. Luzi, L.; Hailemikael, S.; Bindi, D.; Pacor, F.; Mele, F.; Sabetta, F. ITACA (ITalian ACcelerometric Archive): A web portal for the dissemination of Italian strong-motion data. *Seismol. Res. Lett.* **2008**, *79*, 716–722. [CrossRef]
56. Stucchi, M.; Akinci, A.; Faccioli, E.; Gasperini, P.; Malagnini, L.; Meletti, C.; Montaldo, V.; Valensise, G. *Redazione della Mappa di Pericolosità Sismica prevista dall'Ordinanza PC del 20 marzo 2003, n. 3274, All. 1 Rapporto Conclusivo*; Istituto Nazionale di Geofisica e Vulcanologia (INGV): Rome, Italy, 2004.

57. Seed, H.B.; Sun, J.H. *Implication of Site Effects in the Mexico City Earthquake of September 19, 1985 for Earthquake-Resistant Design Criteria in the San Francisco Bay Area of California*; Report No. UCB/ EERC-89/03; Earthquake Engineering Research Center: Berkeley, CA, USA, 1989.
58. Candela, M.; Viggiani, C. The effects of the Irpinia earthquake in the ancient centre of Avellino, Italy. In Proceedings of the International Symposium of IAEG, Athens, Greece, 19–23 September 1988; Volume 3: Earthquakes, Vibrations and Other Hazards in Relation to the Study and the Protection of Monuments and Historical Sites Athens, Greece; Marinos, P.G., Koukis, G.C., Eds.; 1988.

geosciences

MDPI

Article

The Role of the Water Level in the Assessment of Seismic Vulnerability for the 23 November 1980 Irpinia–Basilicata Earthquake

Davide Forcellini

Department of Civil and Environmental Engineering, University of Auckland, 20 Symonds Street, Auckland 1010, New Zealand; dfor295@aucklanduni.ac.nz

Received: 30 April 2020; Accepted: 12 June 2020; Published: 13 June 2020

Abstract: The seismic vulnerability of structures is closely related to changes in the degree of soil saturation that may cause significant changes in volume and shear strength, and consequently, bearing capacity. This paper aims to consider this issue during the strong earthquake that struck Southern Italy on 23 November 1980 (Ms = 6.9) and affected the Campania and Basilicata regions. Several 3D numerical finite element models were performed in order to consider the effects of soil–structure interaction (SSI) on a representative benchmark structure. In particular, the role of the water level depth is herein considered as one of the most significant parameters to control the shear deformations inside the soil, and thus the performance of the superstructure. Results show the importance of considering the water level for buildings on shallow foundations in terms of settlements, base shear forces and floor displacements.

Keywords: Irpinia–Basilicata earthquake; seismic assessment; soil–structure interaction; numerical simulations; OpenSees

1. Background

The historical 23 November 1980 Irpinia–Basilicata (Southern Italy) earthquake (Ms = 6.9) showed the importance of assessing seismic risk for Italian communities. In particular, earthquake vulnerability depends on the mechanisms occurring inside the superficial layers that filter the input motion from the bedrock. Many parameters may drive these mechanisms, such as soil properties, bedrock characteristics, layers depth, stratification and water level. In particular, the effects caused by water level may modify the vertical pressures on the soil layers, as observed and discussed by [1,2]. Other contributions showed the importance of controlling the water table as a soil improvement [3,4]. In addition, the correlation between shallow groundwater levels and liquefaction occurrence is proposed by [5] for the May 2006 earthquake at Yogyakarta (Indonesia). In addition, [6] showed the effects of water level on analytical indexes for liquefaction susceptibility, while [7] investigated the effects of water table level during the recent Emilia Romagna earthquake, where the oscillations were shown to be limited to less than 2 m.

In this regard, there are many approaches to measure the depth of the water table. One of the most common consists of applying piezometers that are typically inserted into the soil at different depths below the surface with measurements made manually or automatically with a continuous registration [8]. As shown in [9], other interesting issues are the density, the frequency of measurement sites, the spatially interpolating point data and the extrapolating of the water table depth procedures. In this regard, interpreting redoximorphic features [10] is another common methodology in routine soil surveys. More extended approaches that allow for the prediction of the variation of shallow water table are based on climatic records, field evidences from soil morphology and properties, outputs of physically based water balance models and combinations of the different approaches. Soil morphology is particularly valuable for interpreting water table dynamics in the soil profile [11], and simulation models were recently proposed (e.g., [12]).

After the 23 November 1980 Irpinia–Basilicata earthquake, several contributions considered the various mechanisms of rupture [13–17] and estimated damages [18], which focused on the effects of spectral accelerations [19–21] or quantified the consequences [22–24], and proposed several models [25,26]. Even with such extended literature, information regarding the level of the water table was not registered; this paper proposes numerical simulations of 3D models with different water levels in order to investigate the role of this parameter on the seismic vulnerability of the structure, which was previously investigated [27]. In particular, the effects of water level are relevant especially in the case of shallow foundations, since they are more sensitive to the change of the vertical stresses due to changing water level. For example, [28] showed that structures with shallow foundations on soft soils are expected to experience big damages and losses due to soil–structure interaction (SSI), which can be beneficial, detrimental or uninfluential on the seismic vulnerability. In addition, [29] showed that considering SSI for unreinforced masonry (URM) buildings with shallow foundations is non-negligible and that taking into account the inelastic behaviour of the soil foundation system may lead to smaller structural displacements. Furthermore, [30] focused on the mechanisms at the base of the permanent deformation of the soil to conclude that buildings founded on shallow foundations are particularly vulnerable to seismically induced settlements.

With this background, the present paper aims firstly at covering the lack of information regarding the position of the water level during the Irpinia–Basilicata earthquake by presenting a case study to assess the most detrimental water level conditions under which the earthquake may have occurred. Secondly, this paper proposes an attempt to generalize the outcomes presented to other seismic assessments by accounting the water level position, which is an important source of uncertainty, and sometimes underestimated or even neglected. The methodology applied in this paper consists of performing advanced 3D finite element models of the entire system (soil + foundation + structure) that allow consideration of the soil nonlinear mechanisms of shear deformation. In particular, different positions of water level were considered in order to assess several saturated conditions among the homogenous 20 m layer selected in the previous work [27]. The structural configuration performed is representative of the Italian residential buildings that were mainly damaged during the Irpinia–Basilicata earthquake, and consists of a reinforced concrete (RC) structure with infilled masonry walls. In particular, the documentary sources are based on two main typologies of technical data preserved in local archives: the "Scheda A" and "Scheda B" (Figure 1), which report the damages to the buildings and were fundamental tools to detect the level of damage to the buildings, [31,32]. These were used to summarize the data from the surveys that were carried out in two phases: the first, to evaluate the conditions of the entire building (Scheda A) and the second, to verify the effects on each housing unit (Scheda B) [18]. Other important documents are the recovery plans (named "Piani di Recupero") of the historical centres, and sources used to analyse the outcomes of the earthquake at the urban scale. For an extended literature review, please refer to [27]. Results in terms of settlements, accelerations, base shear forces and floor displacements are calculated in order to assess the most detrimental positions of the water level.

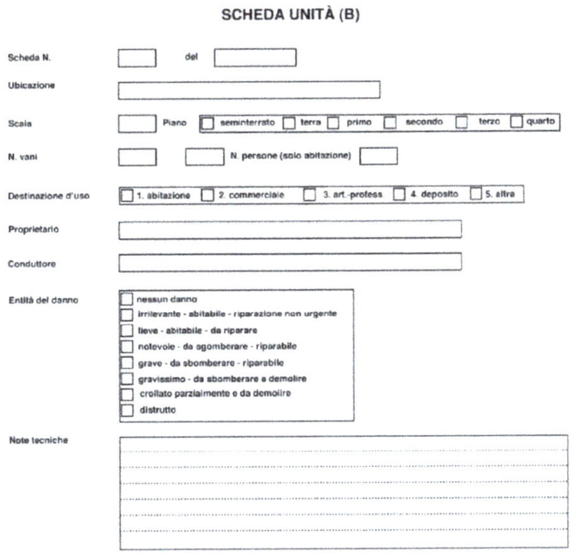

SCHEDA UNITÀ (B)

Scheda N. del

Ubicazione

Scala Piano seminterrato terra primo secondo terzo quarto

N. vani N. persone (solo abitazione)

Destinazione d'uso 1. abitazione 2. commerciale 3. art.-profess 4. deposito 5. altra

Proprietario

Conduttore

Entità del danno
- nessun danno
- irrilevante - abitabile - riparazione non urgente
- lieve - abitabile - da riparare
- notevole - da sgomberare - riparabile
- grave - da sbomberare - riparabile
- gravissimo - da sbomberare e demolire
- crollato parzialmente e da demolire
- distrutto

Note tecniche

Figure 1. Example of Scheda B [31].

2. The Irpinia–Basilicata Earthquake

The 23 November 1980 Irpinia–Basilicata earthquake (Figure 2) may be considered one of the strongest earthquakes recorded in Italy [18,24,33–37], causing profound changes in communities and extended damages to civil structures and infrastructures. In particular, [18] conducted an extended literature review showing that the knowledge of the damages focused on some of Campania's and Basilicata's most heavily damaged towns, but without a systematic study. In particular, estimations assessed that approximately 1.85 million buildings were involved in the event, with 75,000 destroyed, 275,000 seriously damaged and 480,000 slightly damaged. In addition, the environmental effects induced by the 23 November 1980 Irpinia–Basilicata earthquake were also significantly severe since primary and secondary effects (such as hydrological variations and liquefaction-induced mechanism) brought important natural modifications to the natural conditions [27], such as ground deformations, slopes, landslides, lateral displacements and settlements, which were extensively documented by several contributions [37–43].

The damaged buildings consisted mostly of reinforced concrete (RC) structures characterised by infilled masonry walls (IMWs), the object of this paper (see next section). Specific studies were carried on as recovery plans (named "Piani di Recupero") and described the procedures to recover the historical centres. These plans are useful as sources of information regarding the seismic effects at the urban scale [27,44–49].

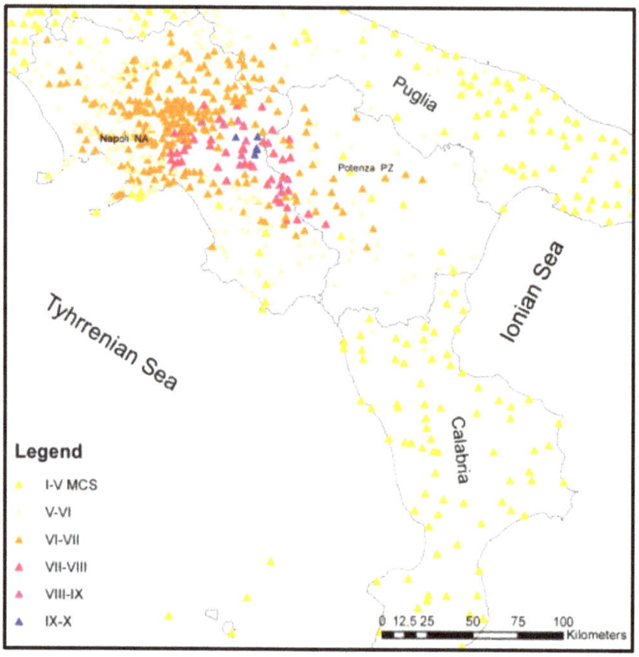

Figure 2. Macroseismic field of the 23 November 1980 Irpinia–Basilicata earthquake [18] (MCS = Mercalli Cancani Sieberg scale).

3. Methodology

SSI consists of several mechanisms that define the mutual behaviour of the soil, the foundation and the building, and consequently affect the structural performance. The water level depth modifies the vertical pressures in the soil and thus the shear mechanisms that cause soil deformations and settlements at foundation levels, and, consequently, stresses in the superstructures that affect the whole behaviour of the system. The proposed 3D finite element models (Figure 3) aim to reproduce these complex nonlinear mechanisms to represent realistically the behaviour of the entire system (soil + foundation + structure). This goal is particularly challenging because of the mutual effects of two sources of nonlinearity: the shear mechanisms in the soil and the structural behaviours. The study proposed here is based on the previous contribution [27], here extended to consider the role of water level depth.

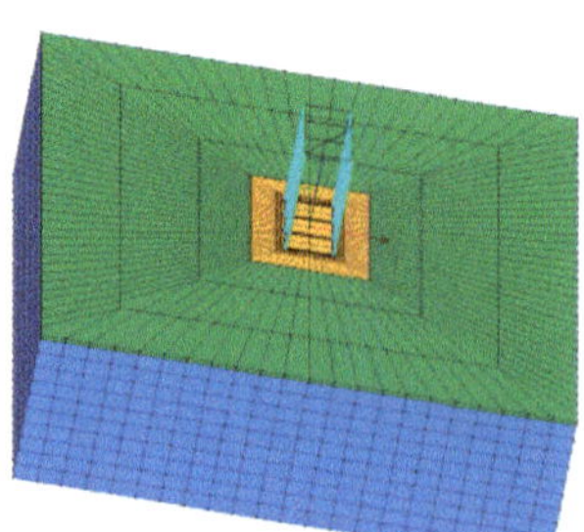

Figure 3. Mesh 2: soil–structure interaction (SSI) model; uniform soil layer (blue), infill (green) and foundation (yellow) [27].

The input motion performed (Figure 4), selected from the Italian Accelerometric Archive [50], consists of the registration at the Sturno (STN) station (latitude: 41.0183°, longitude: 15.1117°) in Avellino, Campania, which is defined on soil B (following the Eurocode 8 classification); more details are available in [27].

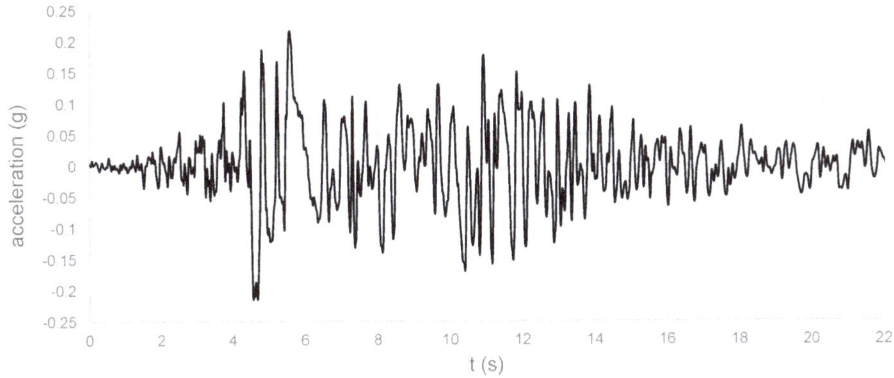

Figure 4. Selected input motion (longitudinal direction) [27].

The tridimensional (3D) models performed are based on the *u–p* formulation (*u* is the displacement of the soil skeleton and *p* is the pore pressure), defined in detail in [51] and [52]. The *Pressure Depend Multi Yield02* model [53,54], selected to represent the soil layer, consists of a multi-yield-surface plasticity framework that may reproduce the mechanism of cycle-by-cycle permanent shear strain accumulation in incoherent soil materials, as shown in [55]. The model is based on the definition of several parameters, such as the low-strain shear modulus, the friction angle, shear wave velocity and permeability, shown in Table 1. In particular, an equivalent uniform linear layer was adopted in order to calculate the soil fundamental periods [56]. Note that the paper aims to reproduce a historical event and geotechnical parameters for superficial layers are currently not available for the Irpinia–Basilicata earthquake, as specified in [27]. The data that are necessary to be implemented in a numerical model require material characterizations with in situ tests that are impossible 40 years after the earthquake occurred, since soil materials and hydraulic conditions have changed significantly. Therefore, the values in Table 1 were defined on the available information (mainly [20]) and the free-field study that was carried out in [27].

Table 1. Soil characteristics.

Soil	S1	S2	S3	S4
Density (Mg/m^3)	1.7	1.9	1.9	2.1
Reference shear modulus (kPa)	3.83×10^4	4.28×10^4	5.50×10^4	1.32×10^5
Reference bulk modulus (kPa)	1.50×10^5	2.00×10^5	2.00×10^5	3.90×10^5
Shear wave velocity (m/s)	150	150	170	250
Soil fundamental period (s)	0.53	0.53	0.47	0.32
Cohesion (kPa)	5	5	5	5
Friction angle (°)	27	29	35	40
Horizontal permeability (m/s)	1.0×10^{-7}	1.0×10^{-7}	1.0×10^{-7}	1.0×10^{-7}
Vertical permeability (m/s)	1.0×10^{-7}	1.0×10^{-7}	1.0×10^{-7}	1.0×10^{-7}

Figure 5 shows the water levels performed, corresponding to the six numerical models that were performed in this study. The dimensions of the mesh were calibrated with a calibration procedure [27] and the final ones were 118.4 × 124.4 × 20.5 m. A total of 31,860 nodes and 35,868 20-node *BrickUP* elements were applied, following the previous contributions [57–59]. As explained in [27], S2 is implemented amongst the soil materials that were considered in Table 1.

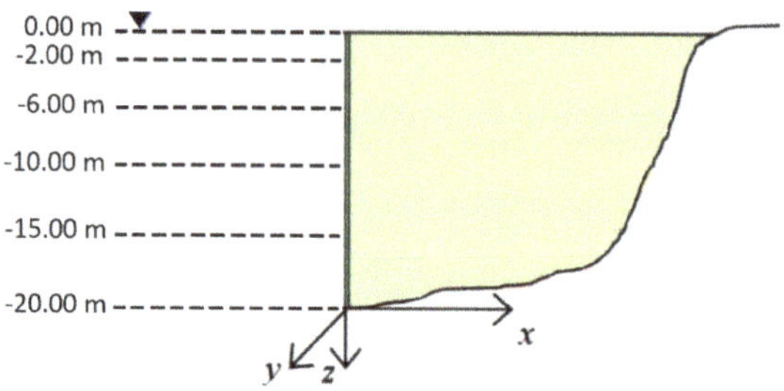

Figure 5. Performed water levels.

Boundary conditions are fundamental when SSI analyses are carried out, especially to represent correctly the development of pore pressures. The penalty method with a tolerance value of 10^{-4} was chosen to ensure strong constrain conditions without computational problems associated with conditioning of the system of equations (more details in [60,61]). Base boundaries (20.5 m depth) were considered rigid and all the directions were constrained. In order to allow shear deformations, longitudinal and transversal directions were left unconstrained in correspondence with the lateral boundaries, while the vertical direction (described by the third Degree of Freedom (DOF) of the nodes) was constrained. Hydraulic conditions were defined on the fourth DOF of each node that were blocked during the gravity analyses (step 1) and then released when hydraulic conditions were applied (more details in [60,61]).

The structure consists of an RC building with infill masonry walls (RCIMW) as a benchmark, in order to represent the Italian residential buildings that were mostly damaged during the 1980 Irpinia–Basilicata earthquake, and already studied in [27]. Figure 6 shows the structural scheme consisting of 4 × 2 columns (4 columns in the transversal direction (spaced 8 m apart) and 2 columns in the longitudinal direction (spaced 10 m apart) and 3 floors (a 3.4 m-storey height, 10.2 m total height). RCMIW is a typical Italian system based on two schemes (Figure 6): RC concrete columns and beams are superposed on vertical masonry walls. It is worth noting that the structural configuration performed here is a simplified version of frame-type buildings because detailed information should have required material characterizations with in situ tests that are almost impossible after 40 years from the earthquake, since the conditions of the materials could have changed significantly. Both vertical and horizontal elements are composed by RC concrete columns and beams, respectively, and characterized by fibre section models. *Concrete02* material (Figure 7a,b [27]) is chosen to model the core and the cover portions. A total of 30 bars are used and represented by *Steel02* material (Figure 7c [27]). The masonry walls are modelled as equivalent diagonal *elastic Beam Column* elements [53,54], in both longitudinal and transversal directions. The masonry walls properties are selected based on the Italian code provisions, as shown in [27]. Table 2 shows the vibration periods of the structure with and without the infill masonry walls, related to the fixed-base condition (no SSI effects included). It is worth seeing that the masonry walls affect the structural natural period (from 0.3012 to 0.2085 s), since they increase the lateral stiffness of the whole structure (as shown in [27]).

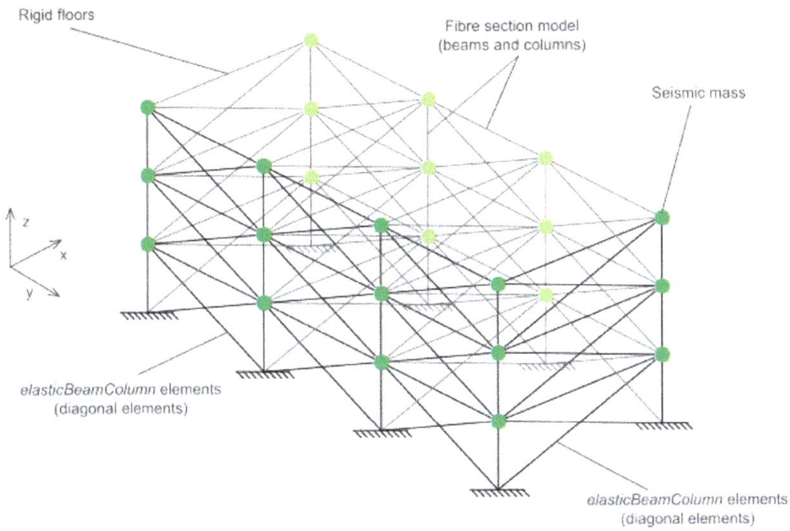

Figure 6. Structural 3D model [27].

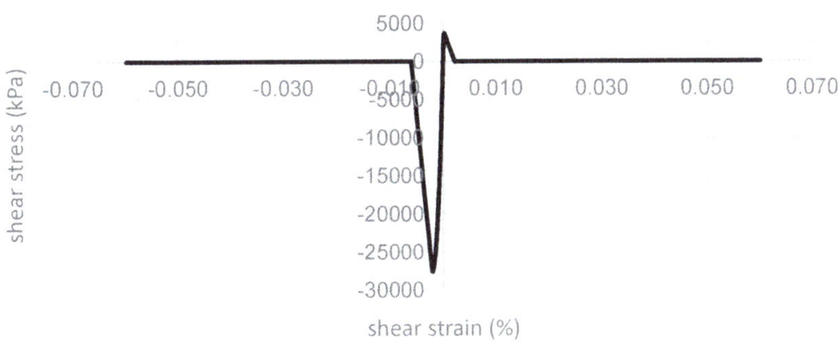

(a)

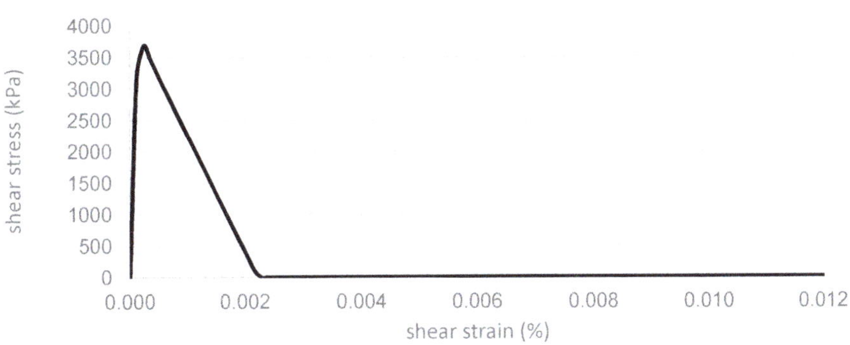

(b)

Figure 7. *Cont.*

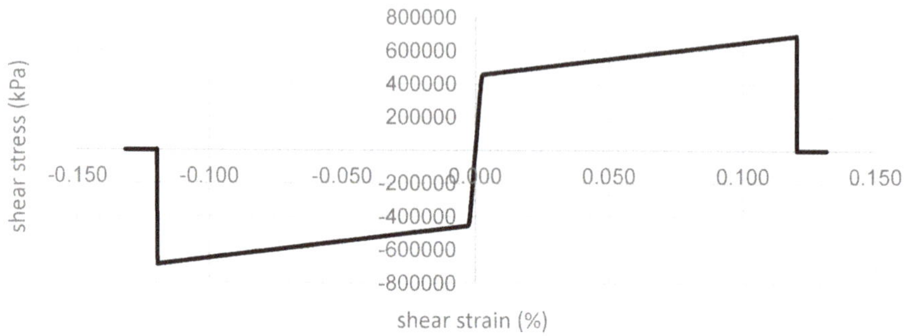

(**c**)

Figure 7. Shear stress vs. shear strain relationship for Concrete02: core (**a**) and cover (**b**) and Steel02 (**c**) [27].

Table 2. Structural periods; reinforced concrete (RC), infill masonry walls (IMWs).

Models	T1 (s)	T2 (s)	T3 (s)
RC	0.301	0.107	0.073
RC with IMWs	0.209	0.074	0.049

The foundation is modelled as a 0.50 m-deep rectangular concrete raft foundation (28.4 × 34.4 m) in order to represent recurring shallow foundation typologies for residential buildings. The foundation considered is assumed to be rigid, by tying all the columns base nodes together with those of the soil domain surface, using *equalDOF* [53,54]. The foundation is modelled with an equivalent concrete material, by applying the *Pressure Independent Multi-Yield* model [53,54] (Table 3). The first 0.5 m-deep soil layer around the foundation is modelled with a backfill defined by the *Pressure Depend Multi Yield* model [53,54] and Table 4 shows the adopted parameters, such as the low-strain shear modulus, the friction angle and the permeability. The number of yield surfaces is equal to 20.

Table 3. Foundation.

Parameters	Concrete
Density (Mg/m^3)	2.4
Reference Shear Modulus (kPa)	1.25×10^7
Reference Bulk Modulus (kPa)	1.67×10^7

Table 4. Infill soil characteristics.

Soil	S1
Density (Mg/m^3)	1.7
Reference shear modulus (kPa)	3.83×10^4
Reference bulk modulus (kPa)	1.50×10^5
Shear wave velocity (m/s)	150
Cohesion (kPa)	5
Friction angle (°)	27
Horizontal permeability (m/s)	1.0×10^{-7}
Vertical permeability (m/s)	1.0×10^{-7}

4. Results

This section shows the results of the models performed, focusing on the performance of the soil, the foundation and the structure. The vertical stresses in the soil depend on the weight of the water for those models where the material is submerged and saturated conditions occur. It is worth noting that the effects connected with the water level can potentially be related with liquefaction occurrence, which is not considered in this paper because no historical relevancies of such phenomenon were found in the literature.

4.1. Soil Results

Significant values of the main soil parameters are herein compared in correspondence with different positions and different water depths, with particular attention to the role of water level. Figures 8–12 show the relationship between the effective confinement pressures and the shear stress at various depths in correspondence with model-10 m that was chosen herein to underline the two different conditions of saturated and dry soil. It is worth noticing the role of the water in generating several levels of confinement and the increase of the shear stress with the soil depth. In order to ensure that liquefaction did not occurred, the pore pressure ratio (ru), defined as the ratio between the total pore pressure and the total overburden pressure [60,61], was calculated and it was verified that the maximum value (0.57) was significantly smaller than 1, the value that is considered for liquefaction occurrence. Figures 7 and 8 show the conditions of dry soil and that the effective confinement pressures depend on the vertical stresses. The role of the water in increasing the effective confinement is shown in Figures 10–12, where it is possible to see the maximum level reached for the several depths: approximately 170, 220 and 250 kPa corresponding to depths of 11.50, 15.50 and 19.50 m, respectively.

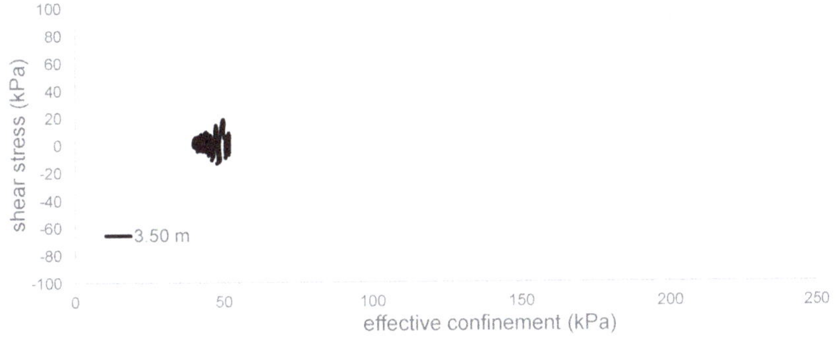

Figure 8. Model-10 m: effective confinement vs. shear stress at 3.50 m depth.

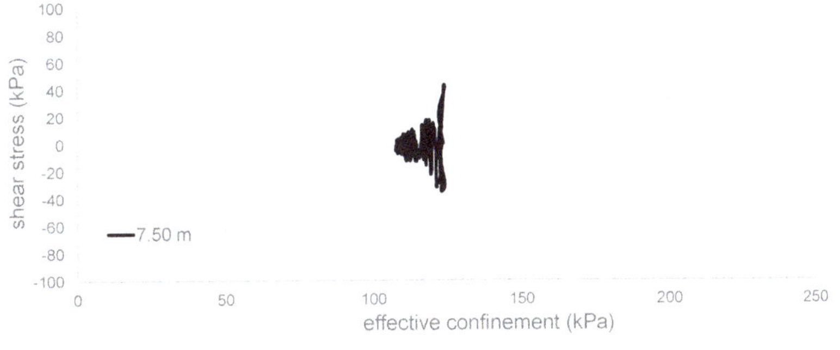

Figure 9. Model-10 m: effective confinement vs. shear stress at 7.50 m depth.

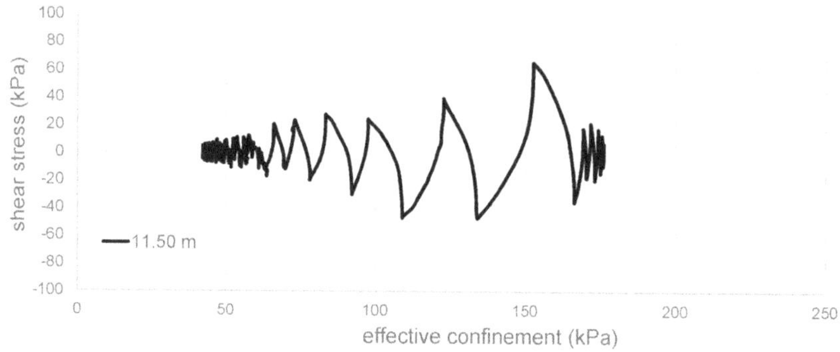

Figure 10. Model-10 m: effective confinement vs. shear stress at 11.50 m depth.

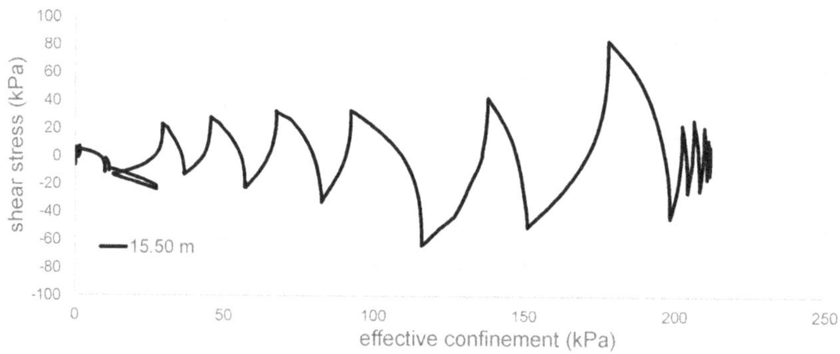

Figure 11. Model-10 m: effective confinement vs. shear stress at 15.50 m depth.

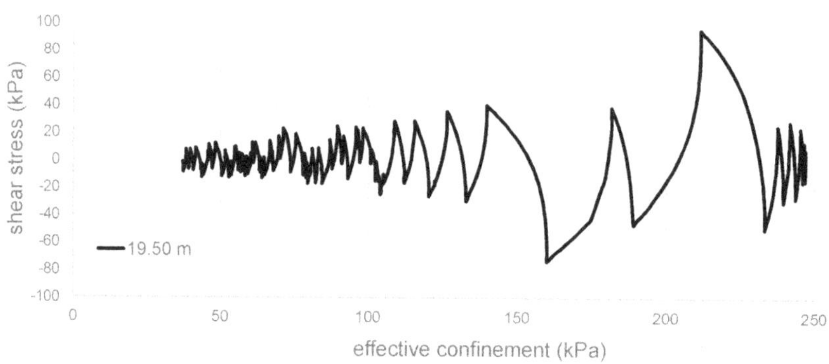

Figure 12. Model-10 m: Effective confinement vs. shear stress at 19.50 m depth.

Figures 13 and 14 show the acceleration time histories in correspondence with the two extreme conditions (fully saturated and dry) at the surface and at −16.5 m depth. It is worth seeing that in the lower layers, dry conditions (water level at −20 m, Figure 14) are shown to be more detrimental while at the surface fully saturated conditions show large acceleration values (Figure 13).

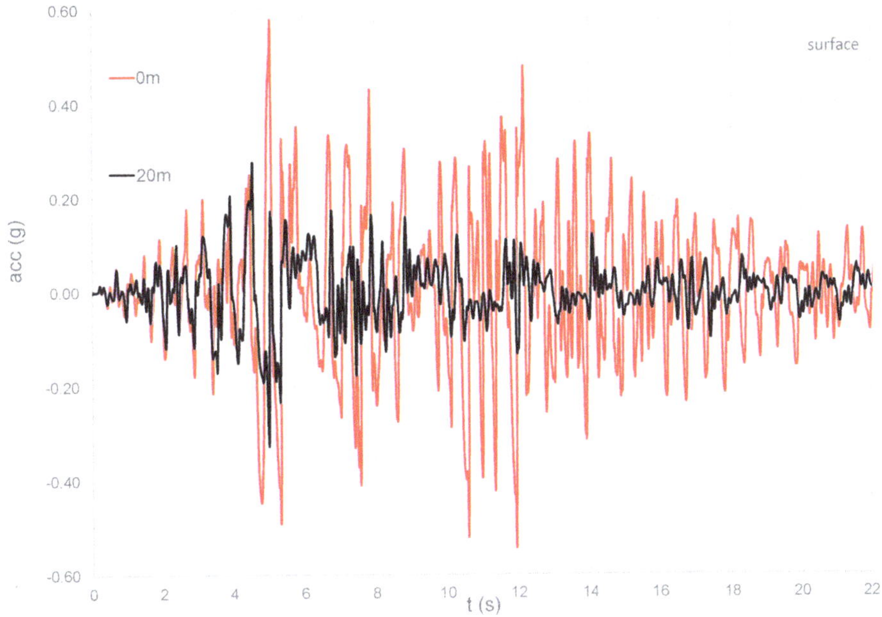

Figure 13. Acceleration time histories at the surface for model-0 m and model-20 m.

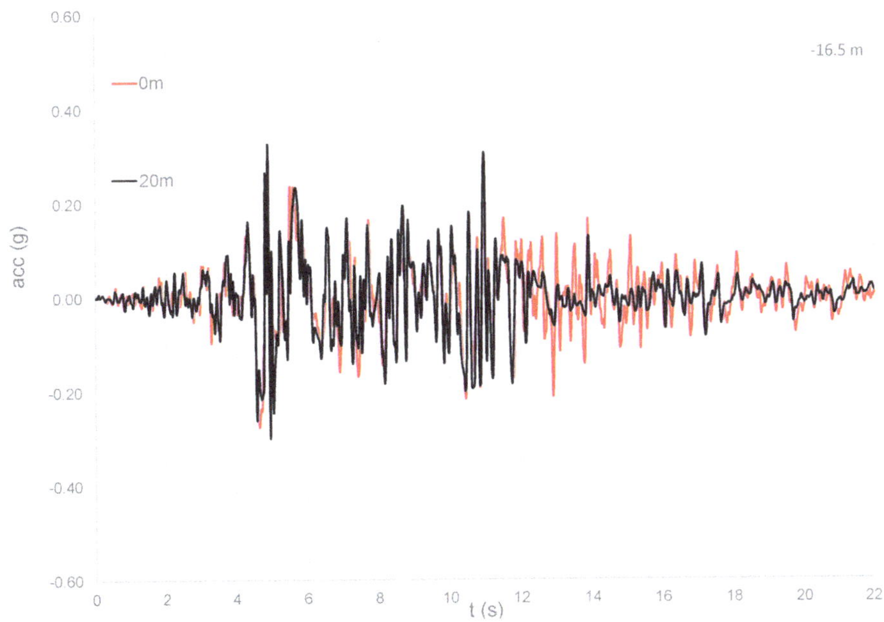

Figure 14. Acceleration time histories at −16.5 m depth model-0 m and model-20 m.

4.2. Foundation Results

Foundation settlements (mainly differential) are one of the most significant causes of damage, depending on several parameters, like material quality, geometry of the structure, amount of openings, type of foundation or the actual state of preservation. Figure 15 shows the time history of absolute settlements (in the centre of the foundation) for the various models, demonstrating the role of the water level depth in the development of settlements. It is worth noting that the settlements start in correspondence with the peak of the input motion (compare with Figure 4); they increase and then they remain stable around the residual values (Table 5). These values were verified to be lower than the condition defined as serviceability level SLS1 (1/25) in the New Zealand code (NZS 1170.0:2002). It is worth considering that the calculated values are not comparable with liquefaction occurrence (compare with the values assessed in [60,61]).

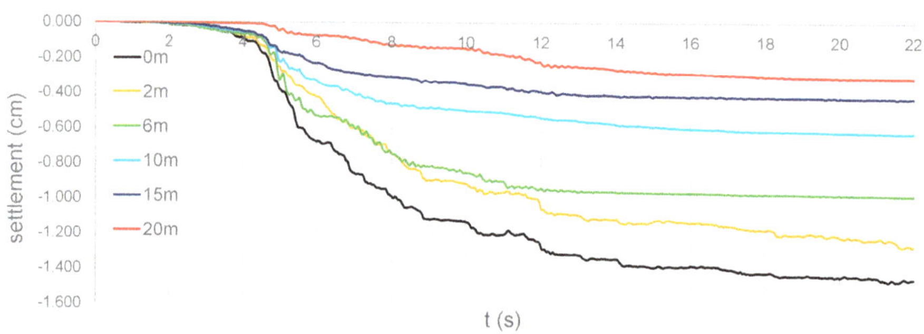

Figure 15. Foundation settlements.

Table 5. Foundation settlements.

Models	Settlement (cm)
Model-0 m	1.49
Model-2 m	1.29
Model-6 m	0.99
Model-10 m	0.64
Model-15 m	0.43
Model-20 m	0.33

Figure 16 shows the normalized foundation tilt time histories, defined as the ratio between the differential foundation settlements and the width (along the longitudinal direction: 15 m). It is worth seeing that the tilts are related with the water level depths (maximum values: 1.20% and 0.60% for water levels at the surface and at −2 m depth, respectively). In addition, the main tilts are shown between 5–12 s, where the water level at the surface is shown to be at the most detrimental condition. At the end of the motion, the values of the tilts (particularly for 0 and −2 m) increase due to shear resistance and associated accumulation of permanent deformations in the soil [52]. It is important to notice that these nonlinear mechanisms were reproduced thanks to the 3D numerical models performed, which implemented advanced material models and procedures developed inside the OpenSees platform [53,54]. The permanent deformations in the soil and the foundation tilts are both responsible for transferring significant stress to the superstructure and thus contribute to the SSI effects (shown in [27,60,61]) and described in the next section.

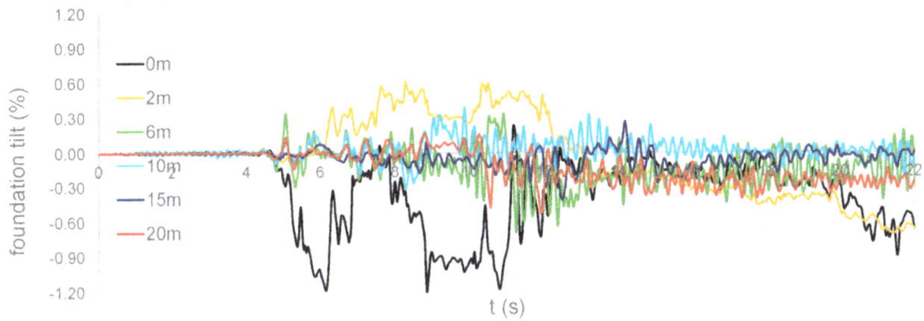

Figure 16. Foundation tilts.

4.3. Structural Results

This section investigates the role of the water level on the structural performance in terms of shear forces at the base of the column and floor displacements at the top of the structure. Figures 17 and 18 show the relationship between the longitudinal displacements and base shear forces in the fully saturated case (0 m) and the dry case (20 m). It is worth noticing the different values of the displacements, due to the levels of deformations that occur in the soil, in correspondence with the two different water levels and that depend on the local soil effects, as discussed in [27]. The fully saturated case shows values of displacement that are more than two times the ones that occur for the dry case. The values of the shear forces are different as well, with maximum values of 390 kN for the case of 0 m (Figure 17 and [27]) and 307 kN for the case of 20 m (Figure 18). In particular, the maximum tensile stresses are approximately 65 kPa (for Model-0 m) and 53 kPa (for model-20 m), corresponding to two different levels of damage and thus the potential collapse mechanisms of the masonry wall [27].

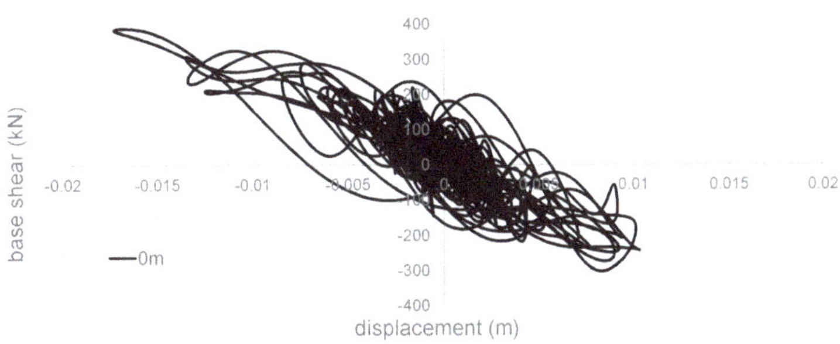

Figure 17. Model-0 m: base shear.

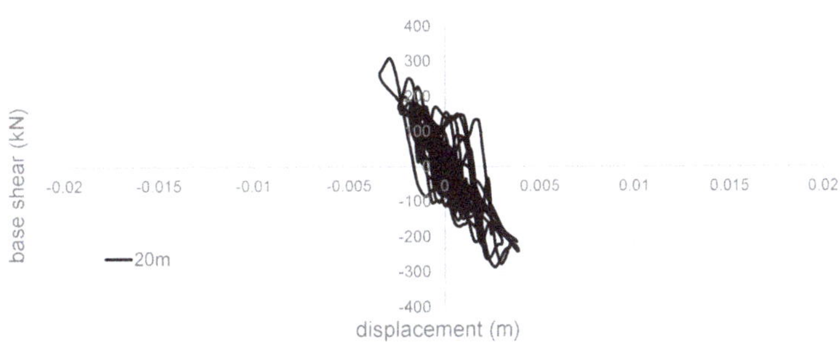

Figure 18. Model-20 m: base shear.

To consider the effects of the structural performance, Figure 19 shows the time histories of the floor displacements (for all the performed models) in correspondence with the top of the building. Table 6 shows the maximum displacements, demonstrating that in the case of fully saturated conditions (0 m), maximum displacements are more than four times the values reached for the dry conditions (20 m). Comparing the results in correspondence with the different levels, it is possible to see that the peak values (around 6 and 12 s) of the displacements are reached for the highest positions of the water level (at the surface and −2 m). Regarding the values of the displacements at the end of the transient, intermediate water level positions (−6, −10 and −15 m) are shown to be the most detrimental conditions. These results may depend on the delay in developing permanent deformations due to the highly nonlinear mechanisms, both in the soil and in the structure. The results also show the importance of performing the 3D advanced models to represent realistically the interaction between soil deformability and structural flexibility.

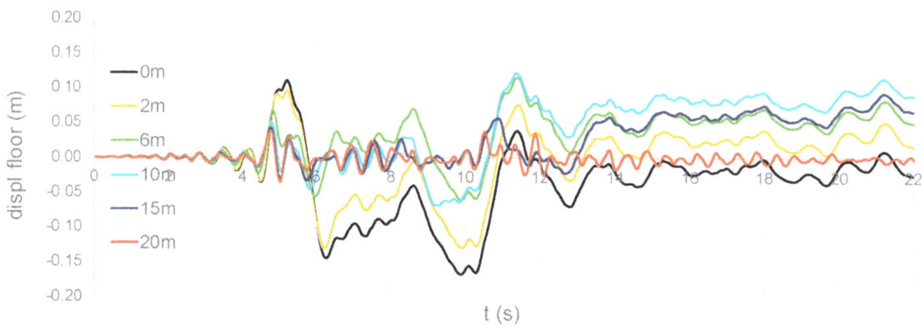

Figure 19. Top displacement time histories.

Table 6. Maximum top displacements.

Models	Max. Displacement (cm)
Model-0 m	1.67
Model-2 m	1.30
Model-6 m	1.15
Model-10 m	1.22
Model-15 m	0.93
Model-20 m	0.39

5. Discussion

The results demonstrated the role of the water level in modifying the response of the entire system (ground–foundation–structure), with particular attention to the consequences on the structure. In this regard, as shown in [62], the presence of the masonry infill walls increases the lateral stiffness and introduces different mechanisms that significantly modify the seismic behaviour, and thus the structural vulnerability of the structure. In particular, the role of the shallow foundation is important in affecting the global stability of the system, being particularly sensitive to the nonlinear soil behaviour (such as gapping, sliding and uplift) that may lead to an unconservative prediction of the superstructure response, as shown in [29]. The role of the water level was demonstrated to be relevant, especially in correspondence with the base floors, which are characterized by large displacements that contribute to the failure of the diagonal elements (representing the masonry infill walls). In this regard, the global behaviour of the system is affected by the complex of mutual nonlinear effects of the soil and of the structure that is generally recognized as SSI, and that depends on the dynamic characteristics of the structure and the foundation soil. By this way, performing 3D advanced nonlinear numerical models was fundamental in order to assess realistic estimations of the structural vulnerability.

6. Conclusions

The paper investigates the role of the water level on a typical Italian building during the 23 November 1980 Irpinia–Basilicata earthquake by performing several 3D numerical models of the entire system (soil–foundation–structure). The finite element models were built with the advanced computational framework OpenSeesPL in order to assess the several mechanisms known globally as soil–structure interaction (SSI). The results show that the shear mechanisms and the consequent permanent deformations inside the soil are driven by the presence of the water. Therefore, knowing the position of the water level (and eventually how it changes during the seasons) is fundamental in order to assess the seismic vulnerability of structural configurations. In the case of masonry buildings, the water level may affect significantly the whole stability of the buildings, both in terms of settlements (absolute or tilts) and the failure mechanisms in correspondence with the structural wall. Although the findings are limited to the specified conditions performed, they may potentially be useful to propose formulations within code provisions.

Funding: This research received no external funding.

Conflicts of Interest: The author declares no conflict of interest.

References

1. Wang, C.Y.; Dreger, D.S.; Wang, C.H.; Mayeri, D.; Berryman, J.G. Field relations among coseismic ground motion, water level change and liquefaction for the 1999 Chi-Chi (Mw 7.5) earthquake, Taiwan. *Geophys. Res. Lett.* **2003**, *30*. [CrossRef]
2. Charlie, W.A.; Doehring, D.O. Groundwater table mounding, pore pressure, and liquefaction induced by explosions: Energy-distance relations. *Rev. Geophys.* **2007**, 45. [CrossRef]
3. Nishiwaki, K.; Yasuhara, K.; Komine, H.; Murakami, S. Displacement and countermeasures for existing structure with rising groundwater levels. In Proceedings of the 8th Japan National Symposium on Environmental Geotechnology, Shanghai, China, 2–5 June 2009; pp. 381–384.
4. Yasuda, S.; Ishikawa, K. Effect of lowering the ground water table as the countermeasure against liquefaction-induced damage to houses. *J. Jpn. Assoc. Earthq. Eng.* **2015**, *15*, 205–219.
5. Hartantyo, E.; Brotopuspito, S.K. Sismanto; Waluyo. Correlation of shallow groundwater levels with the liquefaction occurrence cause by May 2006 earthquake in the south volcanic-clastic sediments Yogyakarta, Indonesia. *Int. J. Appl. Sci.* **2014**, *5*, 1–8.
6. Chung, J.W.; Rogers, J.D. Influence of assumed groundwater depth on mapping liquefaction potential. *Environ. Eng. Geosci.* **2013**, *19*, 377–389. [CrossRef]
7. Santisi d'Avila, M.P.; Lenti, L.; Martino, S.; Romeo, R.W. Nonlinear Numerical Simulation of the Soil Seismic Response to the 2012 Mw 5.9 Emilia Earthquake Considering the Variability of the Water Table Position. *Bull. Seismol. Soc. Am.* **2019**, *109*, 505–524. [CrossRef]
8. Morgan, C.P.; Stolt, M.H. A comparison of several approaches to monitor water-table fluctuation. *Soil Sci. Soc. Am. J.* **2004**, *68*, 562–566. [CrossRef]
9. Calzolari, C.; Ungaro, F. Predicting shallow water table depth at regional scale from rainfall and soil data. *J. Hydrol.* **2012**, *414*, 374–387. [CrossRef]
10. Simonson, G.H.; Boersma, L. Soil morphology and water table relations: II. Correlaion between annual water table fluctuations and profile features. *Soil Sci. Soc. Am. J.* **1972**, *36*, 649–653. [CrossRef]
11. Morgan, C.P.; Stolt, M.H. Soil morphology-water table cumulative duration relationships in Southern New England. *Soil Sci. Soc. Am. J.* **2006**, *70*, 816–823. [CrossRef]
12. He, X.; Vepraskas, M.J.; Skaggs, R.W.; Lindbo, D.L. Adapting a drainage model to simulate water table levels in coastal plain soils. *Soil Sci. Soc. Am. J.* **2002**, *67*, 1722–1731. [CrossRef]
13. Westaway, R.; Jackson, J. Surface faulting in the Southern Italian Campania-Basilicata earthquake of 23 November 1980. *Nature* **1984**, *312*, 436–438. [CrossRef]
14. Cotecchia, V. Ground deformations and slope instability produced by the earthquake of 23 November 1980 in Campania and Basilicata. In Proceedings of the International Symposium Engineering Geology Problems in Seismic Areas, Bari, Italy, 13–19 April 1986; Volume 5, pp. 31–100.

15. Del Prete, M. Examples of mudslides hazard in Southern Apennines (Italy). *Ann. Geofisica* **1993**, *36*, 71–80.
16. Esposito, E.; Gargiulo, A.; Iaccarino, G.; Porfido, S. Distribuzione dei fenomeni franosi riattivati dai terremoti dell'Appennino meridionale. Censimento delle frane del terremoto del 1980. In Proceedings of the International Convention on Prevention of Hydrogeological Hazards, Torino, Italy, 1998; pp. 409–429.
17. Pantosti, D.; Valensise, G. Source geometry and long-term behavior of the 1980 fault based on field geologic observations. *Ann. Geofisica* **1993**, *36*, 41–49.
18. Gizzi, F.T.; Potenza, M.R.; Zotta, C. 23 November 1980 Irpinia–Basilicata earthquake (Southern Italy): Towards a full knowledge of the seismic effects. *Bull. Earthq. Eng.* **2012**, *10*, 1109–1131. [CrossRef]
19. Nunziata, C.; Costa, G.; Marrara, F.; Panza, F. Validted Estimation of Response Spectra for the 1980 Irpinia Earthquake in the Eastern Area of Naples. *Earthquake Spectra* **2000**, *16*. [CrossRef]
20. Ameri, G.; Emolo, A.; Pacor, F.; Gallovič, F. Ground-Motion Simulations for the 1980 M 6.9 Irpinia Earthquake (Southern Italy) and Scenario Events. *Bull. Seismol. Soc. Am.* **2011**, *101*, 1136–1151. [CrossRef]
21. Westaway, R.; Jackson, J. The earthquake of 1980 November 23 in Campania–Basilicata (Southern Italy). *Geophys. J. R. Astron. Soc.* **1987**, *90*, 375–443. [CrossRef]
22. Bernard, P.; Zollo, A. The Irpinia (Italy) 1980 earthquake: Detailed analysis of a complex normal faulting. *J. Geophys. Res.* **1989**, *94*, 1631–1648. [CrossRef]
23. Porfido, S.; Esposito, E.; Michetti, A.M.; Blumetti, A.M.; Vittori, E.; Tranfaglia, G.; Guerrieri, L.; Ferreli, L.; Serva, L. Areal distribution of ground effects induced by strong earthquakes in the Southern Apennines (Italy). *Surv. Geophys.* **2002**, *23*, 529–562. [CrossRef]
24. Porfido, S.; Alessio, G.; Gaudiosi, G.; Nappi, R.; Spiga, E. The resilience of some villages 36 years after the Irpinia-Basilicata (Southern Italy) 1980 earthquake. In Proceedings of the 4th WLF, Ljubljana, Slovenia, 29 May–2 June 2017; Mikoš, M., Vilímek, V., Yin, Y., Sassa, K., Eds.; Springer: Cham, Switzerland, 2017; pp. 121–133. [CrossRef]
25. Pingue, F.; De Natale, G. Fault mechanism of the 40 seconds subevent of the 1980 Irpinia (Southern Italy) earthquake from levelling data. *Geophys. Res. Lett.* **1993**, *20*, 911–914. [CrossRef]
26. Ascione, A.; Mazzoli, S.; Petrosino, P.; Valente, E. A decoupled kinematic model for active normal faults: Insights from the 1980, MS = 6.9 Irpinia earthquake, Southern Italy. *Geol. Soc. Am. Bull.* **2013**, *125*, 1239–1259. [CrossRef]
27. Mina, D.; Forcellini, D. Soil–Structure Interaction Assessment of the 23 November 1980 Irpinia Basilicata Earthquake. *Geosciences* **2020**, *10*, 152. [CrossRef]
28. Khosravikia, F.; Mahsuli, M.; Ghannad, M.A. The effect of soil—Structure interaction on the seismic risk to buildings. *Bull. Earthq. Eng* **2018**, *16*, 3653–3673. [CrossRef]
29. Cavalieri, F.; Correia, A.A.; Crowley, H.; Pinho, R. Dynamic soil-structure interaction models for fragility characterisation of buildings with shallow foundations. *Soil Dyn. Earthq. Eng.* **2020**, *132*, 106004. [CrossRef]
30. Dashti, S.; Bray, J.D.; Pestana, J.M.; Riemer, M.; Wilson, D. Mechanisms of seismically induced settlement of buildings with shallow foundations on liquefiable soil. *J. Geotech. Geoenviron. Eng. ASCE* **2010**, *136*, 151–164. [CrossRef]
31. Gizzi, F.; Masini, N. Il terremoto irpino del 23 Novembre 1980 a Tursi: Dal dato parametrico a quello descrittivo. In *Tursi, La Rabatana*; Fonseca, C.D., Ed.; Fondazione Sassi Matera: Matera, Italy, 2004; pp. 305–316.
32. Benedetti, D.; Petrini, V. On seismic vulnerability of masonry buildings: Proposal of an evaluation procedure. *L'Industria Costr.* **1984**, *18*, 66–78.
33. Postpischl, D.; Branno, A.; Esposito, E.; Ferrari, G.; Marturano, A.; Porfido, S.; Rinaldis, V.; Stucchi, M. The Irpinia earthquake of November 23, 1980. *Atlas Isoseismal Maps Ital. Earthq.* **1985**, *114*, 152–157.
34. Porfido, S.; Esposito, E.; Vittori, E.; Tranfaglia, G.; Michetti, A.M.; Blumetti, M.; Ferreli, L.; Guerrieri, L.; Serva, L. The geological evidence for earthquakes induced effects in the Southern Apennines (Italy). *Surv. Geophys.* **2002**, *23*, 529–562. [CrossRef]
35. Locati, M.; Camassi, R.; Rovida, A.; Ercolani, E.; Bernardini, F.; Castelli, V.; Caracciolo, C.H.; Tertulliani, A.; Rossi, A.; Azzaro, R.; et al. *DBMI15, the 2015 Version of the Italian Macroseismic Database*; INGV: Rome, Italy, 2016. [CrossRef]
36. Porfido, S.; Alessio, G.; Avallone, P.; Gaudiosi, G.; Lombardi, G.; Nappi, R.; Salvemini, R.; Spiga, E. The 1980 Irpinia-Basilicata earthquake: The environmental phenomena and the choices of reconstruction. In Proceedings of the EGU General Assembly 2016, Vienna, Austria, 17–22 April 2016.

37. Serva, L.; Esposito, E.; Guerrieri, L.; Porfido, S.; Vittori, E.; Comerci, V. Environmental Effects from some historical earthquakes in Southern Apennines (Italy) and macroseismic intensity assessment. Contribution to INQUA EEE scale project. *Quat. Int.* **2007**, *173*, 30–44. [CrossRef]
38. Porfido, S.; Esposito, E.; Guerrieri, L.; Vittori, E.; Tranfaglia, G.; Pece, R. Seismically induced ground effects of the 1805, 1930 and 1980 earthquakes in the Southern Apennines, Italy. *Ital. J. Geosci.* **2007**, *126*, 333–346.
39. Blumetti, A.M.; Esposito, E.; Ferreli, L.; Michetti, A.M.; Porfido, S.; Serva, L.; Vittori, E. New data on the novembre 23, 1980, M 6.9, Irpinia-Lucania earthquake (Southern Apennine) coseismic surface effects. *Studi Geol. Camerti* **2002**, *2002*, 19–27.
40. Carulli, G.B.; Migliacci, A.; Onofri, R.; Porfido, S. Indagini geologiche ed ingegneristiche in prospettiva sismica a S. Michele di Serino (AV). *Rend. Della Soc. Geol. Ital.* **1981**, *4*, 161–164.
41. Cotecchia, V.; Nuzzo, G. Hydrological study of the upper valley of the Sele and Ofanto River struck by the Novembre 23, 1980 earthquake. *Geol. Appl. Idrogeol.* **1986**, *21*, 65–95.
42. Esposito, E.; Pece, R.; Porfido, S.; Tanfaglia, G. Hydrological anomalies connected to earthquakes in Southern Apennines (Italy). *Nat. Hazards Earth Syst. Sci.* **2001**, *1*, 137–144. [CrossRef]
43. Galli, P. New empirical relationships between magnitude and distance for liquefaction. *Tectonophysics* **2000**, *324*, 169–187. [CrossRef]
44. Rossi, M. *Situazione, Problemi e Prospettive Dell'area più Colpita dal Terremoto del 23/11/1980*; Einaudi: Torino, Italy, 1981.
45. Verderosa, A. Legge 219/81 Ovvero la Distruzione Legalizzata Dei Centri Storici. In Civiltà Altirpina; Trimestrale, Anno I/lug-dic.90; 1990. Available online: http://verderosa.files.wordpress.com/2008/01/civilto-altirpinia1990.pdf (accessed on 13 June 2020).
46. AAVV. *Indagini di Microzonazione Sismica CNR-PFG*; CNR-PFG: Rome, Italy, 1983; p. 221.
47. Gimma, M.G. *I Piani di Recupero nei Centri Storici, in Atti del Convegno e Mostra Roma*; Bema Editrice: Roma, Italy, 1987.
48. Giuffrè, A. Centri storici in Zona Sismica, Analisi Tipologica Della Danneggiabilità e tecnichedi Intervento Conservativo, Castelvetere sul Calore; in Studi e ricerche. Sulla sicurezza sismica dei Monumenti. Master's Thesis, Università La Sapienza, Facoltà di Architettura, Roma, Italy, 1988.
49. Ordine Architetti Avellino. *Immagini di architettura 1980–1990*; Jacelli: Avellino, Italy, 1991.
50. ITACA. ITalian ACcelerometric Archive (1972–2011); Version1.1; 2011. Available online: http://itaca.mi.ingv.it/ItacaNet/ (accessed on 13 June 2020).
51. Zienkiewicz, O.C.; Chan, A.H.C.; Pastor, M.; Paul, D.K.; Shiomi, T. Static and dynamic behavior of soils: A rational approach to quantitative solutions: I. Fully saturated problems. *Proc. R. Soc. Lond. Ser. A* **1990**, *429*, 285–309.
52. Elgamal, A.; Lu, J.; Forcellini, D. Mitigation of Liquefaction-Induced lateral deformation in sloping stratum: Three-dimensional Numerical Simulation. *J. Geotech. Geoenvironmental Eng.* **2009**, *135*, 1672–1682. [CrossRef]
53. Lu, J.; Elgamal, A.; Yang, Z. OpenSeesPL: 3D Lateral Pile-Ground Interaction, User Manual, Beta 1.0; 2011. Available online: http://soilquake.net/openseespl/ (accessed on 13 June 2020).
54. Mazzoni, S.; McKenna, F.; Scott, M.H.; Fenves, G.L. *Open System for Earthquake Engineering Simulation, User Command-Language Manual*; OpenSees Version 2.0; Pacific Earthquake Engineering Research Center, University of California, Berkeley: Berkeley, CA, USA, 2009; Available online: http://opensees.berkeley.edu/OpenSees/manuals/usermanual (accessed on 13 June 2020).
55. Yang, Z.; Elgamal, A.; Parra, E. A computational model for cyclic mobility and associated shear deformation. *J. Geotech. Geoenviron. Eng.* **2003**, *129*, 1119–1127. [CrossRef]
56. Kramer, S.L. *Geotechnical Earthquake Engineering*; International Series in Civil Engineering and Engineering Mechanics; William, J., Ed.; Prentice-Hall: Upper Saddle River, NJ, USA, 1996.
57. Forcellini, D.; Gobbi, S.; Mina, D. Numerical Simulations of Ordinary Buildings with Soil Structure Interaction. In Proceedings of the SEMC 2016, Alphose Zingoni, Insights and Innovations in Structural Engineering, Mechanics and Computation, Cape Town, South Africa, 5–7 September 2016; pp. 364–369, ISBN 978-1-138-02927-9.
58. Forcellini, D. Cost Assessment of isolation technique applied to a benchmark bridge with soil structure interaction. *Bull. Earthq. Eng.* **2017**. [CrossRef]
59. Forcellini, D. Seismic Assessment of a benchmark based isolated ordinary building with soil structure interaction. *Bull. Earthq. Eng.* **2018**. [CrossRef]

60. Forcellini, D. Numerical simulations of liquefaction on an ordinary building during Italian (20 May 2012) earthquake. *Bull. Earthq. Eng.* **2019**. [CrossRef]
61. Forcellini, D. Soil-structure interaction analyses of shallow-founded structures on a potential-liquefiable soil deposit. *Soil Dyn. Earthq. Eng.* **2020**, *133*, 106108. [CrossRef]
62. Furtado, A.T.; Rodrigues, H.; Arede, A.; Varum, H.; Grubisic, M.; Sipos, T.K. Prediction of the earthquake response of a three-storey infilled RC structure. *Eng. Struct.* **2018**, *171*, 214–235. [CrossRef]

geosciences

MDPI

Article

Soil–Structure Interaction Assessment of the 23 November 1980 Irpinia-Basilicata Earthquake

Daniele Mina [1] and Davide Forcellini [2,*]

[1] School of Engineering and Built Environment, Griffith University, Engineering Building (G09), Engineering Drive, Southport 4222, Australia; minad@live.it

[2] Department of Civil and Environmental Engineering, Te Whare Wānanga o Tāmaki Makaurau-University of Auckland, 20 Symonds Street, Auckland 1010, New Zealand

* Correspondence: dfor295@aucklanduni.ac.nz

Received: 17 March 2020; Accepted: 20 April 2020; Published: 22 April 2020

Abstract: This paper aimed to present a systematic study of the effects caused by the strong earthquake that struck southern Italy on 23 November 1980 (Ms = 6.9) and affected the Campania and Basilicata regions. Two aspects are discussed here: The broadening of the knowledge of the response site effects by considering several soil free-field conditions and the assessment of the role of the soil–structure interaction (SSI) on a representative benchmark structure. This research study, based on the state-of-the-art knowledge, may be applied to assess future seismic events and to propose new original code provisions. The numerical simulations were herein performed with the advanced platform OpenSees, which can consider non-linear models for both the structure and the soil. The results show the importance of considering the SSI in the seismic assessment of soil amplifications and its consequences on the structural performance.

Keywords: Irpinia-Basilicata earthquake; seismic assessment; soil structure interaction; numerical simulations; OpenSees

1. Background

The 23 November 1980 Irpinia–Basilicata (Southern Italy) earthquake (Ms = 6.9) caused deep changes in the urban socio-economic layout, and primary and secondary effects that brought about changes to the natural environment, such as landslides (e.g., Senerchia, Buoninventre, Caposele, Calitri, San Giorgio La Molara, and Grassano) [1–4]. It consisted of several rupture episodes, which occurred at 0.18 and 40 s from the foreshock, and it was assigned a surface-wave magnitude of Ms = 6.9 [5,6]. A wide area (about 3500 km^2) recorded serious damage, many casualties, and 15 localities were almost destroyed, including Sant'Angelo dei Lombardi, Laviano, Lione, Santomenna, Senerchia, Pescopagano, and Balvano. It was estimated that of a total of approximately 1.85 million buildings involved in the event, 75,000 were destroyed, 275,000 seriously damaged, and 480,000 slightly damaged [6].

With respect to this event, the documentary sources are based on two main typologies of technical data preserved in local archives: The "Scheda A" and "Scheda B", which report the damages to the buildings, consisting mostly of reinforced concrete (RC) structures characterized by infill masonry walls (IMWs), which are representative of the Italian residential buildings. Eight damage levels were defined by considering the action to be undertaken, such as repairing works, evacuation, or demolition [6]. Other important documents are the recovery plans (named "Piani di Recupero") of the historical centers, the other sources used to analyze the outcomes of the earthquake at the urban scale. An important study regarding the effects of spectral accelerations was proposed by [7], who analyzed the effects of the soil on the accelerations in several locations, with particular attention to the Naples area. In addition, [8] simulated the recorded strong-motion data by computing spectral accelerations and peak amplitude residual distributions in order to investigate the influence of site effects and compute synthetic ground motions around the fault. They simulated the expected ground motions varying the hypocenters, the rupture velocities, and the slip distributions and compared the median ground motions and related standard deviations from all scenario events with empirical ground-motion prediction equations (GMPEs). Recent earthquakes, such as the Athens (Greece, 1999) [9], the Kocaeli (Turkey, 1999) [10], the Haiti (2010) [11], and the Gorkha (Nepal, 2015) [12–14] earthquakes, showed the importance of taking into account soil amplifications. In the literature, several approaches have been applied to perform ground motion analyses including site effects: hybrid analyses that consist of a combination of probabilistic and deterministic methods (e.g., [15,16]), convolution approaches that provide modifications of the rocking hazard (e.g., [17,18]), and 1D seismic site response analyses (e.g., [19,20]).

Even if the 1980 Irpinia–Basilicata (southern Italy) earthquake is well documented with several contributions (e.g., [21–24]) and models proposed [25,26], the assessment of the role of the soil on the structural damage is still a relatively unexplored issue and this paper aimed to fill this gap. In particular, the principal aim was to propose numerical simulations of different soil conditions and assess the effects of the soil–structure interaction (SSI), which can significantly affect the seismic vulnerability of structures [27–29]. In this regard, when the superficial deposits overlie the bedrock, amplifications of the surface seismic accelerations may not be conservatively predicted by the codes. The so-called site effects consist of a combination of soil and topographical effects, which can modify (amplify and attenuate) the characteristics (amplitude, frequency content, and duration) of the incoming wave field and are primarily based on the geotechnical properties of the subsurface materials [30]. In particular, the response of the superficial layers is strongly influenced by the uncertainty associated to the definition of the soil properties and model parameters that are fundamental to assess the well-known mechanism of seismic amplifications of ground motion [31]. Therefore, accounting for the amplification effects of superficial layers has become critically important in seismic design [32] and widely adopted in many codes' prescriptions, such as Eurocode 8 [33], ASCE (American Association of Civil Engineering) standards 7-05 [34], and 4-98 [35]. These codes provide soil parameters, generally determined through geological investigations [36–40], that can largely vary even within the same area [41,42]. The methodology followed in this paper consists of a first step, where free-field (FF) analyses were computed on several layers of soil, and secondly, an SSI (Soil-Structure Interaction) analysis was performed on a selected structural configuration that is representative of the buildings that were damaged during the Irpinia-Basilicata earthquake.

2. Case Study

SSI analyses require the definition of geomechanical parameters that are fundamental to describe the dynamic soil behavior, such as the modulus reduction and damping curves (see [43,44]). According to the current state of knowledge on the Irpinia-Basilicata earthquake, strength parameters for superficial layers are not available. Therefore, it was necessary to select representative values based on available information, such as [8], for a preliminary study. These values are herein determined with free-field analyses since the actual values at each building site will slightly differ when the building characteristics are considered. In particular, the present paper aimed to model a low-rise building based on a relatively shallow foundation assuming that the ground motion amplitude, which decreases at the foundation level with respect to the free field, may be negligible [42].

The study here proposed was divided into two steps. First of all, several FF models with different soil conditions were considered (Figure 1), in order to study the effects of soil deformability on the amplification of the motion. In particular, four incoherent soils were performed on the basis of the contributions that were found in the literature. Then, a complete 3D numerical model with the soil-foundation-structure system was performed (Figure 2). The FF soil models consist of a one-layer 20-m-deep homogenous incoherent material with a 3D mesh (Figure 1). The penalty method was adopted for the boundary conditions (tolerance of 10^{-4}), chosen as a compromise for the soil domain definition, which was modelled large enough to ensure strong constraint conditions but not too large in order to avoid problems associated with the equations system conditions. Base boundaries (depth of 20 m) were considered as rigid. Base and lateral boundaries vertical direction (described by the third degree of freedom (DOF)) were constrained, while longitudinal and transversal directions were left unconstrained on the lateral boundaries, in order to allow shear deformations of the soil. The definition of the mesh elements dimension follows the approach already adopted [45–48] and, in order to verify proper simulation of FF conditions, accelerations at the top of the mesh were compared with the FF ones, which were found to be identical, confirming the effective performance of the mesh. The benchmark structure was calibrated in order to be representative of the buildings that were present in Irpinia in 1980. In this regard, a 3-storey concrete building with masonry walls was considered.

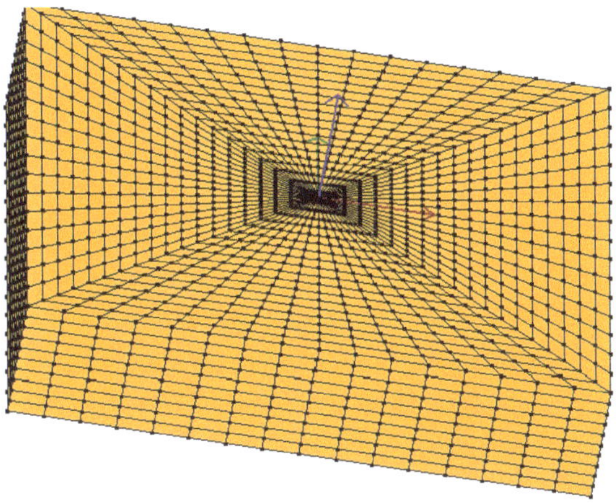

Figure 1. Mesh 1: free-field analysis.

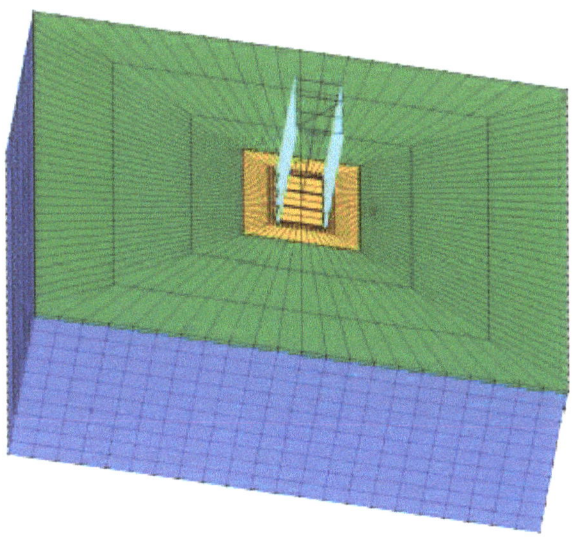

Figure 2. Mesh 2: SSI model; uniform soil layer (blue), infill (green), and foundation (yellow).

Dynamic analyses were performed with OpenSeesPL. The selected input motion was chosen from the Italian Accelerometric Archive [49] and it represents the acceleration registered in Sturno (STN) station (lat: 41.0183°, long: 15.1117°) in Avellino, Campania (Figure 3), and located less than 5 km from the fault and 33 km from the epicenter (41.76°N, 15.31°E). For more details, see [8]. The input was defined on soil B, as classified by Eurocode 8, and applied at the base of the model along the longitudinal direction.

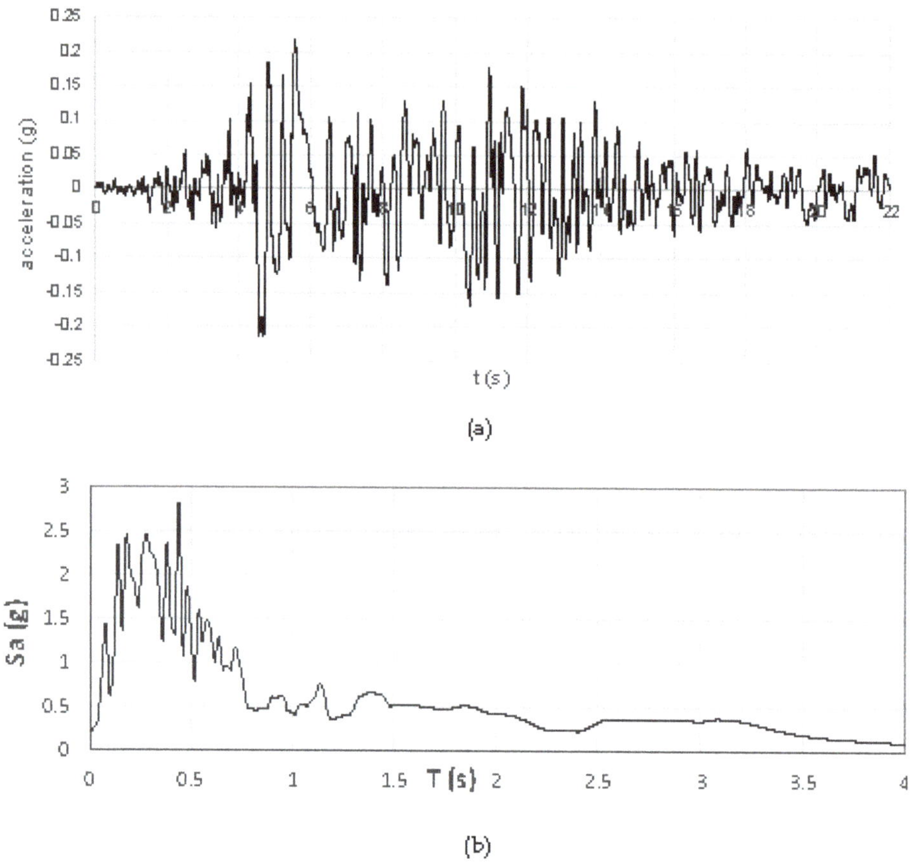

Figure 3. Selected input motion: acceleration (**a**) and spectrum (**b**).

2.1. Step 1: FF Analyses

The soil models were built up on a two-phase material following the *u-p* formulation [50], where *u* is the displacement of the soil skeleton and *p* is the pore pressure. The soil material was based on the following assumptions: (1) Small deformations and rotations, as well as solid and fluid densities remain constant in both time and space; (2) porosity is locally homogeneous and constant with time; (3) soil grains are incompressible; and (4) solid and fluid phases are accelerated equally [51]. The 20-m-deep soil layer was defined by the *PressureDependMultiYield02* model [52,53], based on the multi-yield-surface plasticity framework developed by [54], in order to reproduce the mechanism of cycle-by-cycle permanent shear strain accumulation in clean sands (Figure 4). Table 1 shows the adopted parameters, such as the low-strain shear modulus and friction angle, as well as the shear wave velocities and permeability. Soil fundamental periods were estimated considering an equivalent uniform linear layer, following [55]. The number of yield surfaces was equal to 20 for all soil models. Figure 5 shows the backbone curves for all the selected soil models.

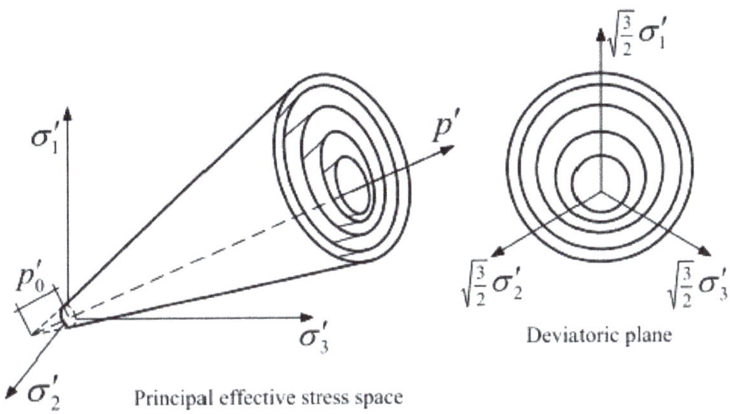

Figure 4. Multi-yield surfaces in principal stress space and deviatoric plane [53].

Table 1. Soil characteristics.

Soil	S1	S2	S3	S4
Density [Mg/m^3]	1.7	1.9	1.9	2.1
Reference Shear Modulus [kPa]	3.83×10^4	4.28×10^4	5.50×10^4	1.32×10^5
Reference Bulk Modulus [kPa]	1.50×10^5	2.00×10^5	2.00×10^5	3.90×10^5
Shear wave velocity [m/s]	150	150	170	250
Soil fundamental period [s]	0.53	0.53	0.47	0.32
Cohesion [kPa]	5	5	5	5
Friction angle [°]	27	29	35	40
Hor. Permeability [m/s]	1.0×10^{-7}	1.0×10^{-7}	1.0×10^{-7}	1.0×10^{-7}
Ver. Permeability [m/s]	1.0×10^{-7}	1.0×10^{-7}	1.0×10^{-7}	1.0×10^{-7}

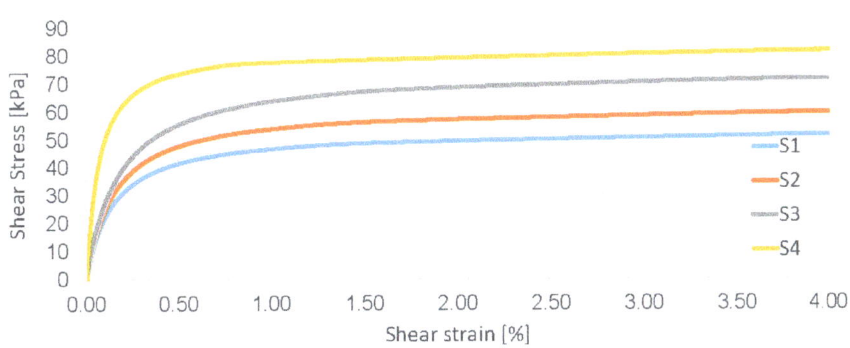

Figure 5. Backbone curves (selected soils).

The 3D soil models consist of a 100 m × 100 m × 20 m mesh, built up with 8000 20-node *BrickUP* elements and 9163 nodes to simulate the dynamic response of solid-fluid fully coupled material [52,53]. For each *BrickUP* element, 20 nodes describe the solid translational degrees of freedom, while the eight nodes on the corners represent the fluid pressure 4 degrees of freedom. For each node, Degree of Freedom (DOF)s 1, 2, and 3 represent solid displacement (u) and DOF 4 describes fluid pressure (p), which were recorded using OpenSees Node Recorder [52,53] at the corresponding integration points. The element dimension increases from the structure (center of the model) to the lateral boundaries, which were modelled to behave in pure shear and located far away from the center of the mesh.

2.2. Step 2: SSI Analyses

The study considered an RC structure with infill masonry walls as a benchmark, in order to represent the Italian residential buildings that were mostly damaged during the 1980 Irpinia–Basilicata earthquake. The benchmark structure was built with a 4 × 2 column scheme (4 columns in the transversal direction (8 m spaced) and 2 in the longitudinal direction (10 m spaced)) and modelled to have periods in the range of those of residential buildings, considering 3 floors (a 3.4 m storey height, with a total structure height of 10.2 m). The structure was modelled as a superposition of two schemes (Figure 6). Both vertical and horizontal elements were composed by RC concrete columns and beams, respectively, and characterized by fiber section models. *Concrete02* material [56,57] was chosen to model the core and the cover portions (Figure 7a,b, respectively) of the section (0.40 m × 0.40 m) and with the parameters defined in Table 2. The ratio between the unloading slope (related to the maximum strength) and the initial slope was taken as equal to 0.1. A total of 30 bars were used and represented by *Steel02* material [58], with the properties shown in Table 3 and the ratio between the post-yield tangent and initial elastic tangent equal to 0.01 (Figure 8). The parameters that control the transition from elastic to plastic branches were assumed R0 = 15, CR1 = 0.925 and CR2 = 0.15, as suggested by [53]. The masonry walls were modelled as equivalent diagonal *elasticBeamColumn* elements [52,53], in both the longitudinal and transversal directions. The masonry walls' properties were selected based on the Italian code provisions, with low-to-medium mechanical characteristics (Table C8A.2.1 [59]), as shown in Table 4. Table 5 shows the vibration periods of the structure with and without the infill masonry walls. It is worth noting that the masonry walls affect the structural natural period (from 0.3012 s to 0.2085 s), since they increase the lateral stiffness of the whole structure (as shown in [60]). In particular, the infill masonry walls introduce different mechanisms that may significantly modify the seismic behavior of the structure. The foundation was modelled as a 0.50-m-deep rectangular concrete raft foundation (28.4 m × 34.4 m) in order to represent the recurring shallow foundation typologies for residential buildings. These types of foundation can be particularly vulnerable due to their bearing capacity, which depends only on the contact pressure and not on the frictional mechanisms (as in the case of deep foundations). The considered foundation was assumed to be rigid, by tying all the columns base nodes together with those of the soil domain surface, using *equalDOF* [52,53]. Horizontal rigid beam-column links were set normal to the column longitudinal axis to simulate the interface between the column and the foundation. The foundation was designed by calculating the eccentricity (the ratio between the overturning bending moment at the foundation level and the vertical forces) in the most detrimental condition of the minimum vertical loads (gravity and seismic loads) and maximum bending moments. The foundation was modelled with an equivalent concrete material, by applying the *Pressure Independent Multi-Yield* model [52,53] (Table 6). This model consists of a non-linear hysteretic material with a Von Mises multi-surface kinematic plasticity model, which can simulate a monotonic or cyclic response of materials whose shear behavior is insensitive to the confinement change. The nonlinear shear stress-strain backbone curve is represented by the hyperbolic relation, defined by the two material constants (low-strain shear modulus and ultimate shear strength) [52,53].

Table 2. Concrete02 (core and cover).

Concrete02	Core	Cover
Compressive strength at 28 days [kPa]	-4.630423×10^4	-2.757900×10^4
Strain at maximum strength [%]	-3.484000×10^{-3}	-2.00000×10^{-3}
Crushing strength [kPa]	-4.466062×10^4	0
Strain at crushing strength [%]	-3.572200×10^{-3}	-6.0000×10^{-3}
Tensile strength [kPa]	6.482592×10^3	3.861060×10^3
Tension softening stiffness [kPa]	1.860438×10^6	1.930530×10^6

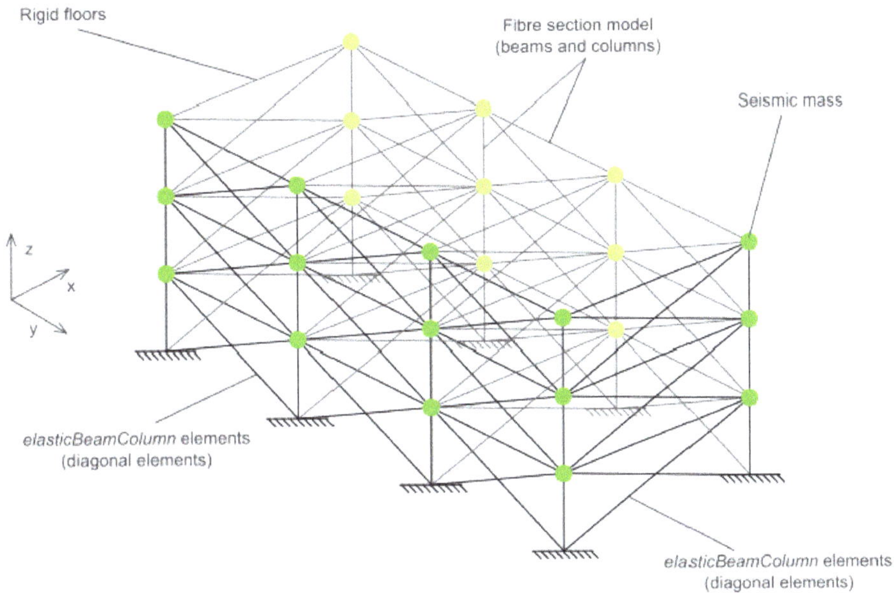

Figure 6. Structural 3D model.

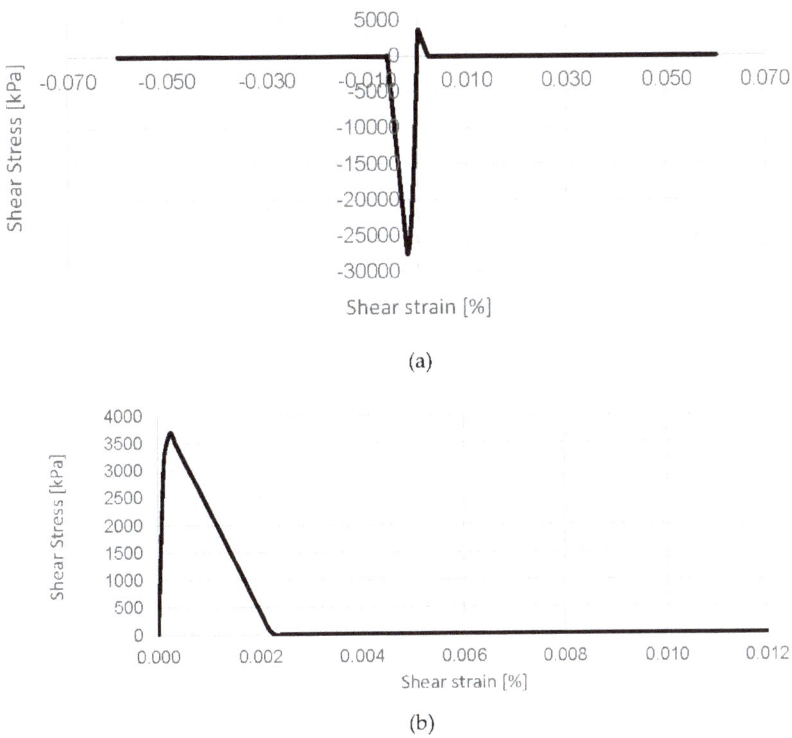

Figure 7. Shear stress vs. shear strain relationship for Concrete02: core (**a**) and cover (**b**).

Table 3. Steel02 (bars).

Steel02	Core	Cover
Yield Strength [kPa]	4.55054×10^5	-2.757900×10^4
Initial elastic tangent [kPa]	2.00000×10^8	-2.00000×10^{-3}

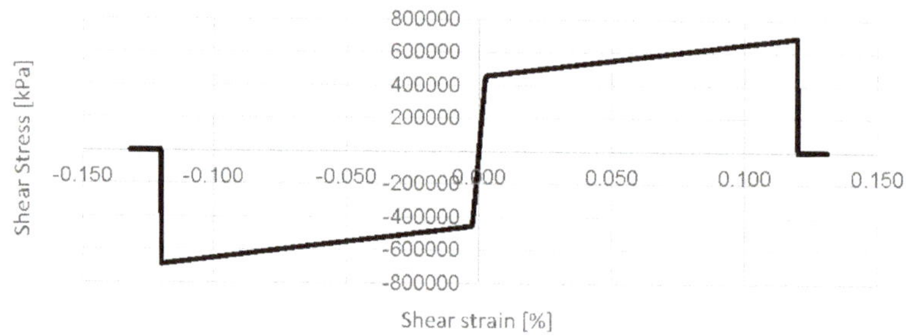

Figure 8. Shear stress vs. shear strain relationship for Steel02 (bars).

Table 4. Masonry characteristics [53].

Diagonals	Masonry
Density [Mg/m^3]	1.8
Young Modulus [kPa]	1.50×10^6
Shear Modulus [kPa]	5.00×10^5
Compressive strength [kPa]	3.00×10^3
Shear strength [kPa]	70

Table 5. Structural periods.

Models	T1 [s]	T2 [s]	T3 [s]
RC	0.301	0.107	0.073
RC with IMWs	0.209	0.074	0.049

Table 6. Foundation.

Parameters	Concrete
Density [Mg/m^3]	2.4
Reference Shear Modulus [kPa]	1.25×10^7
Reference Bulk Modulus [kPa]	1.67×10^7

The 3D soil models consist of a 118.4 m × 124.4 m (20.5 m thick) mesh, built up with 31,860 nodes and 35,868 20-node *BrickUP* elements to simulate the dynamic response of solid-fluid fully coupled material [52,53] and with the same assumptions considered for the free-field models (Section 2.1). As explained in Section 3.1, S2 was implemented amongst the soil materials that were considered in step 1. The first 0.5-m-deep soil layer around the foundation was modelled with a backfill defined by the *PressureDependMultiYield* [52,53] model, based on the multi-yield-surface plasticity framework developed by [54]. Table 7 shows the adopted parameters, such as the low-strain shear modulus, the friction angle, and the permeability. The number of yield surfaces was equal to 20. Figure 9 shows the backbone curves.

Table 7. Infill soil characteristics.

Soil	S1
Density [Mg/m^3]	1.7
Reference Shear Modulus [kPa]	3.83×10^4
Reference Bulk Modulus [kPa]	1.50×10^5
Soil fundamental period [s]	150
Cohesion [kPa]	0.53
Friction angle [°]	5
Hor. Permeability [m/s]	27
Ver. Permeability [m/s]	1.0×10^{-7}

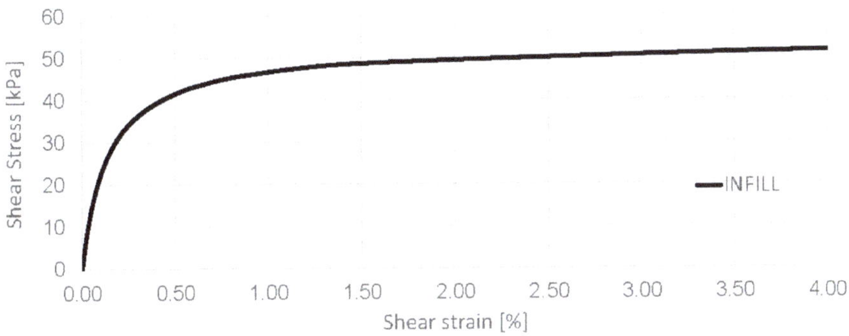

Figure 9. Backbone curve (infill soil).

3. Results

In this section, the results are discussed on the basis of the assumptions made so far. In particular, it is important to state that the findings are limited to the conditions considered herein, especially to those regarding the selected soils.

3.1. FF Analyses

The selected soil profiles were considered under the assumption that the superficial layers are characterized by sand deposits with shear wave velocities in the range of 150–250 m/s. In this regard, the four materials were selected to be representative of real soil conditions from low to medium-low stiffness. Saturated and dry conditions were chosen in order to perform dynamic analyses. The position of the water table is fundamental in order to assess the performance of the system (soil + structure). However, it is extremely difficult to know this parameter in real situations. In this study, the water table depth was set at a depth of 2 m from the ground surface for the saturated condition and for all soil models. For each soil condition, transfer functions (TFs) were calculated as the ratio between the acceleration at the ground surface (depth of 0 m) and the one at the base of the soil domain (depth of 20 m), considering the selected input motion (Figure 3) along the longitudinal axis.

Figure 10 compares the behavior of S1, S2, S3, and S4 for both saturated and dry conditions for the range of periods between 0 and 1 s. It is worth considering that the role of soil deformability in the mechanism of amplification inside the range of periods of the selected structure is paramount (Table 5). In particular, S2 with the saturated condition is shown to be the most detrimental soil above which the structure can be founded (maximum amplification equal to 1.66), since the TF peak occurs in conjunction with the fundamental period of the structure (Table 5), and thus S2 is applied in the SSI model (see Section 3.2). Moreover, dry conditions are noticeable for those periods that are far from the structural ones. In general, it is possible to state that S2, S3, and S4 with the saturated condition are the most detrimental cases. On the other hand, S1 seems to behave differently from the other cases, since dry conditions are more detrimental for the structural configurations that were herein considered.

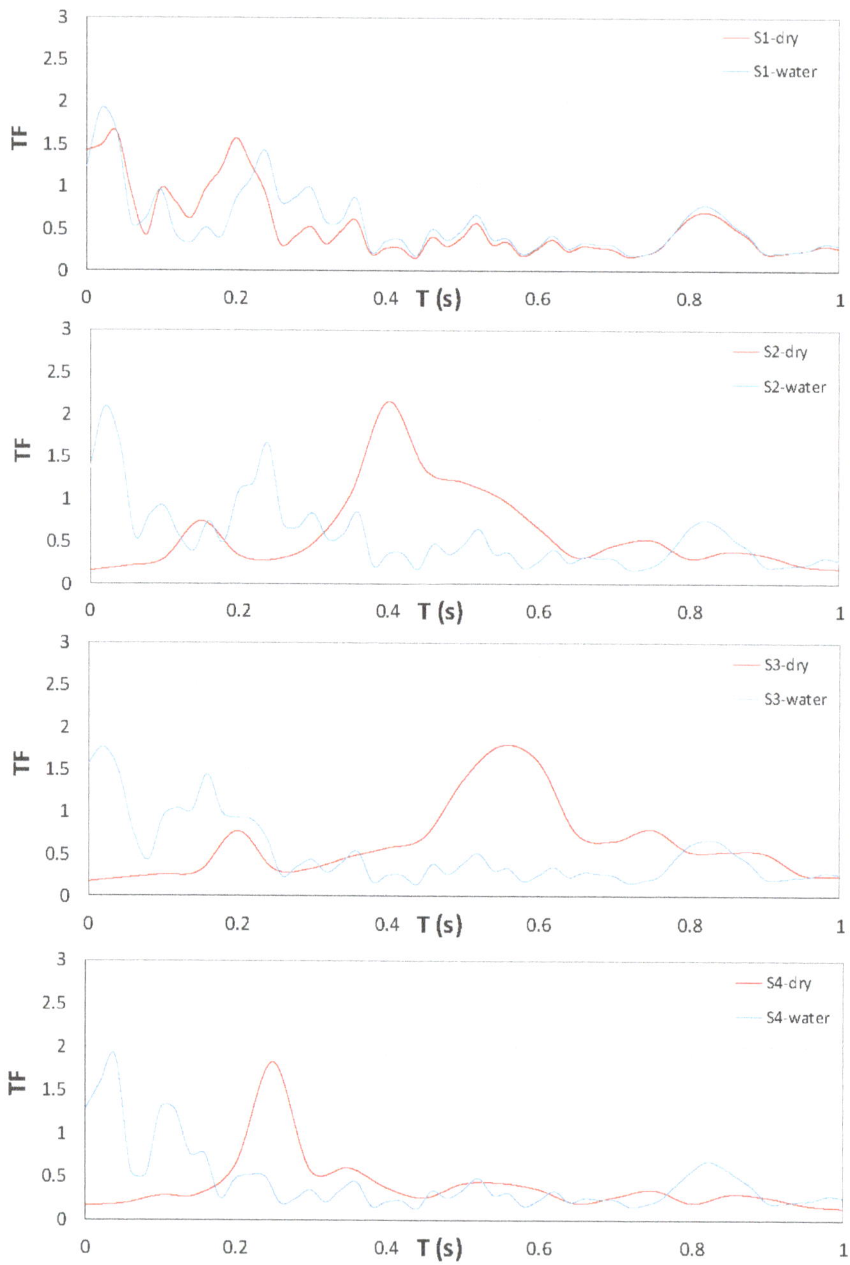

Figure 10. Transfer functions: S1, S2, S3, and S4 with dry and saturated conditions.

3.2. *SSI Analyses*

With the recent development of the performance-based earthquake engineering (PBEE) methodology [61–63], there has been an increasing attention in the new engineering demand parameter (EDP) to assess the structural performance of buildings, such as the floor accelerations. In particular, many codes [64–67] are implementing new provisions based on floor performance. In this regard, the paper aimed to move in the direction of this new approach by calculating not only the peak values of the aforementioned EDP but also the top floor accelerations. This section presents the structural performance at the foundation level and along the height of the building, considering the soil S2 under saturated conditions, which was found to be the most detrimental (Section 3.1).

Figure 11 represents the rotation versus bending moment related to an RC column at the base of the structure. It is possible to see that the diagram presents the typical hysteretic mechanism registered during a seismic event. In particular, the values of the rotations are not significant, meaning that the rocking component of the foundation does not relevantly affect the performance of the structure, which is tied to the ground. The settlement of the foundation is not substantial as well (maximum 1.8 mm) and this is the reason for the low level of overturning moments and interstorey drifts (Figure 12), calculated as the ratio between the relative longitudinal displacement and the height of the floor from the foundation level.

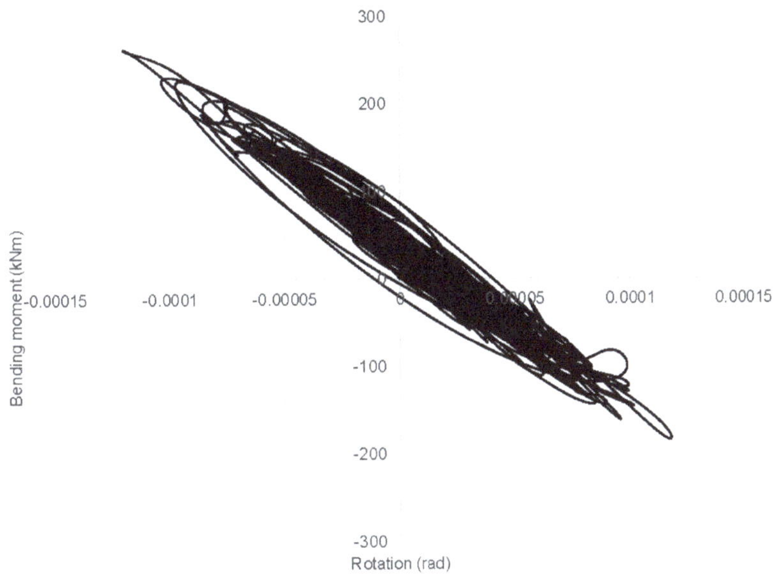

Figure 11. Rotation vs. bending moment.

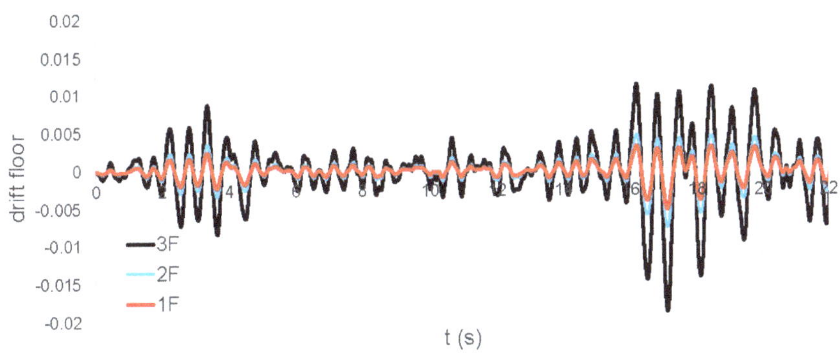

Figure 12. Drift floor time histories.

The floor spectra (5% damping), which were considered in order to assess the amplifications of the longitudinal accelerations, and thus the seismic performance of the structure, are represented in Figure 13. It is worth noticing that the peaks correspond to the fundamental period of the structure (as expected), demonstrating that the numerical model performed properly. Moreover, the accelerations increase with the height of the structure (0.426, 0.628, and 0.776 g, respectively, for floor 1, floor 2, and floor 3), with amplifications for floor 2 and floor 3 of 23.5% and 82.1% greater than those resulted for floor 1. Additionally, the significant peak related to the period of 0.6 s is noticeable, where all the structures show the same level of amplification (nearly 0.9 g). This peak corresponds to a period that is close to the fundamental period of the S2 soil, and thus may be a consequence of the mutual behavior of the soil and the structure [68].

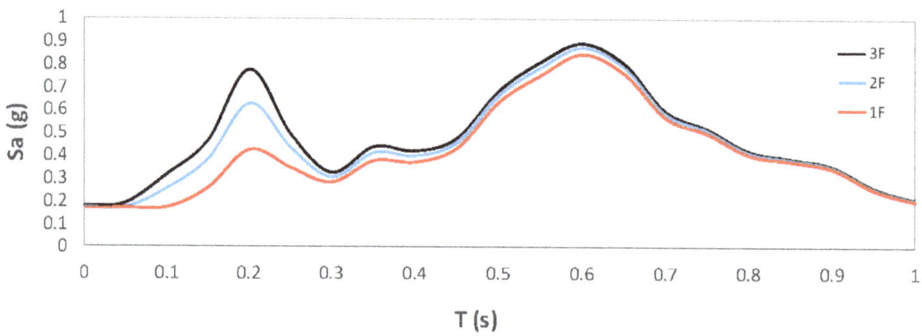

Figure 13. Spectra at various floors (damping ratio = 5%).

Figure 14 shows the maximum longitudinal displacements along the height of the structure. It is noticeable that the foundation maximum displacement is 1.725 cm, which means that a considerable translation occurred. In addition, since the structure is a low-rise building, the structural stiffness drives the increase in displacements along the height of the buildings and in relation to the various floors. The maximum displacement at the top of the structure is 1.885 cm, which is significant for a three-storey building in terms of structural performance.

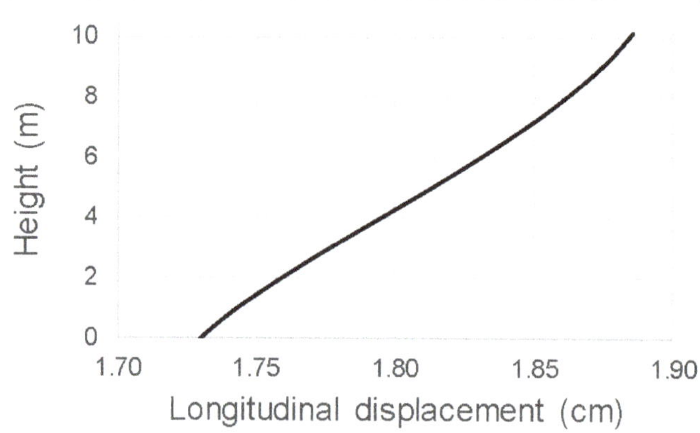

Figure 14. Maximum longitudinal displacements.

Figure 15 represents the shear forces versus longitudinal displacements for the masonry walls. These outputs are fundamental in order to define the damage conditions of the wall and to determine the potential collapse mechanism. It is worth noticing that the results are somewhat significant, demonstrating that the masonry walls are the weakest elements in the structure, as expected by the historical evidence during the 23 November 1980 Irpinia–Basilicata (southern Italy) earthquake. In particular, the maximum base shear was 390 kN and the corresponding maximum tensile stresses were approximately 65 kPa, which is close to the ultimate tensile stress. This aspect suggests that the potential damage is primarily due to the shear failure of the masonry walls.

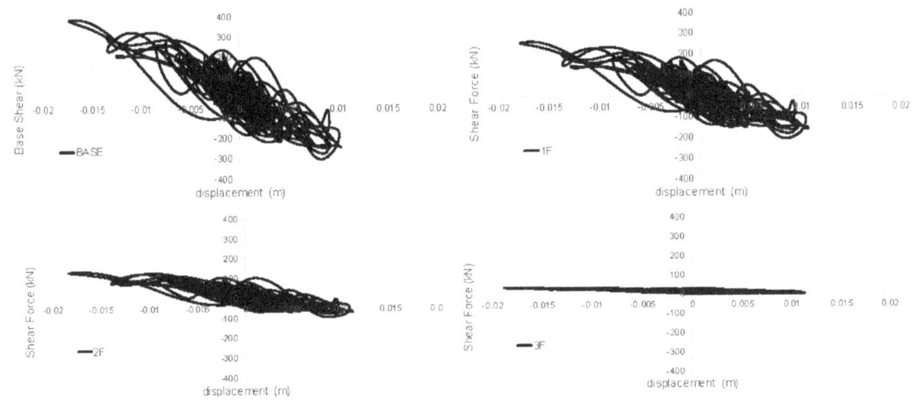

Figure 15. Shear forces vs. longitudinal displacements at various floors.

Overall, the results demonstrate that the numerical model is in good agreement with the assumptions made so far, and thus the ground-foundation-structure system was simulated properly. The paper assessed the role of soil deformability in the amplification of accelerations, and thus its consequences on the structure. In particular, the outputs were chosen in accordance with the new approach proposed by the recent code provisions and the top floor accelerations and other significant EDP were calculated for the performed structure. In this regard, the values of the shear forces that occurred in the masonry elements show that shear failure may potentially occur in these elements, as expected and as proved by the damage that occurred during the 23 November 1980 Irpinia–Basilicata (southern Italy) earthquake.

4. Conclusions

The paper investigated the effects of soil deformability on a typical structural configuration by analyzing a 3-D soil–structure model built up with OpenSeesPL. The results are the consequence of several mechanisms known globally as the soil–structure interaction (SSI). The principal novelty of the paper consisted of proposing a model that performs detailed 3-D simulations of both the soil and the structure and assessing the structural performance in terms of displacements, drifts, and accelerations at various floors. Overall, the paper demonstrated that the soil may cause several spectral amplifications under free-field conditions (maximum amplifications: 1.66) and that a rigid low-rise building is sensitive to SSI effects, which need to be considered. Although the findings were limited to the specified conditions, they may potentially be useful to propose formulations that include SSI effects within code provisions. In this regard, future parametric numerical simulations on the response of other structural typologies and soil characteristics (e.g., water-level depth) will be performed.

Author Contributions: Please consider this paragraph that specifies the individual contributions of the authors: Conceptualization, D.F.; methodology, D.F.; software, D.M.; validation, D.M.; formal analysis, D.M. and D.F.; investigation, D.M. and D.F.; resources, D.M. and D.F.; data curation, D.M. and D.F.; writing—original draft preparation, D.M. and D.F.; writing—review and editing, D.M. and D.F.; visualization, D.M. and D.F.; supervision, D.F.; project administration, D.M. and D.F.; funding acquisition, D.M. and D.F. All authors have read and agreed to the published version of the manuscript.

Funding: This research received no external funding.

Conflicts of Interest: The authors declare no conflict of interest.

References

1. Westaway, R.; Jackson, J. Surface faulting in the Southern Italian Campania-Basilicata earthquake of 23 November 1980. *Nature* **1984**, *312*, 436–438. [CrossRef]
2. Cotecchia, V. Ground deformations and slope instability produced by the earthquake of 23 November 1980 in Campania and Basilicata. In *Proceedings of the International Symposium Engineering Geology Problems in Seismic Areas*; IAEG: Bari, Italy, 1986; Volume 5, pp. 31–100.
3. Del Prete, M. Examples of mudslides hazard in Southern Apennines (Italy). *Ann. Di Geofis.* **1993**, *36*, 71–80.
4. Esposito, E.; Gargiulo, A.; Iaccarino, G.; Porfido, S. Distribuzione dei fenomeni franosi riattivati dai terremoti dell'Appennino meridionale. Censimento delle frane del terremoto del 1980. In *Proceedings of the Conversion International Prevention of Hydrogeological Hazards*; CNR-IRPI: Torino, Italy, 1998; Volume 1, pp. 409–429.
5. Pantosti, D.; Valensise, G. Source geometry and long-term behavior of the 1980 fault based on field geologic observations. *Ann. Di Geofis.* **1993**, *36*, 41–49.
6. Gizzi, F.T.; Potenza, M.R.; Zotta, C. 23 November 1980 Irpinia–Basilicata earthquake (Southern Italy): Towards a full knowledge of the seismic effects. *Bull. Earthq. Eng.* **2012**, *10*, 1109–1131. [CrossRef]
7. Nunziata, C.; Costa, G.; Marrara, F.; Panza, F. Validted Estimation of Response Spectra for the 1980 Irpinia Earthquake in the Eastern Area of Naples. *Earthq. Spectra* **2000**, *16*, 643–660. [CrossRef]
8. Ameri, G.; Emolo, A.; Pacor, F.; Gallovič, F. Ground-Motion Simulations for the 1980 M 6.9 Irpinia Earthquake (Southern Italy) and Scenario Events. *Bull. Seismol. Soc. Ameirca* **2011**, *101*, 1136–1151. [CrossRef]
9. Assimaki, D.; Kausel, E.; Gazetas, G. Oil-Dependent Topographic Effects: A Case Study from the 1999 Athens Earthquake. *Earthq. Spectra* **2005**, *21*, 929–966. [CrossRef]
10. Seyhan, F.; Nihat, I.; Hasan, A.; Mesut, D.; Isa, V. Investigation of the soil amplification factor in the Adapari region. *Bull. Eng. Geol. Environ.* **2016**, *75*, 141–152.
11. Assimaki, D.; Jeong, S. Ground-Motion observations at Hotel Montana during the 7.0 2010 Haiti Earthquake: Topography or Soil Amplification? *Bull. Seismol. Soc. Ameirca* **2013**, *103*, 2577–2590. [CrossRef]
12. Varum, H.; Furtado, A.; Rodrigues, H.; Dias-Oliveira, J.; Vila-Pouca, N.; Arêde, A. Seismic performance of the infill masonry walls and ambient vibration tests after the Ghorka 2015, Nepal earthquake. *Bull. Earthq. Eng.* **2017**, *15*, 1185–1212. [CrossRef]
13. Gautam, D.; Rodrigues, H.; Bhetwal, K.K.; Neupane, P.; Sanada, Y. Common structural and construction deficiencies of Nepalese buildings. *Innov. Infrastruct. Solut.* **2016**, *1*, 1. [CrossRef]
14. Varum, H.; Dumaru, R.; Furtado, A.; Barbosa, A.R.; Gautam, D.; Rodrigues, H. Seismic Performance of Buildings in Nepal After the Gorkha Earthquake. In *Impacts and Insights of the Gorkha Earthquake*; Gautam, D., Rodrigues, H., Eds.; Elsevier: Amsterdam, The Netherlands, 2018; Chapter 3, pp. 47–63.
15. Cramer, C.H. Site-specific seismic-hazard analysis that is completely probabilistic. *Bull. Seismol. Soc. Am.* **2003**, *93*, 1841–1846. [CrossRef]
16. Goulet, C.A.; Stewart, J.P. Pitfalls of deterministic application of nonlinear site factors in probabilistic assessment of ground motions. *Earthq. Spectra* **2009**, *25*, 541–555. [CrossRef]
17. Bazzurro, P.; Cornell, C.A. Nonlinear soil-site effects in probabilistic seismic-hazard analysis. *Bull. Seismol. Soc. Am.* **2004**, *94*, 2110–2123. [CrossRef]
18. Rathje, E.M.; Pehlivan, M.; Gilbert, R.; Rodriguez-Marek, A. *Incorporating Site Response into Seismic Hazard. Assessments for Critical Facilities: A Probabilistic Approach, in Perspectives on Earthquake Geotechnical Engineering*; Springer International Publishing: New York, NY, USA, 2015; pp. 93–111.
19. Régnier, J.; Bonilla, L.-F.; Bard, P.-Y.; Bertrand, E.; Hollender, F.; Kawase, H.; Sicilia, D.; Arduino, P.; Amorosi, A.; Asimaki, D. International benchmark on numerical simulations for 1D, nonlinear site response (PRENOLIN): Verification phase based on canonical cases. *Bull. Seismol. Soc. Am.* **2016**, *106*, 2112–2135. [CrossRef]
20. Groholski, D.R.; Hashash, Y.M.; Kim, B.; Musgrove, M.; Harmon, J.; Stewart, J.P. Simplified model for small-strain nonlinearity and strength in 1D seismic site response analysis. *J. Geotech. Geoenviron. Eng.* **2016**, *142*, 4016042. [CrossRef]

21. Westaway, R.; Jackson, J. The earthquake of 1980 November 23 in Campania–Basilicata (Southern Italy). *Geophys. J. R. Astron. Soc.* **1987**, *90*, 375–443. [CrossRef]
22. Bernard, P.; Zollo, A. The Irpinia (Italy) 1980 earthquake: Detailed analysis of a complex normal faulting. *J. Geophys. Res.* **1989**, *94*, 1631–1648. [CrossRef]
23. Porfido, S.; Esposito, E.; Michetti, A.M.; Blumetti, A.M.; Vittori, E.; Tranfaglia, G.; Guerrieri, L.; Ferreli, L.; Serva, L. Areal distribution of ground effects induced by strong earthquakes in the Southern Apennines (Italy). *Surv. Geophys.* **2002**, *23*, 529–562. [CrossRef]
24. Porfido, S.; Alessio, G.; Gaudiosi, G.; Nappi, R.; Spiga, E. The resilience of some villages 36 years after the Irpinia-Basilicata (Southern Italy) 1980 earthquake. In Proceedings of the 4th Workshop on World Landslide Forum, Ljubljana, Slovenia, 29 May–2 June 2017.
25. Pingue, F.; De Natale, G. Fault mechanism of the 40 seconds subevent of the 1980 Irpinia (Southern Italy) earthquake from levelling data. *Geophys. Res. Lett.* **1993**, *20*, 911–914. [CrossRef]
26. Ascione, A.; Mazzoli, S.; Petrosino, P.; Valente, E. A decoupled kinematic model for active normal faults: Insights from the 1980, MS = 6.9 Irpinia earthquake, Southern Italy. *Geol. Soc. Am. Bull.* **2013**, *125*, 1239–1259. [CrossRef]
27. Cavalieri, F.; Correia, A.A.; Crowley, H.; Pinho, R. Dynamic soil-structure interaction models for fragility characterisation of buildings with shallow foundations. *Soil Dyn. Earthq. Eng.* **2020**, 106004. [CrossRef]
28. Pitilakis, K.D.; Karapetrou, S.T.; Fotopoulou, S.D. Consideration of aging and SSI effects on seismic vulnerability assessment of RC buildings. *Bull. Earthq. Eng.* **2014**, *12*, 1755–1776. [CrossRef]
29. Rajeev, P.; Tesfamariam, S. Seismic fragilities of non-ductile reinforced concrete frames with consideration of soil structure interaction. *Soil Dyn. Earthq. Eng.* **2012**, *40*, 78–86. [CrossRef]
30. Anbazhagan, P.; Aditya, P.; Rashmi, H.N. Amplification based on shear wave velocity for seismic zonation: Comparison of empirical relations and site response results for shallow engineering bedrock sites. *Geomech. Eng.* **2011**, *3*, 189–206. [CrossRef]
31. Rota, M.; Lai, C.G.; Strobbia, C.L. Stochastic 1D site response analysis at a site in central Italy. *Soil Dyn. Earthq. Eng.* **2011**, *31*, 626–639. [CrossRef]
32. Shahri, A.A.; Esfandiyari, B.; Hamzeloo, H. Evaluation of a nonlinear seismic geotechnical site response analysis method subjected to earthquake vibrations (case study: Kerman Province, Iran). *Arab. J. Geosci.* **2011**, *4*, 1103–1116. [CrossRef]
33. EC-8-3. *Design of Structures for Earthquake Resistance, Part. 3: Strengthening and Repair of Buildings*; European standard EN 1998-3; European Committee for Standardization (CEN): Brussels, Belgium, 2005.
34. ASCE. *Minimum Design Load for Buildings and Other Structures*; ASCE standard no. 007-05; American Society of Civil Engineering: Reston, VA, USA, 2006.
35. ASCE. *Seismic Analysis of Safety-Related Nuclear Structures and Commentary*; ASCE standard no. 004-98; American Society of Civil Engineering: Reston, VA, USA, 2000.
36. Foti, S.; Parolai, S.; Albarello, D.; Picozzi, M. Application of Surface-Wave Methods for Seismic Site Characterization. *Surv. Geophys.* **2011**, *32*, 777–825. [CrossRef]
37. Stokoe, K.H.; Nazarian, S.; Rix, G.J.; Sánchez-Salinero, I.; Sheu, J.C.; Mok, Y.J. *In Situ Seismic Testing of Hard-to-Sample Soils by Surface Wave Method, Earthquake Engineering and Soil Dynamics II - Recent Advances in Ground Motion Evaluation*; Von Thun, J.L., Ed.; ASCE Geotechnical Special Publication No. 20; ACSE: Reston, VA, USA, 1988; pp. 264–278.
38. Amorosi, A.; Castellaro, S.; Mulargia, F. Single-Station Passive Seismic Stratigraphy: An inexpensive tool for quick subsurface investigations. *GEOACTA* **2008**, *7*, 29–39.
39. Hobiger, M.; Wegler, U.; Shiomi, K.; Nakahara, H. Single-station cross-correlation analysis of ambient seismic noise: Application to stations in the surroundings of the 2008 Iwate-Miyagi Nairiku earthquake. *Geophys. J. Int.* **2014**, *198*, 90–109. [CrossRef]
40. Wair, B.R.; DeJong, J.T. *Guidelines for Estimation of Shear Wave Velocity Profiles*; PEER Report 2012/08; Pacific Earthquake Engineering Research Center: Berkeley, CA, USA, 2012.
41. Tanganelli, M.; Viti, S.; Forcellini, D.; D'Intisonante, V.; Baglione, M. Effect of soil modeling on Site Response Analysis (SRA). Alphose Zingoni, Insights and Innovations in Structural Engineering, Mechanics and Computation. In Proceedings of the SEMC 2016, Cape Town, South Africa, 5–7 September 2016; pp. 364–369, ISBN 978-1-138-02927-9.

42. Viti, S.; Tanganelli, M.; D'Intinosante, V.; Baglione, M. Effects of soil characterization on the seismic input. *J. Earthq. Eng.* **2017**, *21*. [CrossRef]
43. Rodriguez-Marek, A.; Kruiver, P.P.; Meijers, P.; Bommer, J.J.; Dost, B.; van Elk, J.; Doornhof, D. A regional site-response model for the Groningen gas field. *Bull. Seism. Soc. Am.* **2017**, *107*, 2067–2077. [CrossRef]
44. Kruiver, P.P.; van Dedem, E.; Romijn, R.; de Lange, G.; Korff, M.; Stafleu, J.; Gunnink, J.L.; Rodriguez-Marek, A.; Bommer, J.J.; van Elk, J.; et al. An integrated shear-wave velocity model for the Groningen gas field, The Netherlands. *Bull. Earthq. Eng.* **2017**, *15*, 3555–3580. [CrossRef]
45. Forcellini, D. Cost Assessment of isolation technique applied to a benchmark bridge with soil structure interaction. *Bull. Earthq. Eng.* **2017**. [CrossRef]
46. Forcellini, D. Seismic Assessment of a benchmark based isolated ordinary building with soil structure interaction. *Bull. Earthq. Eng.* **2018**. [CrossRef]
47. Forcellini, D. Numerical simulations of liquefaction on an ordinary building during Italian (20 May 2012) earthquake. *Bull. Earthq. Eng.* **2019**. [CrossRef]
48. Forcellini, D. Soil-structure interaction analyses of shallow-founded structures on a potential-liquefiable soil deposit. *Soil Dynamics Earthq. Eng.* **2020**, *133*, 106108. [CrossRef]
49. ITACA 1.1, 2011, ITalian ACcelerometric Archive (1972-2011) version1.1. Available online: http://itaca.mi.ingv.it/ItacaNet/ (accessed on 21 April 2020).
50. Zienkiewicz, O.C.; Chan, A.H.C.; Pastor, M.; Paul, D.K.; Shiomi, T. Static and dynamic behavior of soils: A rational approach to quantitative solutions: I. Fully saturated problems. *Proc. R. Soc. Lond. Ser. A* **1990**, *429*, 285–309.
51. Elgamal, A.; Lu, J.; Forcellini, D. Mitigation of Liquefaction-Induced lateral deformation in sloping stratum: Three-dimensional Numerical Simulation. *J. Geotech. Geoenvironmental Eng.* **2009**, *135*, 1672–1682. [CrossRef]
52. Lu, J.; Elgamal, A.; Yang, Z. OpenSeesPL: 3D Lateral Pile-Ground Interaction, User Manual, Beta 1.0. Available online: http://soilquake.net/openseespl/ (accessed on 21 April 2020).
53. Mazzoni, S.; McKenna, F.; Scott, M.H.; Fenves, G.L. *Open System for Earthquake Engineering Simulation, User Command-Language Manual*, OpenSees version 2.0; Pacific Earthquake Engineering Research Center, University of California: Berkeley, CA, USA, 2009. Available online: http://opensees.berkeley.edu/OpenSees/manuals/usermanual (accessed on 21 April 2020).
54. Yang, Z.; Elgamal, A.; Parra, E. A computational model for cyclic mobility and associated shear deformation. *J. Geotech. Geoenviron. Eng. (ASCE)* **2003**, *129*, 1119–1127. [CrossRef]
55. Kramer, S.L. *Geotechnical Earthquake Engineering*; William, J., Ed.; International Series in Civil engineering and engineering Mechanics; Prentice-Hall: Upper Saddle River, NJ, USA, 1996.
56. Madas, P.; Elnashai, A. A new passive confinement model for the analysis of concrete structures subjected to cyclic and transient dynamic loading. *Earthq. Eng. Struct Dyn.* **1992**, *21*, 409–431. [CrossRef]
57. Mander, J.; Priestley, M.; Parks, R. Theoretical stress-strain model for confined concrete. *J. Struct. Eng.* **1988**, *114*, 1804–1826. [CrossRef]
58. Menegotto, M.; Pinto, P. Method of analysis for cyclically loaded RC plane frames including changes in geometry and non-elastic behaviour of elements under combined normal force and bending. In Proceedings of the Symposium on Resistance and Ultimate Deformability of Structures Acted on by Well Defined Repeated Loads, Zurich, Switzerland, 1973.
59. CIRCOLARE 2 febbraio 2009, n. 617 Istruzioni per l'applicazione delle «Nuove norme tecniche per le costruzioni» di cui al decreto ministeriale 14 gennaio 2008. Available online: http://cslp.mit.gov.it/ (accessed on 21 April 2020).
60. Furtado, A.; Rodrigues, H.; Arêde, A.; Varum, H.; Grubišić, M.; Šipoš, T.K. Prediction of the earthquake response of a three-storey infilled RC structure. *Eng. Struct.* **2018**, *171*, 214–235. [CrossRef]
61. Garcia, R.; Sullivan, T.J.; Della Corte, G. Development of a displacement-based design method for steel frame-RC wall buildings. *J. Earthq. Eng.* **2010**, *14*, 252–277. [CrossRef]
62. Maley, T.J.; Sullivan, T.J.; Della Corte, G. Development of a displacement-based design method for steel dual systems with buckling-restrained braces and moment resisting frames. *J. Earthq. Eng.* **2010**, *14*, 106–140. [CrossRef]
63. Priestley, M.J.N.; Calvi, G.M.; Kowalsky, M.J. *Direct Displacement-Based Seismic Design*; IUSS Press: Pavia, Italy, 2007; p. 720.

64. ASCE. *ASCE Standard, Minimum Design Loads for Buildings and Other Structures*; ASCE/SEI 7-10; American Society of Civil Engineers: Reston, VA, USA, 2010.
65. ATC-3-06. *Amended Tentative Provisions for the Development of Seismic Regulations for Buildings*; ATC Publication ATC 3-06, NBS Special Publication 510, NSF Publication 78-8; Applied Technology Council, US Government Printing Office: Washington, DC, USA, 1978.
66. CEN. *Eurocode 8: Design of Structures for Earthquake Resistance. Part. 1: General Rules, Seismic Actions and Rules for Buildings*; Final Draft prEN 1998; European Committee for Standardization: Brussels, Belgium, 2003.
67. CEN. *Eurocode 8 Design of Structures for Earthquake Resistance. Part. 1: General Rules, Seismic Actions and Rules for Buildings*; European standard EN 1998-1; European Committee for Standardization: Brussels, Belgium, 2004.
68. Forcellini, D.; Gobbi, S.; Mina, D. Numerical Simulations of Ordinary Buildings with Soil Structure Interaction. Alphose Zingoni, vol. Insights and Innovations in Structural Engineering, Mechanics and Computation. In Proceedings of the SEMC 2016, Cape Town, South Africa, 5–7 September 2016; pp. 364–369, ISBN 978-1-138-02927-9.

geosciences

MDPI

Article

Near-Real-Time Loss Estimates for Future Italian Earthquakes Based on the M6.9 Irpinia Example

Max Wyss * and Philippe Rosset

International Centre for Earth Simulation Foundation, 1223 Cologny, Switzerland; rossetp@orange.fr
* Correspondence: max@maxwyss.ch

Received: 17 March 2020; Accepted: 30 April 2020; Published: 3 May 2020

Abstract: The number of fatalities and injured was calculated, using the computer code QLARM and its data set and assuming information about the Irpinia 1980 earthquake became available in near-real-time. The casualties calculated for a point source, an approximate line source and a well-defined line source would have become available about 30 min, 60 min and years after the main shock, respectively. The first estimate would have been satisfactory, indicating the seriousness of the disaster. The subsequent loss estimate after 60 min would have defined the human losses accurately, and the ultimate estimate was most accurate. In 2009, QLARM issued a correct estimate of the number of fatalities within 22 min of the M6.3 L'Aquila main shock. These two results show that the number of casualties and injuries in large and major earthquakes in Italy can be estimated correctly within less than an hour by using QLARM.

Keywords: earthquake risk; earthquake fatalities; Italian earthquakes; Irpinia earthquake

1. Introduction

At the time of the M6.9 Irpinia earthquake of 1980, near-real-time loss estimates for earthquakes did not exist. Today however, two teams distribute these estimates for major earthquakes worldwide. The PAGER (USGS) and the QLARM (Quake Loss Assessment for Response and Mitigation) teams distribute loss alerts within 25 min and 29 min of potential earthquake disasters, respectively [1].

The question we ask in this article is how reliable are theoretical fast loss assessments in cases of major and large earthquakes in Italy? The M6.9 Irpinia earthquake is used as an example. Calculations are presented of what the estimates of fatalities and injured would have been immediately after the Irpinia earthquake. The results are an indication of the quality of loss estimates by QLARM within less than an hour of the next large earthquake neighboring the Irpinia or L'Aquila areas and, by implication, in all of Italy.

The routine by which QLARM alerts are issued in near-real-time is described by Wyss [1]. The details of the program and data sets in QLARM are given by Trendafiloski et al. [2]. Different aspects and uses of QLARM have been discussed in several articles [3–6]. Here, we do not repeat these explanations; rather, we focus on the Irpinia earthquake and the quality of near-real-time earthquake loss assessments in Italy.

Our aim in loss calculations is to estimate the total sum of fatalities that are likely as a measure of the extent of earthquake disasters. The unknown parameters within minutes after an earthquake are numerous, as all initial earthquake parameters are afflicted by significant uncertainties [7]. The direction of rupture and consequential focusing of radiated energy becomes known only later, if ever. The local wave attenuation was not known, so we used an average. Soil conditions that might amplify accelerations are not known for any specific location. Building resistance to shaking was only known approximately in an average sense, so we used an approximate model for the building types. Finally, the occupancy rate was also largely unknown, so we used 99% occupancy at night and 26% during the day.

All these unknowns mean that we cannot attempt to calculate what happens to a specific building. Instead, we rely on an average. We only have confidence in the overall sum of casualties, not even in those within one settlement, and much less those within a building. It is this overall average estimate of fatalities that we distribute minutes after earthquakes, and it is this value that would have been calculated by QLARM after the Irpinia earthquake that we are testing here.

2. Building Stock Used

An aspect of QLARM application to Italy that has not previously been explained is the distribution of buildings into vulnerability classes and their occupants. As QLARM is a tool operated pro-bono for worldwide application, we did not have the information or the resources necessary to construct detailed information on the specific buildings in each settlement. This information can be developed by special projects [8,9], but, on a worldwide basis, it was available to us only for Greece. The Greek 2001 population and building census includes information about the construction material, age and number of floors for each building (Hellenic Statistical Authority). This information allowed the classification of all building types in each settlement according to the European Macroseismic Scale 1998 (EMS98) [9] for Greece only. For other regions, like southern Europe, we used averages with three size categories. The population limits for building models we used for Italy were pop1 < 2000 ≤ pop2 < 20,000 ≤ pop3 (Table 1).

Table 1. Percentage of building types and percentage of the population in buildings in the three class sizes of settlements used by QLARM for Italy.

Cities Above 20,000 Inhabitants			Towns			Villages Below 2000 Inhabitants		
Type	Buildings (%)	People (%)	Type	Buildings (%)	People (%)	Type	Buildings (%)	People (%)
A	0	0	A	13	11	A	27	22
B	70	27	B	69	41	B	68	55
C	27	63	C	16	43	C	5	23
D	3	10	D	2	5	D	0	0
E	0	0	E	0	0	E	0	0

The building distribution used for Italy and people in them are listed in Table 1 and shown in Figure 1. We are not advocating these approximations to the built environment as the best option for a specific local environment, we simply needed models for all countries of the word that would yield correct results. For several countries, we verified the appropriateness of the models for building stock by determining if the theoretical estimates of the total numbers of fatalities agreed with the reported numbers in large earthquakes [5,6,10].

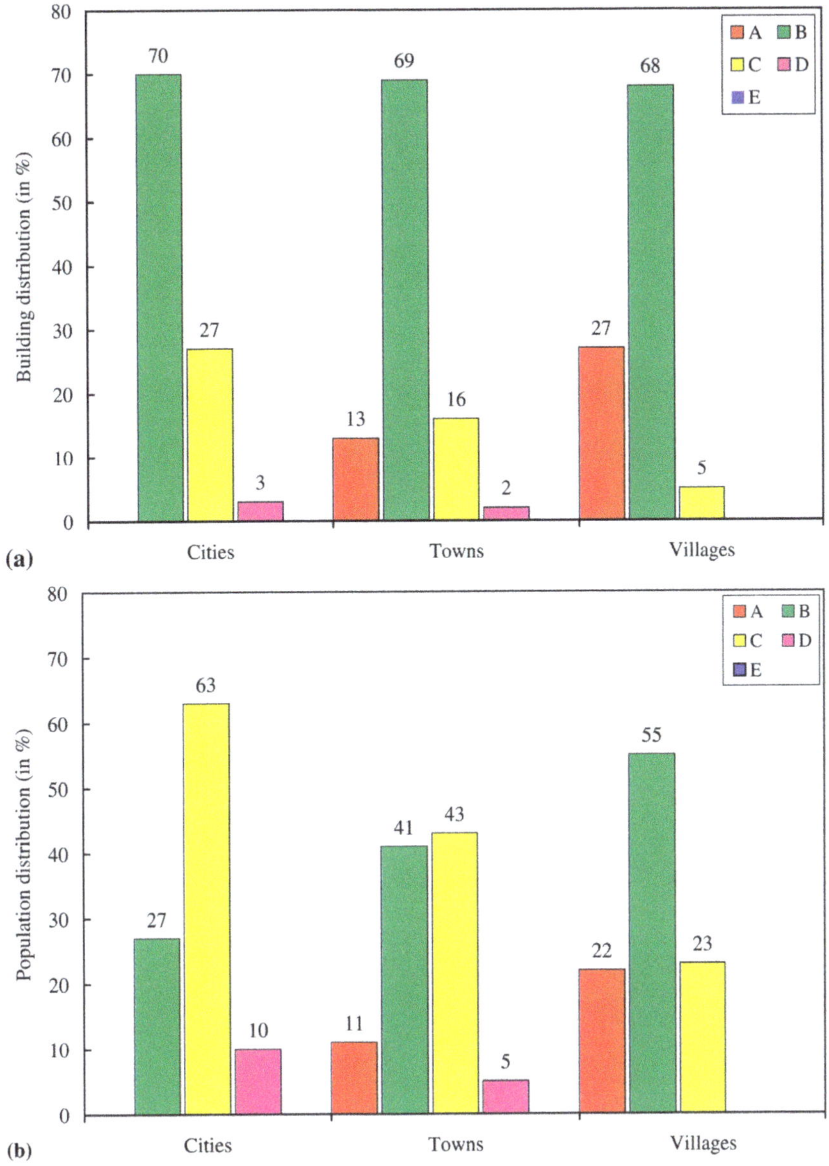

Figure 1. Percentage of building types (**a**) and percentage of the population in them (**b**) in the three class sizes of settlements used by QLARM for Italy.

3. The M6.3 L'Aquila Earthquake

In the case of the L'Aquila earthquake, the QLARM alert was correct (Table 2). This type of estimate can serve as a guide for first responders at a time when no information is available from the devastated area.

Table 2. Copy of the estimate of fatalities and injuries distributed by email 22 min after the M6.3 earthquake of L'Aquila in 2009 to interested parties.

From	wapmerr@sed.ethz.ch
To	wapmerr@sed.ethz.ch
Date	Monday, April 06, 2009 03:54 am
Subject	Earthquake in Italy
The Following Earthquake has been Reported: Date: 2009/04/04 01:31:57.6 Region: Italy Magnitude: M 6.3 Latitude: 42.42 N Longitude: 13.39 E Depth (km): 5.0 Source: GHZ Injured Exp. min/max: 100/1300 Fatalities Exp. min/max: 50/500 L'Aquila is the nearest town at 5 km distance	

4. Loss Estimates for the M6.9 Irpinia Earthquake

The loss estimates for the Irpinia event presented here were calculated as if the earthquake's parameters became available in near-real-time by SMS and no other information existed, as in the aforementioned L'Aquila earthquake. The hypocenters and magnitudes of significant earthquakes worldwide are received by users, such as QLARM, from the GFZ (GeoForschungsZentrum) and the USGS (United States Geological Survey) within 7 and 18 min, respectively. After receiving this message, shaking intensities, damage to buildings and the impact on occupants are immediately calculated. At first, only the hypocenter is known, not the extent and direction of the rupture. Also, initial hypocenter and magnitude values tend to be inaccurate. In the case of the Irpinia earthquake, it is not known by what errors the first parameter estimates were afflicted, so those parameters currently listed by the USGS (M6.9, 1980-11-23 18:34:53 (UTC) 40.914° N 15.366° E) must be used.

The map showing settlements with color-coded damage due to an M6.9 earthquake at the Irpinia epicenter modeled as a point source is shown in Figure 2. The mean damage is given on a scale from 0 to 5. The calculated pattern is circular because the direction of rupture would not have been known. This map and the casualties (both fatalities and injured) given in the first row of Table 3 would have been distributed by the QLARM team about 30 min after the rupture.

Table 3. Casualties calculated for a hypothetical repeat of the M6.9 Irpinia earthquake. The results are presented for three source models that would have become available with increasing accuracy as a function of time and would have been distributed with the delays given in the first column. The calculated casualties are good estimates of the reported ones and become better with the assumed increase of information.

Delay (Minute)	Longitude (Degree)	Latitude (Degree)	Source Model	Depth (km)	Magnitude	Fatalities (Calculated)		Fatalities (Reported)	Injured (Calculated)		Injured (Reported)
						Min.	Max.		Min.	Max.	
30	15.37	40.91	Point	10	6.9	300	1840	2483	1480	9'064	**7700**
60	15.28	41.11	Line end N	15	6.9	410	3120	2483	2110	14'790	**7700**
	15.61	40.67	Line end S								
Ultimate	15.00	41.00	Line end N	15	6.9	780	5840	2483	3860	26'660	**7700**
	15.515	40.355	Line end S								

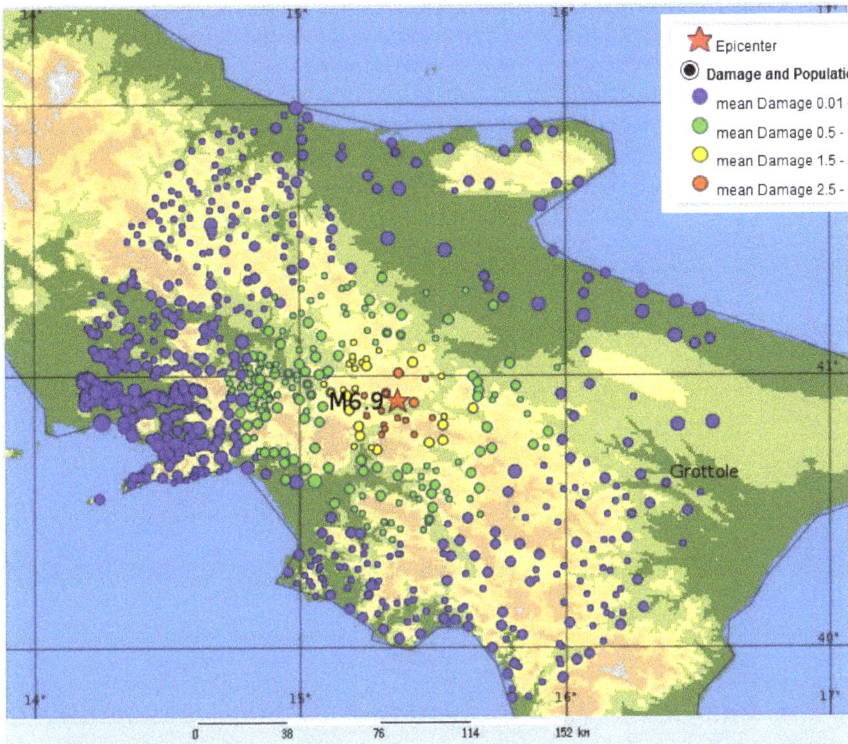

Figure 2. Map of estimated mean damage in the 1980 Irpinia epicenter area, calculated for a point source with M6.9 located at 40.914° N 15.366° E and occurring at 19:35 local time. Each dot is a settlement with size proportional to population.

The location of an aftershock can provide hints about the direction and length of the rupture. In November 1980, the USGS calculated the aftershock hypocenters given in Table 4 during the first 32 min after the initiation of the rupture. Since QLARM distributes loss estimates about 30 min after a given earthquake, estimates of losses based on a line source (defined by the aftershocks), would have been distributed within about one hour of the main shock.

Table 4. Parameters listed by the USGS for the Irpinia main shock and the first three aftershocks that occurred within 32 min of the main shock (delay in third to last column). The distance of each aftershock from the initial rupture point (second to last column) allows an early estimate of the rupture length (last column) as the separation of the most distant aftershocks from each other.

Year	Month	Day	Hour	Minute	Second.	Latitude	Longitude	Depth	Magnitude	Time Difference (Min.)	Distance (km)	Length Estimate (km)
1980	11	23	18	34	53	40.91	15.37	10	6.9			
1980	11	23	18	52	6.3	41.11	15.28	10	4.6	17	24	
1980	11	23	19	4	2.7	40.67	15.61	10	4.5	29	37	61
1980	11	23	19	6	45	40.72	15.59	10	4.4	32	32	

The three aftershocks listed in Table 4 define an approximate direction of the rupture NW to SE. Therefore, in real time, the QLARM operator would have made the usual assumption that the aftershocks most distant from the initiation of the rupture give an approximate indication of the rupture length and direction. In this case, aftershocks 1 and 2 (origin times 18:52 and 19:04) were separated by 61 km. This length agrees with an M6.9 rupture [11,12]. That means it supports the hypothesis that after 30 min the full length of the rupture was approximately defined as 61 km. Assuming that this was the case, a second estimate of the casualties, as given in the second row of Table 3, would have been distributed by QLARM within about an hour of the earthquake.

The final estimate of casualties (Figure 3) is based on a line source connecting the endpoints of the surface rupture (row three of Table 3) and the aftershock distribution as published years after the event [13–15]. This estimate would not have been available in real time but is given here as the best estimate of the reliability of QLARM alerts for losses in Italian earthquakes.

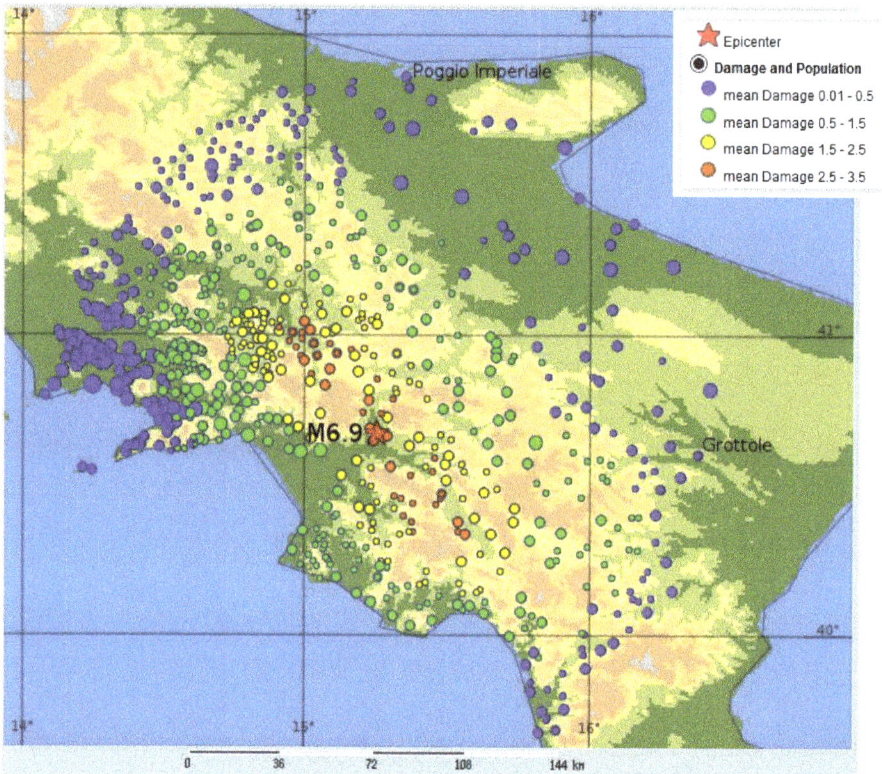

Figure 3. Map of estimated mean damage in the 1980 Irpinia epicenter area, calculated for a line source with M6.9, end points at 41° N/15° E and 40.355° N/15.515° E with occurrence at 19:35 local time. Each dot marks a settlement. The distribution of strong damage is seen better when the rupture is modelled as a line rather than a point (Figure 2).

QLARM also estimates numbers of affected people. We defined strongly affected people as those living in the area of intensity VIII+ because serious damage occurs in this area. The number of people in the area of intensity VI + VII were considered moderately affected because some damage occurs, and some casualties may result. The total number of affected people was defined as the sum of these two categories, that is, all in the area of intensities VI+ (Table 5).

Table 5. Estimates of numbers of affected people for the three types of earthquake source models of Table 3, and in three categories of shaking (intensities VI + VII, VIII + IX, and VI+).

Delay	Source Model	Magnitude	Intensities VI + VII	Intensities VIII + IX	Total Affected VI+
Min.			Million	Million	Million
30	Point	7	2.2	0.1	2.3
60	Approximate line	7	2.4	0.3	2.7
Ultimate	Final line	7	5.4	0.6	6.0

5. Discussion

The loss estimates we have distributed over the last 17 years and presented here are not intended to be highly accurate and applicable to single settlements. They are intended to be order of magnitude assessments of the extent of disasters. Unknowns, such as local soil conditions, tend to average out when many settlements receive strong shaking. Therefore, loss estimates for large earthquakes are more stable than for small ones. Loss estimates for small earthquakes with relatively few fatalities are less reliable because the collapse of a single apartment building or school can kill 100 people, possibly doubling the number of fatalities. For example, on 31 October, 2002, a school collapsed in an M5.9 earthquake in Molise, Italy, killing 27 out of 28 reported fatalities.

The number of fatalities is taken as the best measure of the extent of an earthquake disaster, especially soon after the event. It is a number that is relatively accurate for most earthquakes after months once all the information has been gathered. The number of injured is more uncertain and often not given. Even more nebulous are economic losses.

Comparison of the reported numbers of fatalities with the theoretical ones in Table 3 is a measure of the quality of QLARM performance. It shows the estimates of the number of fatalities at 30 min, 60 min and after a year compared with final reported numbers. The result of the initial point source calculation (distribution with 30 min delay) is already acceptable, given that the maximum fatality estimate is 74% of the ultimate count (row 1 in Table 3). Based on the aftershock locations that would have become known within an additional 30 min, the approximate line source model for the rupture yields a range of numbers of fatalities that encompasses the observed number (row 2 in Table 3). This good agreement is achieved, even though rapidly calculated epicenters can be wrong by about 10 km [4], which means that the preliminary estimate of the rupture line is poorly defined.

The final and best source model of a line is defined by the aftershock distribution and surface ruptures published years later [13–15]. This final model gives the best agreement between calculated and observed fatalities, but it does not become available within an hour after the shock. It is, however, a measure of QLARM performance for Italy with final earthquake source parameters known. The three agreements in Table 3 are most encouraging because they indicate that the program QLARM yields correct estimates of fatalities in Italian earthquakes.

The hypothetically calculated number of injured matches the reported numbers surprisingly well (Table 3). In all models (three rows in Table 3), the minimum and maximum estimates of injuries encompass the reported number.

The history of casualty underestimates by news media is not known to us in the case of the 1980 Irpinia earthquake, but we assume it was similar to the cases documented [16]. Figure 4 shows the delay of assessing the size of the disaster in the case of the L'Aquila M6.3 earthquake on 6 April, 2009, compared to the mean value estimates by QLARM 22 min after the earthquake. The slow reporting of correct fatality numbers by media is not surprising because no information is flowing from regions of earthquake disasters for hours or even days. Even in the case of a small earthquake like in L'Aquila, it takes some time for the extent of the losses to become clear (Figure 4). This is more pronounced in larger disasters [16] as in the Wenchuan M8 and Kashmir M7.6 earthquakes that killed more than 87,000 people. In these instances, more than a week passed before the extent of the disaster became apparent. Given that the Irpinia earthquake's magnitude was intermediary between that of L'Aquila,

Wenchuan and Kashmir, one has to assume that in a future Italian earthquake of M7±, there will be several days of fatality underestimates by the news media. The reports by PAGER and QLARM could correct this misunderstanding by highlighting the need of major rescue efforts.

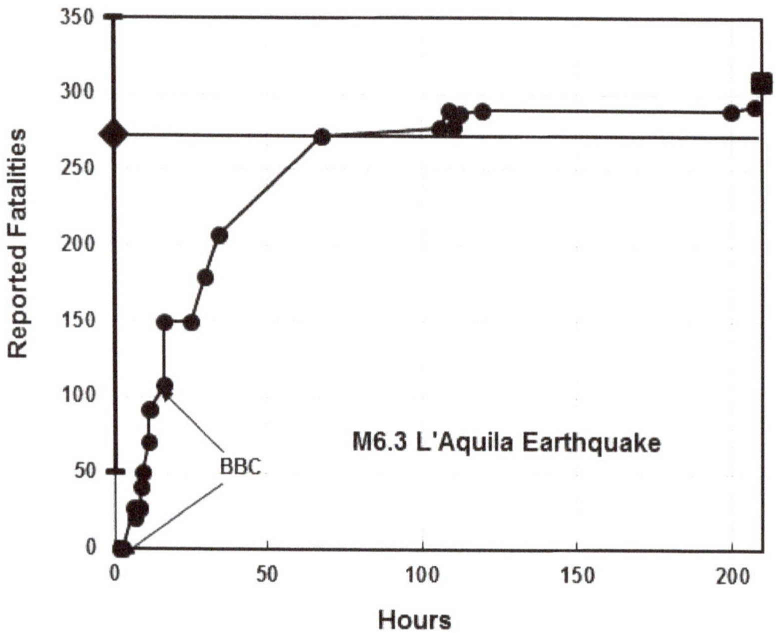

Figure 4. Reports of fatalities as a function of time after the L'Aquila M6.3 earthquake. The average estimate by QLARM was correct and given after 22 min.

Initial epicenter errors can map into large errors in fatality estimates when the population present is concentrated in one spot [7]. However, in the center of Italy, this is a less severe problem because the population is distributed. The results by three models for the Irpinia source with increasing precision do not yield vastly different fatality estimates (Table 3). Nevertheless, in Italy, loss estimators must pay attention to the possibility of one of the larger metropolitan areas being affected.

Back projections of the rupture line source [17] and estimates of early warning [18] may become available in near-real-time. This will replace the need to rely on aftershocks to estimate the rupture direction and length. After the Irpinia earthquake, relevant aftershocks were recorded within about half an hour, but this is not always the case. Therefore, near-real-time back projections and early warnings will become important to increase the accuracy of fast loss estimates after earthquakes.

The Irpinia source was a complex multiple rupture. It is known that large and great earthquakes tend to be multiple ruptures since this was established for the M9.2 Alaskan earthquake of 1964 [19]. One might therefore ask how strong ground motion should be modelled for multiple ruptures. One answer comes from the strong motion record at right angles and at 43 km distance from the M7.2 Kalapana rupture in 1975. The aftershock area in 1975 was 50 km long and six separate sub-earthquakes were identified [20]. In Harvey, D. 1986 [20], Figure 2 demonstrates that pulses of seismic energy were emitted for more than 60 seconds with intervals as long as 10 seconds of no energy being released. The amplitudes of these pulses were not much larger than those of the M5.9 foreshock (Figure 2 of [20]), which means that, for engineering purposes, the Kalapana seismic radiation consisted of a sequence of pulses corresponding to earthquakes in the range of $6.0 < M < 6.5$. The overall rupture, as measured by long period surface waves, however, indicated an M7.2 earthquake.

The complexity of the Irpinia earthquake must have similarly modified the local radiation of high frequency seismic waves. Nevertheless, the sum of the damage and the human losses is estimated correctly using a ground motion prediction equation for an M6.9 earthquake.

Admittedly, the size of the error bars for estimating casualties (minima and maxima in Table 3) are large. This is because minutes after an earthquake there are many poorly known parameters including location, depth, M, direction of rupture, length of rupture, energy propagation effects, local soil conditions, condition of the built environment and occupation rate of buildings. In spite of these numerous uncertainties, casualties and injured in the Irpinia M6.9 and L'Aquila M6.3 earthquakes are estimated correctly.

Satellite images could provide the strongest refinements of estimates of rupture location and extent. An example of this has been shown for the Bam earthquake [21]. In this case, interferometry could have shown that the rupture went straight through the city of Bam, while early estimates placed the epicenter 5 to 10 km west of the city. The difference in fatality estimates in this case was nearly two orders of magnitude. In Italy, the difference would not be that large because the population is more evenly distributed than in the area of Bam, Iran. Nevertheless, rapidly constructed interferograms that would define the crustal deformation due to an earthquake could help greatly in improving rapid loss assessments. For this to be possible, the necessary satellite passes have to become available without delay.

6. Conclusions

Rupture dimensions of disastrous earthquakes in Italy are not very long, only in the range of 20 to 80 km. Therefore, point source estimates of shaking, damage and casualties are reasonably reliable. Definitions of the direction and length of ruptures improve the loss estimates to a level that can be called excellent in the case of the M6.9 Irpinia earthquake. The correct estimates of casualties presented in this paper for Irpinia calculated in 2020 have also been achieved in near-real-time by QLARM for the M6.3 L'Aquila earthquake. This means that QLARM can be expected to correctly estimate casualties after Italian earthquakes within less than an hour. Since two groups, QLARM and PAGER, issue earthquake loss alerts that are similar and reliable, media, government and the general population should base their response to earthquake disasters on these estimates, which become available within less than an hour, instead of relying on the notorious underestimates by media that can last for weeks.

Author Contributions: M.W. has developed the concept and has performed the calculations. P.R. has researched and implemented the data used and made Figure 3. All authors have read and agreed to the published version of the manuscript.

Funding: This research received no external funding

Acknowledgments: Dataset for population and buildings in QLARM for Italia was jointly developed with colleagues Sandra Hurter and Goran Trendafiloski. We thank S. Tolis for information on buildings in Greece and the reviewers for helpful comments, especially reviewer 1.

Conflicts of Interest: The authors declare there exists no conflict of interest.

References

1. Wyss, M. Ten years of real-time earthquake loss alerts. In *Earthquake Hazard, Risk, and Disasters*; Wyss, M., Ed.; Elsevier: Waltham, MA, USA, 2014; pp. 143–165.
2. Trendafiloski, G.; Wyss, M.; Rosset, P. Loss estimation module in the second generation software QLARM. In *Human Casualties in Earthquakes: Progress in Modeling and Mitigation*; Spence, R., So, E., Scawthorn, C., Eds.; Springer: Dordrecht, The Netherland, 2011; pp. 381–391.
3. Rosset, P.; Bonjour, C.; Wyss, M. QLARM, un outil d'aide à la gestion du risque sismique à échelle variable. In Plans communaux de sauvegarde et outils de gestion de crise. In *Collection «Géorisques»*; Leone, F., Vinet, F., Eds.; Presses Universitaires de la Méditerranée: Montpellier, France, 2015; pp. 91–98.
4. Rosset, P.; Wyss, M. Seismic Loss Assessment in Algeria Using the Tool QLARM. *Civ. Eng. Res. J.* **2017**, 2, 1–3. [CrossRef]

5. Wyss, M.; Wu, Z.L. How Many Lives Were Saved by the Evacuation Before the M7.3 Haicheng Earthquake of 1975? *Seismol. Res. Lett.* **2014**, *85*, 126–129. [CrossRef]
6. Wyss, M.; Gupta, S.; Rosset, P. Casualty estimates in repeat Himalayan earthquakes in India. *Bull. Seismol. Soc. Am.* **2018**, *108*, 2877–2893. [CrossRef]
7. Wyss, M.; Elashvili, M.; Jorjiashvili, N.; Javakhishvili, Z. Uncertainties in teleseismic epicenter estimates: Implications for real-time loss estimates. *Bull. Seismol. Soc. Am.* **2011**, *101*, 1152–1161. [CrossRef]
8. Rosset, P.; Tolis, S.; Speiser, M.; Wyss, M. QLARM, un outil au service de la gestion du risque sismique et des crises; Étude de cas au Kirghizstan. *Georisque* **2020**. In Review.
9. Gruenthal, G. *European Macroseismic Scale 1998*; Conseil de l'Europe: Luxembourg, 1998.
10. Wyss, M.; Chamlagain, D. Estimated Casualties in Possible Future Earthquakes South and West of the M7.8 Gorkha earthquake of 2015. *Acta Geophys.* **2019**, *67*, 423–429. [CrossRef]
11. Wells, D.L.; Coppersmith, K.J. New empirical relationships among magnitude, rupture length, rupture width, rupture area and surface displacement. *Bull. Seismol. Soc. Am.* **1994**, *84*, 974–1002.
12. Wyss, M. Estimating expectable maximum magnitude of earthquakes from fault dimensions. *Geology* **1979**, *7*, 336–340. [CrossRef]
13. Deschamps, A.; King, G.C.P. Aftershocks of the Campania-Lucania (Italy) earthquake of 23 November 1980. *Bull. Seismol. Soc. Am.* **1984**, *74*, 2483–2517.
14. Bernard, P.; Zollo, A. The Irpinia (Italy) 1980 earthquake: Detailed analysis of a complex normal faulting. *J. Geophys. Res.* **1989**, *94*, 1631–1647. [CrossRef]
15. Pantosti, D.; Schwartz, D.P.; Valensise, G. Paleoseismology along the 1980 surface rupture of the Irpinia Fault: Implications for Earthquake recurrence in the Southern Apennines, Italy. *J. Geophys. Res.* **1993**, *98*, 6561–6577. [CrossRef]
16. Wyss, M. Report estimated quake death tolls to save lives. *Nature* **2017**, *545*, 151–153. [CrossRef] [PubMed]
17. Wang, D.; Kawakatsu, H.; Zhuang, J.; Mori, J.; Maeda, T.; Tsuruoka, H.; Zhao, X. Automated determination of magnitude and source length of large earthquakes using backprojection and P wave amplitudes. *Geophys. Res. Lett.* **2017**, *44*, 5447–5456. [CrossRef]
18. Li, J.; Böse, M.; Wyss, M.; Wald, D.J.; Hutchison, A.; Clinton, J.F.; Wu, Z.; Jiang, C.; Zhou, S. Estimating rupture dimensions of three major earthquakes in Sichuan, China, for early warning and rapid loss estimates. *Bull. Seismol. Soc. Am.* **2020**, *110*, 920–936. [CrossRef]
19. Wyss, M.; Brune, J.N. The Alaska earthquake of 28 March 1964: A complex multiple rupture. *Bull. Seismol. Soc. Am.* **1967**, *57*, 1017–1023.
20. Harvey, D.; Wyss, M. Comparison of a complex rupture model with the precursor asperities of the 1975 Hawaii Ms = 7.2 earthquake. *Pure Appl. Geophys.* **1986**, *124*, 957–973. [CrossRef]
21. Wyss, M.; Wang, R.; Zschau, J.; Xia, Y. Earthquake loss estimates in near real-time. *EOS Trans. Am. Geophys. Union* **2006**, *87*, 477–479. [CrossRef]

geosciences

MDPI

Article

Reconstruction as a Long-Term Process. Memory, Experiences and Cultural Heritage in the Irpinia Post-Earthquake (November 23, 1980)

Gabriele Ivo Moscaritolo

Department of Social Science, Federico II University of Naples, 80138 Naples, Italy; moskari@hotmail.it

Received: 26 June 2020; Accepted: 12 August 2020; Published: 15 August 2020

Abstract: Reconstruction after an earthquake is often seen as a material issue, which concerns "objects" such as houses, roofs, and streets. This point of view is supported by the mass media showing the work progress in the disaster areas, especially in conjunction with anniversaries. Rather, we should consider reconstruction as a complex social process in which cultural backgrounds, expectations, and ideas of the future come into play, without neglecting geological, historical, legislative, economic, and political factors. Combining oral history sources and archival records, the article shows the paths taken by two small towns among the most affected by the earthquake of 23rd November 1980 (Mw 6.9). These towns have made opposite reconstruction choices (in situ and ex novo) representing two classical and different ways in which human societies can face their past and think their own future. A careful analysis of these forty-year experiences, with a special focus on cultural heritage, provides useful indications for post-disaster reconstructions in which more attention to the process, and not just to the final product, should be paid.

Keywords: disasters; earthquake; reconstruction; cultural heritage; experiences; oral history; memory; 1980 earthquake

1. Introduction

The earthquake occurred on 23rd November 1980 has been one of the most disastrous seismic events in recent Italian history. It affected a large area in Southern Italy, destroyed dozens of towns, and there were 2735 victims, 9000 injured, and 394,000 homeless [1]. Over the last few decades, the media have recounted this event by emphasizing the central government's unpreparedness in managing relief efforts or by remembering the corruption and the wastefulness during the reconstruction phase. Moreover, many Italian scholars have concentrated their attention on political and economic aspects [2,3], especially in the area around Naples [4–6].

This study is part of another line of research, in which the experiences and memories of the affected populations have been investigated [7–11]. More specifically, it retraces the history of Sant'Angelo dei Lombardi and Conza della Campania (province of Avellino) (Figure 1)—two towns where the X MCS was reached [12,13]—which adopted opposite approaches to reconstruction. A "philological reconstruction" [14] of the old town center was the choice in Sant'Angelo dei Lombardi, whereas Conza della Campania opted for a new settlement rebuilt ex novo near the ancient center, which today has become an archaeological site. These special cases represent both future-oriented choices and two ways in which the past and cultural heritage can be preserved. The aim is to illustrate how a natural phenomenon interacts with human society and how different responses may arise from the same event. Furthermore, another purpose is to show the complex social process of reconstruction and how people after forty years evaluate and rationalize it. From this perspective, the 1980 earthquake is an extremely interesting case study, because the reconstruction act of law (Act. 219/81) granted much autonomy to the local municipality and, after four decades, a broad spectrum of different choices and outcomes is observable.

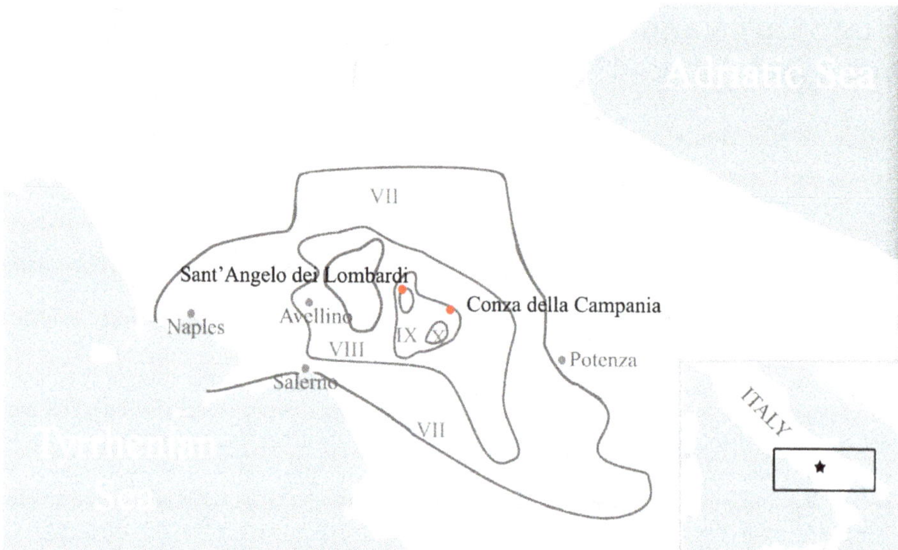

Figure 1. Isoseismal lines of the 1980 earthquake and localization of the two studied towns (modified after Postpischl et al. [13]).

To fully understand the impact of an earthquake on human societies an ecological perspective is necessary. "The natural world and human societies are more easily understandable when they are considered as two systemic and complex realities, fully interactive with each other. They are the most strongly interactive with each other because they rest on the same material, physical, chemical and biological base" [15] (p. vi). Accordingly, "disasters occur at the intersection of nature and culture and illustrate, often dramatically, the mutuality of each in the constitution of the other" [16] (p. 24). Time is also a fundamental factor in understanding catastrophes, because the complex relationship between man and nature is historically constructed and it is based on short or long-duration social processes whereby human beings adapt to their environment [17]. Therefore, an approach capable of encompassing environment, culture, and history becomes essential. In this complex interaction, it is useful to consider the notion of "resilience", which is widely used in disaster studies. The term has received some criticism because diverse actors infuse the concept with diverse meanings [18] but an agreement on its definition may be very productive in this field [19]. Today, the most common definition is provided by United Nations Office for Disaster Risk Reduction (UNDRR): "The ability of a system, community or society exposed to hazards to resist, absorb, accommodate, adapt to, transform and recover from the effects of a hazard in a timely and efficient manner, including through the preservation and restoration of its essential basic structures and functions through risk management" [20].

From this perspective, it is interesting to consider how disasters affect tangible cultural heritage and, consequently, how people try (or do not try) to preserve it. In any area that has been settled for centuries, historic buildings and monuments tend to be a highly visible part of daily life. They embody the continuity of time between the generations and help define the genius loci, or spirit of place, of a settlement [21]. When a disaster occurs, the tangible legacy inherited from past generations can be damaged and at the same time, the cultural identity of a geographical locality is threatened. As Ian Convery et al. underline: "Disasters and catastrophic events can be seen as 'happenings' that entangle people, place and their heritage, and disasters and displacement can leave people overcome by a 'loss of self' and a 'loss of place' [...]. While the intangible cultural heritage of a community might be considered as less at risk from catastrophe, in extreme cases the loss of culture bearers, or dramatic shifts in society, can result in the loss of these heritage assets" [22] (p. 2).

We will see how two small towns with a thousand-year history have decided to rebuild their settlement and cultural heritage differently. After crucial choices, a long social process started, and the witnesses interpret it in the light of the present.

2. Materials and Methods

This study is based on both archival records and oral interviews. The archives are Archivio Storico Protezione Civile (ASPC), Archivio di Stato di Avellino (ASAV), Comune di Conza della Campania (CCC), and Comune di Sant'Angelo dei Lombardi (CSL). ASPC and ASAV allow us to reconstruct how the central state dealt with the impact of the earthquake. In particular, about 600 documents have been examined, concerning both relief operations and the emergency structure that was organized in the following months [23–40]. The local archives (CSL, CCC) show how local authorities discussed and then planned the material reconstruction of the towns. Here, about 240 documents have been consulted, including scripts of council meetings, reconstruction plans, and maps [41–58]. Additionally, oral testimonies have been used. They provide important insights into the affected populations' point of view [59]. In particular, the study of memory can help us to understand what are the reasons behind important choices, how these are interpreted after 40 years, and how the whole reconstruction process influences both the material circumstances and the lives of populations. All the interviews have been collected by the author between 2014 and 2017 [60,61]. Witnesses belong to different generations and social classes: there are "institutional" people (mayors or municipal administrators), adult residents in 1980, and the new generations (born after 1980, but to whom memories have been transmitted) (Table 1). Contact with the witnesses took place in different ways. In some cases, the support of municipalities and local associations was central, in others, personal knowledge networks have been activated. The use of multiple channels to know new witnesses allowed us to reach different points of view and experiences. In general, people told their stories with pleasure, except in some cases where they did not want to speak about the evening of 23rd November. This is a very significant fact, as it is a sign that for someone the trauma has not yet been worked out. The interviews were videotaped and the transcriptions in original language are available upon request. Some of them are also available on the Multimedia Archive of the Memory (Archivio Multimediale delle Memorie) [62], hosted by the Department of Social Sciences at the University of Naples "Federico II".

Table 1. List of witnesses with description, year of birth, and date of the interview.

Name	Description	Year of Birth	Date of the Interview
	Sant'Angelo dei Lombardi		
Franco A.	Council employee and volunteer in 1980	1957	14 July 2015
Tonino C.	Lawyer	1950	3 June 2016
Angela C.	Council employee	1957	13 June 2016
Giovanna C.	Council employee	1957	13 June 2016
Assunta F.	Student in 1980	1966	12 July 2013
Giulio D.	Journalist	1978	3 February 2016
Angelo F.	Doctor	1951	6 February 2013
Elisa F.	Journalist	1981	6 February 2016
Michele G.	Council employee and volunteer in 1980	1955	2 September 2016
Concetta M.	Housewife	1932	4 October 2014
Luigi M.	Municipal councilor in 1980	1950	17 July 2015
Giuseppe L.	Municipal councilor	1989	1 April 2015
Vincenzo L.	Municipal councilor in 1980	1948	1 September 2016
Tony L.	Hospital employee in 1980	1956	31 May 2016
Romualdo M.	Historian	1943	2 February 2016
Carmine M.	Hairdresser	1950	3 February 2016

Table 1. *Cont.*

Name	Description	Year of Birth	Date of the Interview
Francesco P.	Council employee and volunteer in 1980	1945	19 August 2015
Rosanna R.	Mayor in 1980	1950	3 September 2016
Rosaria S.	Social worker during the emergency	1947	22 September 2015
Michele V.	Teacher	1949	4 February 2016
	Conza della Campania		
Vito C.	Mayor from 2013 to 2017	1959	19 February 2016
Vincenzo C.	Ex miner	1933	19 September 2016
Erberto C.	Architect	1962	2 February 2016
Giuseppe F.	Soldier in 1980	1960	18 February 2016
Vito F.	Mayor from 2003 to 2008	1962	17 November 2015
Antonia G.	Housewife	1937	19 February 2016
Felice I.	Mayor in 1980	1949	17 November 2015
Luigi L.	Historian	1951	17 November 2015
Maria M.	Student in 1980	1971	18 February 2016
Gerardina M.	Housewife	1942	19 September 2016
Antonia P.	President "Pro Loco Compsa"	1981	1 February 2016
Michele P.	Worker	1977	18 February 2016
Domenico T.	Merchant in 1980	1928	18 February 2016
Mario T.	Bricklayer	1958	17 November 2015

3. Results: Impact, Choices and Experiences

3.1. The Earthquake and Its Impact

"I remember a beautiful day... with a hot sun and a crowded square... full of people with children" says Rosanna. The 23rd November 1980 was an unusually sweet day, and this is a leitmotiv in the collective memory of witnesses. After this pleasant picture, many people use terms such as "apocalypse" or "end of the world" to indicate the sudden impact and effects of the earthquake. In the words of Romualdo: "I heard a noise... terrible... of irons... an explosion... and instinctively I headed for the exit... but I realized that the stairs were beginning to writhe ... the building collapsed on the other floors, and then the door collapsed on me... I could not go back into the house and I was in that condition overnight" (Romualdo R.). The interruption of roads and communication lines, the delay in the arrival of reliefs, and the absence of a civil protection plan were the causes that amplified the tragedy [24]. The final toll was 2735 victims, 9000 injured, and 394,000 homeless; 687 municipalities were affected, including 37 declared "devastated", 314 "seriously damaged" and 336 "damaged" [63].

After the first moments of shock, people and communities faced different situations. Before the earthquake, Conza della Campania retained its original medieval configuration of narrow streets and closely-packed houses. These buildings collapsed in a tragic domino effect becoming a pile of stones and sand. Destruction reached 95% of the settlement (Figure 2). There were 184 dead and first aid came from the inhabitants themselves. Hence, the population abandoned the hill on 24th November, found shelter in a construction site located down the valley, and here spent the first months. "Fortunately, that building was able to accommodate many people [...] we were really very crowded, but safe" remembers Luigi. The availability of a safe building to house the survivors allowed people to overcome some initial difficulties, such as the removal of rubble and the construction of temporary lodging. In the following days and months, the situation improved, thanks to the province of Bologna, which provided hundreds of volunteers and means to deal with the emergency [40].

The old city center of Sant'Angelo dei Lombardi also retained a medieval configuration. After the quake, it became a pile of rubble but most of the 432 deaths occurred in the "new" buildings, those that had arisen since the 1960s around the main square. These ones did not always comply with anti-seismic standards and collapsed generating the well-known pancake effect (Figure 3). "You could touch the roof of all of these buildings with your hands... they had become like an accordion" recalls Carmine. Compared to Conza, the situation was complicated, due to both the larger devastated area and the lack of the means to remove the reinforced concrete. There was no immediate availability of facilities to accommodate people, and the survivors spent the first few days in different ways as in their car or hosts by relatives in other towns. Then, the situation was brought under control thanks to the intervention of the volunteers from the Regione Toscana, and from the Provinces of Brescia and Pesaro-Urbino, who set up camps for homeless [36,37].

Figure 2. Effects of the earthquake in Conza della Campania (Courtesy of Pro Loco Compsa).

Figure 3. The earthquake in Sant'Angelo dei Lombardi. On the background, the "pancake effect" (Courtesy of Michele V.).

Thanks to the famous speech by President Sandro Pertini, and to the headlines of the Italian press, the devastating impact of the eartquake had a global echo, and economic aid came from all over Italy and other nations. However, while the central state and rescuers were still dealing with the emergency, local populations, experts, and politicians began to debate the future reconstruction.

3.2. *A Persistent Dilemma: Reform vs Continuity*

As noted by Ian Davis and David Alexander [21], this is the first planning dilemma that authorities had to face after a disaster. Post-seism destruction certainly creates a "window of opportunity". On the one hand, this is an occasion to change what was wrong in the past, on the other it can generate the desire to restore the old world remembered with nostalgia. Likewise, within the post-1980 debate, we can distinguish these two opposite positions. "Reformers"advocated the relocation of whole towns closer to the main roads continuing a trend that had already started before the earthquake. However, "conservers" criticized the relocation, because it was not in harmony with the agricultural vocation of the countryside, and would not have allowed the recovery of the historical-artistic heritage of the old towns. These two options, but also various intermediate solutions, could be realized thanks to both the possibilities offered by the reconstruction law and the huge amount of funds allocated. In May 1981, the government issued Act. 219, based on two keywords: "reconstruction" and "development". The aim was to "modernize" the affected areas, still considered in a state of backwardness. More precisely, the article 27 reads: "Rebuilding takes place in the area of existing settlements and, if there are geological, technical and social reasons for it, in the municipal area as a whole. It can also be carried out by means of extensions, completions and adaptations, technical and functional, or by means of new works deemed necessary for the reorganization of an area and for its economic and social development". All the choices took into account the Progetto Finalizzato Geodinamica—CNR, the first major national project for seismic risk assessment. This programme produced maps of seismicity and seismo-tectonics in Italy before the earthquake, and carried out surveys of damage and the potential risk to the stricken towns from seismic sources afterwards. In particular, the most important activities were focused on the production of the Structural Model of Italy, the Neotectonic Map, the Seismic Catalogue, the Atlas of Isoseismala for the largest historical events, the Seismic Hazard Map, the National Seismic Zoning, the Guidelines for seismic risk mitigation of ancient buildings, and the microzonation investigations in the epicentral areas of the 1976 Friuli and 1980 Campania and Basilicata [64,65].

In Conza della Campania the choice of relocation was shared by a large part of the population. Various reasons emerge from the field research:

- *Archeological*. After the removal of debris, the remains of *Compsa* (the original name of the ancient Roman city) came to light. The local heritage protection institution ('Soprintendenza') intervened to protect the archeological heritage. According to the official communication, no construction work should be undertaken in the historic center [53].
- *Geological*. Investigations by prof. Franco Ortolani (Department of Geology—Federico II University of Naples) underline the phenomenon of seismic site amplification on Conza hill, whereby the destructive effects of the earthquake were greater. Reconstruction in situ was not recommended [54]. More specifically, the town "was built on two small hills made by clay and sandy clay in the lower part, conglomerates with sands and sandstones in the middle part, and conglomerates of middle-low resistance in the upper part" [66] (p. 127) [67].
- *Personal*. For many inhabitants, the old town had become associated with the trauma they had experienced. For example, Maria went back to the hilltop for the first time in thirty years.
- *Pre-existing trends*. From the 1960s, a building and commercial development towards the main roads had already started. As the major in charge remembers: "Our emigrants had invested their savings to build their house there … it was along the Ofantina that commercial and craft businesses sprung up, not on the hilltop of Conza della Campania, so we started to entertain

the notion of building, commercial and artisanal expansion along the slopes going down towards the valley" (Felice I.).

Differently, the townspeople of Sant'Angelo intended to restore the lost past. This choice was possible thanks to the geological investigations which considered the area of the ancient center suitable for reconstruction [68]. Various reasons can be identified:

- *Status of the town*. Compared to other towns of the affected area, Sant'Angelo was perceived as very relevant due to the presence of public facilities such as a prison, a court, a tax office, an ASL (local health districts) and a hospital (inaugurated in 1979). The destruction of the new buildings led also to huge economic losses. As the major in charge remembers: "Sant'Angelo was the leader town of the district [...] we thought to get the court back, to get the hospital back, to continue to playing that role" (Rosanna R.). A municipal resolution fully documents this position [42].
- *Cultural/historical*. In the early aftermath of the earthquake, hasty demolitions of historic buildings occurred in many small towns. In Sant'Angelo, many scholars opposed the destruction of medieval monuments and tried to recover archaeological heritage creating a "Village of cultural heritage" [43]. In Romualdo's words, we note the intent to preserve the genius loci of the place: "We all got together united by the common intent to save identity, because if you destroy the site where a community has lived for centuries, you destroy its identity, we said: 'we don't need industries... let us use local resources, transplanting is useless'" (Romualdo R.).
- *Personal*. Unlike Conza, many people did not want to abandon the places of the tragedy. As Michele remembers: "I intended to rebuild the town where it used to be... with its history and its culture... this intent was largely shared by the townspeople . . . in my opinion, there was a prevalence of personal feelings, of affection, of memory, a wish to stay in the very place where these tragic things had happened . . . what most people wanted was to stay together, to remain at the site of the tragedy... of the memory" (Michele G.).

Therefore, starting from various reasons, these two small towns have decided to rebuild their settlement differently. In both cases, the tangible cultural heritage and the ancient history of the centers have played an important role in the decisions.

3.3. Preserving Cultural Heritage

Historical monuments and buildings, but also modern structure or landscape features, are often elements that embody the spirit of a place, or genius loci. They help to create a sense of belonging and a special connection between people who identify them as part of their own identity [69]. For these reasons, cultural heritage protection and restoration are often among the immediate priorities of recovery from disaster, although they are expensive, complex, and time-consuming processes. For example, the earthquakes that hit Italy in 1997 (in the regions of Umbria and Marche) damaged about 1200 religious buildings, and a large recovery project was set up in the aftermath [70]. In the post-1980 reconstruction debate, the recovery of historical buildings was central and each town headed for different choices [71,72]. Both of the case studies discussed here represent two ways of preserving and celebrating ancient origins, although they pursue two opposite choices for the future.

Sant'Angelo dei Lombardi became the seat of the Plan Office of Soprintendenza (Ufficio di Piano) and a strong collaboration between local scholars and from all over Italy was created. They opposed the demolitions of the old town center and established the "Cultural and Environmental Heritage Service" with the task of coordinating recovery initiatives [43]. The "Earthquake archaeologists", as the press nicknamed the group, created a "Village of cultural heritage" and requested funding for interventions in the historic center. After the approval of the Act. 219, the Technical Commission of Cultural Heritage was formed, and it worked on the preparation of the Recovery Plan [41,46]. The aim was to consider the old town center as a whole, and not as a mere sum of buildings [44]. This conception was not applied in other Recovery Plans and, therefore, in Sant'Angelo, the technical standards were established for the specific case. Moreover, among the aims of the Plan, there were both the resettlement of the population and the start of economic activities. The restoration of the entire old town center took many years compared to the new homes. The overall time was approximately 20 years (Figure 4).

Figure 4. The cathedral of Sant'Angelo dei Lombardi, before (above) and after (below) the earthquake.

In comparison to Sant'Angelo, the recovery of the ancient *Compsa* had different purposes. On the one hand, the resettlement of the population was not planned, as a new town close to the main road would be built. On the other, there were Roman archeological finds to bring to light, and no damaged buildings to restore. Professor Corrado Beguinot (School of Architecture—"Federico II" University of Naples) designed the Recovery Plan which was approved in September 1982. The project considered the old settlement an economic and cultural asset, as supported by the most recent debate on the safeguarding and protection of historic centers. For this reason, interventions and restorations for tourist and cultural activities would have been scheduled. After the removal of debris, excavation and restoration campaigns were planned to create the archaeological park where an antiquarium (a sort of museum) and a seismological center would be built. The park was inaugurated in 2003 (Figure 5) [71].

Figure 5. In the foreground, the "new Conza". On the hilltop, the archeological park (Courtesy of Pro Loco Compsa).

3.4. *After the Choices, Life Continues*

Reconstruction after an earthquake is not a mere material issue concerning "objects" such as houses, roofs, and streets. To fully understand its outcomes, we should consider it as a complex social process, rather than just observing the final product. Thus, in order to illuminate the relationship between people and the environment, we have to study the various stages that the populations go through while waiting for the completion of the work. These phases constituted an important part of their own experience and help us understand how people adapt to their environment, especially after the sudden transformations caused by a major earthquake. As Sara Zizzari has shown in her study—where she combines the L'Aquila post-earthquake urban transformations with oral sources—the path to returning home is a complex social-spatial issue, which has consequences for both individual and collective paths [73]. In the cases presented here, the implementation of the plans was the starting point of a long path.

After the first months spent in the construction site, the inhabitants of Conza lived for twelve years in a valley settlement: "The urbanised area was built by the Province of Bologna according to really valid urban criteria... It really was a model of extraordinary urban cohabitation. All the spaces were well arranged... the prefabricated homes, although small, were comfortable... you got a sense of privacy, of intimacy, in short, of family, not separated, because they were contiguous and therefore in a way they restored the downhome dimension of the old town, of the neighborhood" (Luigi L.). This positive memory is largely shared among townspeople. The settlement allowed people to be as close together as they had been in the old town center. Gerardina, in turn, remembers how the condition of being "all earthquake victims" contributed to strengthening ties: "It was good in the prefabricated homes... this was a good experience... because we lived closer to each other... we were all equal and being all equal is important... there were no rich, there were no poor" (Gerardina M.). Unfortunately, after this good period, the transfer to the new town was traumatic. The "new Conza" had an urban structure very different from the old city (Figure 5). This was a long-lasting cause of disorientation for the population. As Antonia recalls: "This is a dark moment in my mind, the move to the new Conza della Campania, because at that time Conza was not a town, it was a group of houses where there were no facilities for the community... the square was not a square, there were no memories associated with those places, there was nothing for us" (Antonia P.). Since 1992, the appearance of the town has changed a lot, as new urban furniture has made the center more livable. For the younger generations, it is certainly easier to adapt to the new places, but for those who used to live in a different spatial context, the "old" Conza remains a world remembered with nostalgia.

As we have seen (Section 3.1), in Sant'Angelo there was no immediate availability of facilities to accommodate people and even establishing temporary settlements was a complicated matter. The population was larger than Conza, and they did not want to leave the destroyed town center. Thus, temporary housing sprung up in various available areas around the ruins, creating a patchwork spatial pattern. This sudden change led to a sense of displacement among the citizens, who were accustomed to a social life concentrated around the main square and the town center. Tony's words show this lasting sense of disorientation: "For many years, this patchwork layout caused us to lose the centrality of the agora, of the main square ... and perhaps we are still bearing the consequences... the square that is usually the heart of the community, where you meet, where you argue, where you walk, where you discuss... it was empty for many years... it was the ghost of the town's main square... in the evening there was nothing at all... you could meet at one or two bars, located within these settlements or near them... for too long a period... settlement at the margins of the town led to the loss of the sense of community" (Tony L.). If in the case of Conza there are well-defined stages that the population has gone through, the transition to the "new" Sant'Angelo has been more gradual. Here, the intent was to recover the destroyed old town, to preserve its artistic and cultural identity, and to allow the resettlement of the population. Of course, restoration of ancient buildings is a time-consuming process and most of the population, after spending about fifteen years in prefabricated buildings, preferred to go and live in the new buildings at the edge of town. As Tonino underlines: "When the earthquake occurred, the historic center of Sant'Angelo was almost empty. Indeed, there were very few deaths in the historic center. People were resettled in expansion areas... this means that those whose house had collapsed in the expansion area preferred to return to the expansion area and to opt to have their home in the historic center as a second residence, because its construction would take much longer" (Tonino C.). Thus, today we can distinguish between the old part, which is well reconstructed but underpopulated, and the fragmented outskirts, where most of the inhabitants live [51].

4. Discussion

In their book, Christof Mauch and Christian Pfister recall the notion of "societies as weaving daily tapestries" [74] (p. 6). Following this metaphor, a disaster is "a gash or a sharply discordant thread suddenly introduced into the pattern". Accordingly, a historical perspective forces analysts to see "how a society repairs/reweaves itself and moves on. In many cases, the tapestry takes off in a dramatically different direction, with new colors and designs" [75] (pp. 1–2). This fascinating idea underlies the study presented here. Forty years after the 1980 earthquake, it is possible to adopt a long-term perspective, retrace the paths taken by the affected communities, and to show how different responses may arise from the same event. Moreover, from a memory studies perspective, forty years represent a significant time frame because "after forty years those who have witnessed an important event as an adult will leave their future-oriented professional career, and will enter the age group in which memory grows as does the desire to fix it and pass it on" [76] (p. 36). As we have seen, the witnesses' stories allow us to fully understand the upheavals caused by a great earthquake and illuminate how important decisions are made, decisions having an impact both on the environment and on the lives of populations. In other words, the memory perspective helps us to deeply investigate the complex relationship between human beings and their environment.

Sant'Angelo dei Lombardi and Conza della Campania represent two classical and different ways in which, after a great calamity, people can face their past and think their own future. The Italian sociologist Alessandro Cavalli proposed three ways whereby communities deal with the experience of space-time discontinuity. These are "ideal types", an idea-construct of social phenomena that do not fully correspond to reality, but are meant to stress certain elements common to most cases of the given phenomenon [77]. These types are the "re-localization" (the move of the entire town), the "philological reconstruction" (which aims to restore the pre-disaster state), and the "selective reconstruction" (which preserves some symbolic elements of the past) [14]. The model of re-localization, which concerns the case of Conza, is a sort of "year zero model", because it represents a real rebirth for the affected communities. In this case, the breaking event is celebrated in order "not to forget", but also to symbolically mark the start of the new course. In some cases, the past may be removed. However, this does not represent the case of Conza, as the town's ancient history is now preserved in the archaeological park, an open-air museum where the pre-quake memory has been "frozen". The case of Sant'Angelo's old town center corresponds to the "philological reconstruction", which aims to restore the past from where it left off. This choice reflects the desire to reclaim both lost time and space but, in this possibility, there is also an attempt to remove the disastrous event and to delete the element of discontinuity.

Both in the "ideal types" by Cavalli and in our studied cases, tangible cultural heritage plays an important role, as buildings and historical monuments are among the elements that contribute to create the sense of a place. This is Tony's opinion on the reconstruction of Sant'Angelo: "The choice was fundamental, the historic centre where it was, even if some mistakes were made [...] I think so... both the town as a whole and the old town centre; thanks to the choice to rebuild as it was, the mayor earned the Zanotti Bianco prize... I have shared and still share this choice today... I am in love with the historic centre" (Tony L.). However, as pointed out above, the reconstruction process is not a mere material issue. Rather, it is a complex social process involving many aspects of community life. For example, the choice to restore the past may also include the desire for a cohesive community. In the words of Tonino: "The story is interpreted in a certain way... as one of successes... that are possibly measured on the 'material' reconstruction... but as regards the 'spiritual' reconstruction, so to speak... I think that Sant'Angelo stopped existing on that exact day... in the sense that... there are still ruins" (Tonino C.). Thus, in this case, the desire to get back the lost past seems to have been partially fulfilled.

Otherwise, the old center of Conza has changed its meaning, as it has become an open-air museum from an inhabited place. Thus, while a tourist can imagine ancient civilizations by observing the remains of the ancient Roman and medieval *Compsa*, the inhabitants have different sensations. For some, the hilltop has become a place of death and they have had difficulty returning over time. For others, like Domenico, there is the pleasure of being a tourist guide for his friends, but also the sadness of not seeing the places of youth anymore: "Sometimes I go there... friends come and I take them to see... I bring them but when I get there... the heart suffers" (Domenico T.). Finally, Antonia, born after 1980, during her visits imagines life, stories and places transmitted by photos and family stories: "I imagine these narrow streets made of stone, these houses... always full of life, of people with their coming and going because the cars could not get the hilltop, so people in their daily lives gave life to the town because they went back and forth to do the daily chores... I imagine it as a coloured town" (Antonia P.).

What lesson can we learn from these stories? Does studying the experiences of communities affected by disasters have a purely historical interest? Or can we use this knowledge for future experiences? What is the role played by people's memory?

According to Christian Pfister, "natural hazards are of course retained in memory if they recur frequently, and the more frequently they occur, the more likely people are to anticipate them and to try to develop adequate adaptive strategies, which are always the result of learning processes and which can take different forms" [78] (p. 4). Consequently, the memory of disasters would be able to develop a "culture of disasters" whereby human societies adapt to risky environments [79]. However, this adaptation process is not obvious because "the manner, scope and thus benefit of this implicit or explicit 'learning' varies greatly and depends on epoch and culture. This variation reveals the role played by history and culture in the learning process" [80] (p. 76).

Starting from historical knowledge, it may be possible to draw lessons that can be useful for living with environmental risk. In other words, stop building vulnerable settlements, being able to manage future emergencies, and planning reconstructions from a long-term perspective. All these actions should take into account the social dimension of the disaster and not only the material one. As the geographer Robert Geipel pointed out: "Disaster and reconstruction are incisive events in the life of the individual and group [...]. Planning that follows only laws of a technical rationality would endanger the already injured identity" [81] (p. 152). For example, our cases inform us about the importance of maintaining social/spatial relations as similar as possible to the pre-disaster state, in order to favor the social cohesion after a traumatic experience. Furthermore, the importance of the recovery of cultural heritage is certainly a fundamental aspect of the reconstructions, as it allows establishing connection and common reference point between generations. However, a "spiritual reconstruction", which aims to repair the social ties of the affected community, should also be pursued. In conclusion, we should start from people's stories, because "local actors play such a crucial role in the transmission of social memory. We lack stories that can translate knowledge into a renewed sense of place and cultural identity. Local communities [...] teach us about the possibility of living differently on (and with) an unstable earth" [82] (p. 77).

Funding: This research received no external funding.

Acknowledgments: First of all, I would like to thank the people who wanted to tell their story. Without their voice, this work would not have been possible. Sincere thanks go to the archives staff who made the research possible. Finally, I would like to express my deep gratitude to Gabriella Gribaudi for her precious teachings during my studies and research.

Conflicts of Interest: The authors declare no conflict of interest.

References and Notes

1. INGV Terremoti. Available online: http://www.ingvterremoti.com (accessed on 23 June 2020).
2. Osservatorio Permanente sul Doposisma. *Le Macerie Invisibili. Rapporto 2010*; Edizioni Mida: Pertosa, Italy, 2010.
3. Osservatorio Permanente sul Doposisma. *La Fabbrica del Terremoto. Come I Soldi Affamano il Sud*; Edizioni Mida: Pertosa, Italy, 2011.
4. Barbagallo, F. *Napoli Fine Novecento. Politici Camorristi Imprenditori*; Einaudi: Torino, Italy, 1997.
5. Barbagallo, F.; Becchi Collidà, A.; Sales, I. (Eds.) *L'affare Terremoto. Libro Bianco Sulla Ricostruzione*; Sciba: Angri, Italy, 1989.
6. De Seta, C. *Dopo il Terremoto la Ricostruzione*; Laterza: Bari, Italy, 1983.
7. Gribaudi, G. *La Memoria, i Traumi, la Storia. La Guerra e le Catastrofi nel Novecento*; Viella: Roma, Italy, 2020.
8. Gribaudi, G.; Zaccaria, A. (Eds.) Terremoti. Storia, memorie, narrazioni. In *Memoria/Memorie*; Cierre Edizioni: Verona, Italy, 2013; Volume 8.
9. Gribaudi, G. Guerra, catastrofi e memorie del territorio. In *L'Italia e le Sue Regioni (1945–2011)*; Salvati, M., Sciolla, L., Eds.; Istituto dell'Enciclopedia Italiana: Roma, Italy, 2015; pp. 251–273.

10. Zaccaria, A.; Zizzari, S. Spaces of resilience. Irpinia 1980, Abruzzo 2009. *Sociol. Urbana Rural.* **2016**, *111*, 64–83. [CrossRef]
11. Ventura, S. *Non Sembrava Novembre Quella Sera*; Mephite: Atripalda, Italy, 2010.
12. Guidoboni, E.; Ferrari, G.; Mariotti, D.; Comastri, A.; Tarabusi, G.; Sgattoni, G.; Valensise, G. *CFTI5Med, Catalogo dei Forti Terremoti in Italia (461 a.C.-1997) e nell'area Mediterranea (760 a.C.-1500)*; Istituto Nazionale di Geofisica e Vulcanologia (INGV), 2018; Available online: https://www.earth-prints.org/bitstream/2122/11895/1/CFTI5Med_ITA.pdf (accessed on 15 August 2020). [CrossRef]
13. Postpischl, D.; Branno, A.; Esposito, E.; Ferrari, G.; Marturano, A.; Porfido, S.; Rinaldis, V.; Stucchi, M. The Irpinia earthquake of November 23, 1980. In *Atlas of Isoseismal Maps of Italian Earthquakes*; Postpischl, D., Ed.; CNR-PFG: Roma, Italy, 1985.
14. Cavalli, A. Tra spiegazione e comprensione: Lo studio delle discontinuità Socio-Temporali. In *La Spiegazione Sociologica. Metodi, Tendenze, Problemi*; Borlandi, M., Sciolla, L., Eds.; Il Mulino: Bologna, Italy, 2005; pp. 195–218.
15. Agnoletti, M.; Neri Serneri, M. *The Basic Environmental History*; Springer: London, UK, 2014. [CrossRef]
16. Oliver-Smith, A. Theorizing Disasters. Nature, Power and Culture. In *Catastrophe and Culture: The Anthropology of Disaster*; Hofmann, S., Oliver-Smith, A., Eds.; School of American Research Press: Oxford, UK, 2002.
17. Bankoff, G. Comparing vulnerabilities: Toward charting an historical trajectory of disaster. *Hist. Soc. Res.* **2007**, *32*, 103–114.
18. Tierney, K. *Disasters. A Sociological Approach*; Polity Press: Cambridge, UK, 2019.
19. Alexander, D.E. Resilience and disaster risk reduction: An etymological journey. *Nat. Hazards Earth Syst. Sci.* **2013**, *13*, 2707–2716. [CrossRef]
20. UNDRR. United Nations Office for Disaster Risk Reduction. Available online: http://www.undrr.org (accessed on 11 July 2020).
21. Alexander, D.E.; Davis, I. *Recovery from Disaster*; Routledge: London, UK, 2015.
22. Convery, I.; Corsane, G.; Davis, P. (Eds.) *Displaced Heritage: Responses to Disaster, Trauma, and Loss*; The Boydell Press: Woodbridge, UK, 2014.
23. Archivio Storico Protezione Civile—ANSA Notizie Varie.
24. Archivio Storico Protezione Civile—Ministero dell'Interno. Direzione Generale della Protezione Civile, Terremoto del 23/11/1980—Sintesi Cronológica degli Avvenimenti Salienti.
25. Archivio Storico Protezione Civile—Ministero dell'Interno. Direzione Generale della Protezione Civile, Rapporto VV.FF.
26. Archivio Storico Protezione Civile—Ministero dell'Interno. Direzione Generale della Protezione Civile, Sintesi Cronológica degli Avvenimenti Salienti FF. AA.
27. Archivio Storico Protezione Civile—Ministero dell'Interno. Direzione Generale della Protezione Civile, Situazione degli uomini, Mezzi e Materiali Inviati nelle Provincie Terremotate, Roma 24.11.1980, ore 14.45.
28. Archivio Storico Protezione Civile—Ministero dell'Interno. Gabinetto del Ministro, Terremoto in Campania e Lucania del 23 Novembre 1980, ore 19.37, Attività di Soccorso e di Assistenza alle Popolazioni Colpite. «Situazione» n. 5—4 Dicembre 1980—ore 10.00.
29. Archivio Storico Protezione Civile—Ministero dell'Interno. Direzione Generale della Protezione Civile, Roma 25.11.1980, ore 4.00.
30. Archivio di Stato di Avellino—Arretramento, Interventi Assistenziali in Favore Delle Popolazioni del Territorio Dipendente dal COS n°1 di Sant'Angelo dei Lombardi, 24 Gennaio 1981.
31. Archivio di Stato di Avellino—Centri Operativi. Organizzazione e Compiti.
32. Archivio di Stato di Avellino—Comando Settore Avellino. Settori di Responsabilità.
33. Archivio di Stato di Avellino—C.O.P. Avellino presso Caserma Berardi. C.O.S Centri Operativi di Settore.
34. Archivio di Stato di Avellino—Commissariato Straordinario del Governo per la Campania e la Basilicata, Scioglimento dell'organizzazione dei Centri Operativi di Settore, 19 Giugno 1980.
35. Archivio di Stato di Avellino—Riunione con i Coordinator dei CC.OO.SS., 26 Gennaio 1981.
36. Archivio di Stato di Avellino—Provincia di Pesaro-Urbino, Intervento a Sant'Angelo dei Lombardi eseguito dall'Unità Operativa Servizi Speciali dell'Amministrazione Provinciale Coadiuvata da Personale di altri enti e Volontari.

37. Archivio di Stato di Avellino—Centro Operativo «Toscano» Località Sant'Angelo dei Lombardi—Situazione Enti, uomini e mezzi Presenti al 9.XII.1980.
38. Archivio di Stato di Avellino—Situazione Insediamenti Provvisori alla data del 13/06/1981 COS n.1.
39. Archivio di Stato di Avellino—COS n.3 Situazione Insediamenti Provvisori alla data del 10/06/1981.
40. Archivio di Stato di Avellino—Enti locali dell'Emilia Romagna.
41. Comune di Sant'Angelo dei Lombardi—Castellano, R., De Cunzo, M., De Martino, V., Franco, V., Marandino, R., Massarelli, A. Piano di Recupero del Centro Storico, D.C. 29 Giugno 1981.
42. Comune di Sant'Angelo dei Lombardi—D.C. n.20, Centro Storico: Orientamenti, 10 Aprile 1981.
43. Comune di Sant'Angelo dei Lombardi—D.C. n. 1, Servizio Beni Culturali e Ambientali, 3 geNnaio 1981.
44. Comune di Sant'Angelo dei Lombardi—De Cunzo, M., Restauro dei Centri Storici, in Particolare in S. *Angelo dei Lombardi*.
45. Comune di Sant'Angelo dei Lombardi—D.C. n. 133, Adozione Piani ai sensi dell'art. 28 della legge 14.5.1981 n. 219, 16 Settembre 1981.
46. Comune di Sant'Angelo dei Lombardi—D.C. n. 75, Commissione Tecnica Beni Culturali—Proposta della Giunta Municipale, 29 Giugno 1981.
47. Comune di Sant'Angelo dei Lombardi—D.C. n. 2, Individuazione aree Prefabbricati, 12 Gennaio 1981.
48. Comune di Sant'Angelo dei Lombardi—Criteri per L'assegnazione dei Prefabbricati, 20 Maggio 1981.
49. Comune di Sant'Angelo dei Lombardi—Studio di Architettura Corrado Beguinot e Associati, Piano Regolatore Generale, Comune di Sant'Angelodei Lombardi 1983.
50. Comune di Sant'Angelo dei Lombardi—Campisi, M., Corona, F., Losco, G., Piano di Recupero. Progetto di variante, Comune di Sant'Angelo dei Lombardi 1991.
51. Comune di Sant'Angelo dei Lombardi—Dal Piaz, A., Apreda, A., Bruno, G., Piano Urbanistico Comunale. Relazione Generale, Comune di Sant'Angelo dei Lombardi, Aggiornamento Aprile 2019.
52. Comune di Conza della Campania—D.C. n. 1/S, 8 Dicembre 1980.
53. Comune di Conza della Campania—DC. n.112, 25 Settembre 1982.
54. Comune di Conza della Campania—Indagine Geologico Tecnico e Geognostica dell'area del Centro Abitato, prof. Franco Ortolani, Maggio 1982.
55. Comune di Conza della Campania—D.C. n. 29, 24 Maggio 1981.
56. Comune di Conza della Campania—Studio di Architettura Corrado Beguinot e Associati, Criteri Generali di Impostazione del P.R.G.C. Piano per L'edilizia Economica e Popolare, Comune di Conza della Campania 1981.
57. Comune di Conza della Campania—Architettura e Studi urbani Valter Bordini, Progetto di variante al Piano per L'edilizia Economica e Popolare, Comune di Conza della Campania 1984.
58. Comune di Conza della Campania—Ciccone, E., Sposito, T., Ciccone, F., Strazza, G., Piano Particolareggiato di Esecuzione per la «zona B», Comune di Conza della Campania 2002.
59. Portelli, A. *Storie Orali. Racconto, Immaginazione, Dialogo*; Donzelli: Roma, Italy, 2007.
60. Moscaritolo, G.I. Storia Sociale di un Terremoto. Esperienze, Memorie e Trasformazioni nel Cratere Irpino. Sant'Angelodei Lombardi—Conza della Campania (1950–2016). Ph.D. Thesis, Federico II University of Naples, Naples, Italy, 2 May 2017.
61. Moscaritolo, G.I. *Memorie dal Cratere. Storia Sociale del Terremoto in Irpinia*; Editpress: Firenze, Italy, 2020.
62. Archivio Multimediale delle Memorie. Available online: http://www.memoriedalterritorio.it (accessed on 23 June 2020).
63. Senato della Repubblica, Camera dei Deputati. *Commissione Parlamentare di Inchiesta sulla Attuazione degli Interventi per la Ricostruzione e lo Sviluppo dei Territori della Basilicata e della Campania Colpiti dai Terremoti del Novembre 1980 e Febbraio 1981, Relazione Conclusiva e Propositiva*; Tipografia del Senato: Roma, Italy, 1991.
64. Postpischl, D. *Catalogo dei Terremoto Italiani Dall'anno 1000 al 1980*; Quaderni della Ricerca Scientifica; 114, 2B; CNR-PFG: Bologna, Italy, 1985.
65. Siro, L. (Ed.) Indagini di Microzonazione Sismica: Intervento Urgente in 39 Centri Abitati della Campania e Basilicata Colpiti dal Terremoto del 23.11.1980. In *Collaboration with Basilicata, Campania, Emilia-Romagna, Toscana Regions*; 492; CNR-PFG: Bologna, Italy, 1983.
66. Porfido, S.; Alessio, G.; Gaudiosi, G.; Nappi, R.; Spiga, E. The Resilience of Some Villages 36 Years after the Irpinia-Basilicata (Southern Italy) 1980. In *Advancing Culture of Living with Landslides. WLF 2017*; Mikoš, M., Vilímek, V., Yin, Y., Sassa, K., Eds.; Springer: Cham, Switzerland, 2017; pp. 121–133. [CrossRef]

67. Guelfi, F.; Monteforti, B.; Bozzo, E.; Galliani, G.; Plesi, G. Comune di Conza della Campania (AV). In *Indagini di Microzonazione Sismica: Intervento Urgente in 39 Centri Abitati Della Campania e Basilicata Colpiti dal Terremoto del 23.11.1980*; 492; CNR-PFG: Bologna, Italy, 1983.
68. Bosi, C.; Parducci, A.; Rossi-Doria, M. (Eds.) *Comune di S. Angelo dei Lombardi: Carta Geologico-Tecnica*; CNR-PFG: Bologna, Italy, 1986.
69. Violi, P. *Paesaggi Della Memoria. Il Tempo, lo Spazio, la Storia*; Bompiani: Milano, Italy, 2014.
70. Canti, M.; Polichetti, M.L. Marche 1997. Dai danni al restauro: Un virtuoso dialogo istituzionale. *Econ. Della Cult.* **2014**, *24*, 335–346. [CrossRef]
71. Corvigno, V. Terremoto e Ricostruzioni in Irpinia, il Restauro e i Piani di Recupero dei Centri Storici Minori. Ph.D Thesis, Federico II University of Naples, Naples, Italy, 12 April 2013.
72. Mazzoleni, D.; Sepe, M. (Eds.) *Rischio Sismico, Paesaggio, Architettura: L'Irpinia Contributi per un Progetto*; CRdC-AMRA: Naples, Italy, 2005.
73. Zizzari, S. *L'Aquila Oltre i Sigilli. Il Terremoto tra Ricostruzione e Memoria*; Franco Angeli: Milano, Italy, 2019.
74. Mauch, C.; Pfister, C. (Eds.) *Natural Disasters, Cultural Responses. CASE Studies Toward Global Environmental History*; Lexington Books: Plymouth, UK, 2009.
75. Olson, R.S.; Gawronski, V.T. Mexico as a Living Tapestry: The 1985 Disaster in Retrospect. *Nat. Hazards Obs.* **2005**, *30*, 1–24.
76. Assmann, J. *Cultural Memory and Early Civilization: Writing, Remembrance, and Political Imagination*; Cambridge University Press: Cambridge, UK, 2011.
77. Weber, M. *Methodology of Social Sciences*; Routledge: London, UK, 2017.
78. Pfister, C. "The Monster Swallows You". Disaster Memory and Risk Culture in Western Europe, 1500–2000. *RCC Perspect.* **2011**, *1*, 1–23.
79. Bankoff, G. *Cultures of Disaster. Society and Natural Hazard in the Philippines*; Routledge: London, UK, 2003.
80. Schenk, G.J. «Learning from history»? Chances, problems and limits of learning from historical natural disasters. In *Cultures and Disasters: Understanding Cultural Framings in Disaster Risk Reduction*; Kruger, F., Bankoff, G., Cannon, T., Orlowski, B., Lisa, E., Schipper, F., Eds.; Routledge: London, UK, 2015; pp. 72–87.
81. Geipel, R. *Long-Term Consequences of Disasters: The Reconstruction of Friuli, Italy, in Its International Context, 1976–1988*; Springer: New York, NY, USA, 1991.
82. Parrinello, G. To Whom Does the Story Belong? Earthquake Memories, Narratives, and Policy in Italy. In *Sites of Remembering: Landscapes, Lessons, Policies*; Lakhani, V., De Smalen, E., Eds.; RCC Perspectives: Munich, Germany, 2018; pp. 69–77. [CrossRef]

geosciences

MDPI

Communication

Photographic Reportage on the Rebuilding after the Irpinia-Basilicata 1980 Earthquake (Southern Italy)

Sabina Porfido [1,2], Giuliana Alessio [2], Germana Gaudiosi [2], Rosa Nappi [2,*], Alessandro Maria Michetti [3] and Efisio Spiga [4]

1 Consiglio Nazionale delle Ricerche-ISA, Via Roma 64, 80100 Avellino, Italy; sabina.porfido@cnr.it
2 Istituto Nazionale di Geofisica e Vulcanologia, Sezione di Napoli Osservatorio Vesuviano, Via Diocleziano 328, 80124 Napoli, Italy; giuliana.alessio@ingv.it (G.A.); germana.gaudiosi@ingv.it (G.G.)
3 Dipartimento di Scienza e Alta Tecnologia, Università degli Studi dell'Insubria, Via Valleggio 11, 22100 Como, Italy; alessandro.michetti@uninsubria.it
4 Independent Researcher, 83100 Avellino, Italy; Spiga.efisio@gmail.com
* Correspondence: rosa.nappi@ingv.it; Tel.: +39-081-610-8324

Abstract: This paper aims to present, through a photographic reportage, the current state of rebuilding of the most devastated villages by the earthquake that hit the Southern Italy on 23 November 1980, in Irpinia-Basilicata. The earthquake was characterized by magnitude Ml = 6.9 and epicentral intensity I_0 = X MCS. It was felt throughout Italy with the epicenter in the Southern Apennines, between the regions of Campania and Basilicata that were the most damaged areas. About 800 localities were serious damaged; 7,500 houses were completely destroyed and 27,500 seriously damaged. The photographic survey has been done in 23 towns during the last five years: Castelnuovo di Conza, Conza della Campania, Laviano, Lioni, Santomenna, Sant'Angelo dei Lombardi, Balvano, Caposele, Calabritto and the hamlet of Quaglietta, San Mango sul Calore, San Michele di Serino, Pescopagano, Guardia dei Lombardi, Torella dei Lombardi, Colliano, Romagnano al Monte, Salvitelle, Senerchia, Teora, Bisaccia, Calitri and Avellino. Forty years after the 1980 earthquake, the photographs show villages almost completely rebuilt with modern techniques where reinforced concrete prevails. Only in few instances, the reconstruction was carried out trying to recover the pre-existing building heritage, without changing the original urban planning, or modifying it. We argue that this photography collection allows to assess the real understanding of the geological information for urban planning after a major destructive seismic event. Even more than this, documenting the rebuilding process in a large epicentral area reveals the human legacy to the natural landscape, and our ability, or failure, to properly interpret the environmental fate of a site.

Keywords: 1980 Irpinia-Basilicata earthquake; photographic reportage; rebuilding

Citation: Porfido, S.; Alessio, G.; Gaudiosi, G.; Nappi, R.; Michetti, A.M.; Spiga, E. Photographic Reportage on the Rebuilding After the Irpinia-Basilicata 1980 Earthquake (Southern Italy). *Geosciences* **2021**, *11*, 6. https://doi.org/10.3390/geosciences11010006

Received: 17 November 2020
Accepted: 21 December 2020
Published: 25 December 2020

Publisher's Note: MDPI stays neutral with regard to jurisdictional claims in published maps and institutional affiliations.

1. Introduction

The earthquake of 23 November 1980, more commonly known as the Irpinia-Basilicata earthquake, was the strongest seismic event to hit the Southern Apennines in the last 100 years. It was characterized by magnitude Mw = 6.9 and intensity Io = X Mercalli Cancani Sieberg (MCS) scale and/or X Environmental Seismic Intensity 2007 (ESI-07) scale [1–4]. It was felt throughout Italy, from Sicily in the South, to Emilia Romagna and Liguria in the North (Figure 1). It caused devastating effects in over 800 localities distributed in the regions of Campania and Basilicata with a total of 75,000 houses destroyed and 275,000 seriously damaged. The number of victims was about 3000, with 10,000 injured people. Several municipalities distributed in the provinces of Avellino, Salerno and Potenza were almost totally destroyed, since the local felt intensity was I > VIII MCS. In the Campania region, the level of damage of 542 towns was classified as follows: 28 towns destroyed, 250 seriously damaged, 264 damaged; in the Basilicata region, 131 municipalities were classified of which nine were destroyed, 63 seriously damaged and 59 damaged [5–7].

The earthquake also caused several striking effects on the natural environment including extensive coseismic surface faulting, observed between Lioni (Avellino) and San Gregorio Magno (Salerno) [8–18]. Moreover, over 200 landslides occurred, the most devasting hit the territories of Calitri in the urban center [11,19–22], Caposele (Buoninventre landslide) and Senerchia (Serra d'Acquara landslide) (Figure 2) [11,23]; also, widespread soil fracturing was observed, and minor liquefaction effects [24–27]. Furthermore, the coseismic faulting of the regional karst aquifer induced important hydrological variations in the springs of Caposele and Cassano Irpino [25] (Table 1).

Forty years after the 1980 earthquake, we decided to record, through a photographic reportage, the state of the rebuilding, the urban changes that the earthquake had induced mainly in the epicenter and near field areas. This study represents a realistic documentation of what has been achieved over all these years, even with the considerable state economic funding, and the resilience of each community [28]. The study of the various localities portrayed is accompanied by a detailed bibliography starting from 1981 until today.

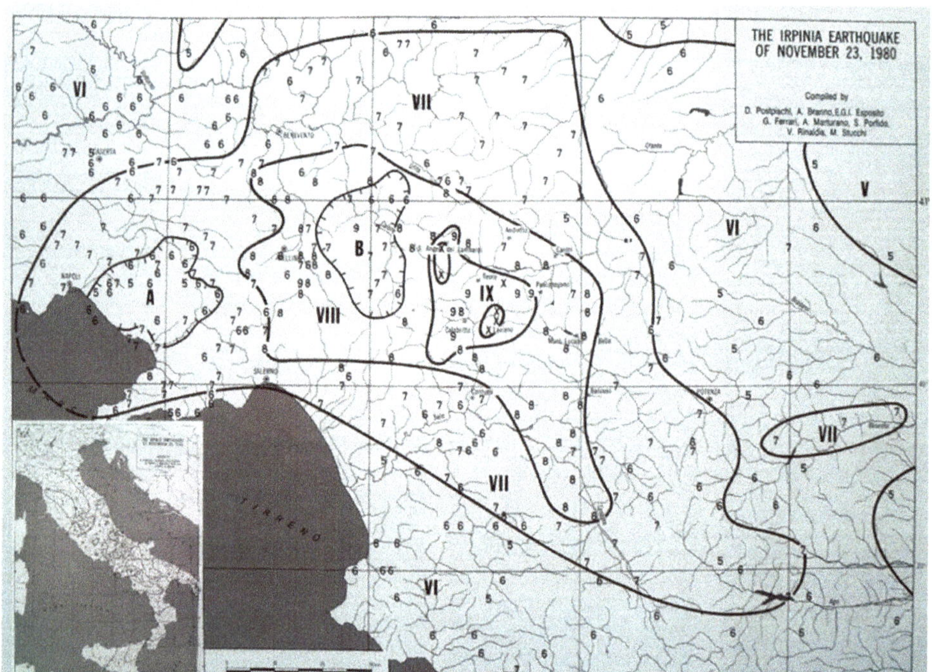

Figure 1. Isoseismal Map of the 23 November 1980 Irpinia-Basilicata earthquake [1].

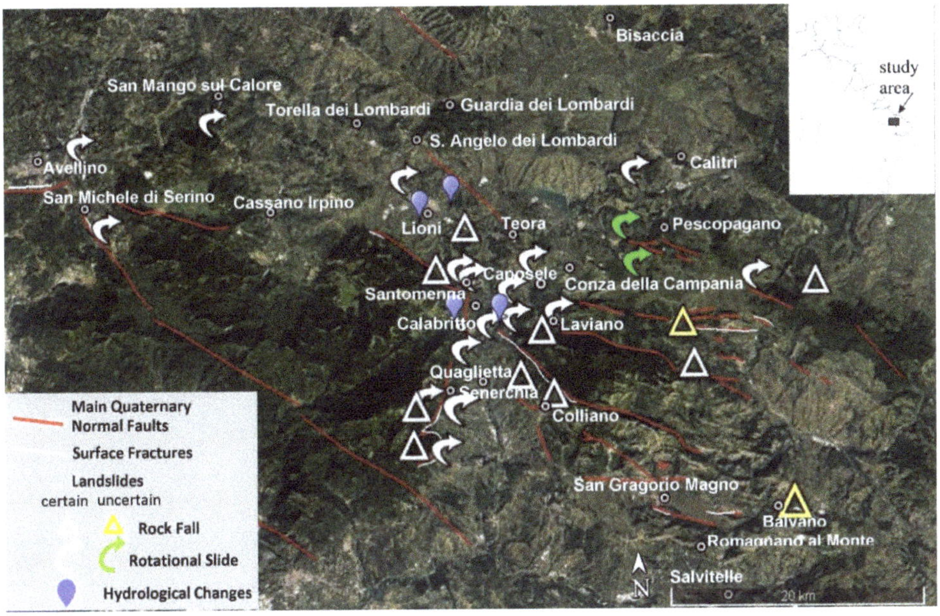

Figure 2. Environmental effects induced in the epicentral area by the 23 November 1980 Irpinia-Basilicata earthquake [21]. The Ortophoto image from Google Earth.

Table 1. The table contains the visited localities, their geographic coordinates and MCS and ESI intensity with the most significant environmental effects observed after the 1980 earthquake (SF surface fault; F fractures, L landslide, SC Soil Compaction, HC Hydrological Changes, LQ Liquefaction) [1–3,11,12,14,19–23,25–27].

Locality	Lat.	Long.	(MCS) Intensity	ESI07 Intensity	Environmental Effects
Castelnuovo di Conza	40.819091°	15.319704°	X	8	SF, L, F
Conza della Campania	40.870562°	15.330818°	X	9	L, F, SC
Laviano	40.786800°	15.309247°	X	9	SF, F, L
Lioni	40.877996°	15.189476°	X	8	L, SF, LQ, SC
S. Angelo dei Lombardi	40.927716°	15.177733°	X	8	L, temporary lake, F
Santomenna	40.808495°	15.321044°	X	-	L, F
Balvano	40.650928°	15.511861°	IX	7	L
Calabritto	40.650928°	15.511861°	IX	9	SF, L, F, HC
Quaglietta	40.745299°	15.235847°	IX	9	L, F
Caposele	40.815372°	15.220236°	IX	10	L, SF, SC, HC
Guardia dei Lombardi	40.953346°	15.209175°	IX	8	L, F

Table 1. *Cont.*

Locality	Lat.	Long.	(MCS) Intensity	ESI07 Intensity	Environmental Effects
Pescopagano	40.835588°	15.399507°	IX	8	L, F
San Mango sul Calore	40.963896°	14.975828°	IX	8	L, F
San Michele di Serino	40.876228°	14.855731°	IX	6	F, LQ
Senerchia	40.740248°	15.203956°	IX	9	SF, L, F, LQ
Teora	40.852752°	15.253197°	IX	6	F
Torella dei Lombardi	40.940058°	15.114519°	IX	8	F
Colliano	40.725860°	15.291517°	VIII-IX	9–10	SF, F, L
Romagnano al Monte	40.628289°	15.456647°	VIII-IX	-	L, F
Salvitelle	40.591196°	15.459109°	VIII-IX	-	F, L
Avellino	40.916753	14.797924°	VIII	-	F, L
Calitri	40.899432°	15.439435°	VIII	8	L, F, LQ
Bisaccia	41.013215°	15.375352°	VIII	8	L, F, SC

2. Discussion

The photographic journey, carried out in the last five years, includes the documentation of the villages almost completely destroyed or seriously damaged (Figure 3), with damage levels evaluated at I $\geq$ VIII MCS, Table 1): Castelnuovo di Conza (Figures 4–9), Conza della Campania (Figures 10–18), Laviano (Figures 19–22), Lioni (Figures 23–28), Santomenna (Figures 29–33), Sant'Angelo dei Lombardi (Figures 34–39), Balvano (Figures 40–43), Caposele (Figures 44–47), Calabritto-Quaglietta (Figures 48–53), San Mango sul Calore (Figures 54–58), San Michele di Serino (Figures 59–65), Pescopagano (Figures 66–70), Guardia dei Lombardi (Figures 71–74), Torella dei Lombardi (Figures 75–79), Colliano (Figures 80–85), Romagnano al Monte (Figures 86–89), Salvitelle (Figures 90–94), Senerchia (Figures 95–99) and Teora (Figures 100–104) [1,2,4,7,29]. In some cases, we also took into account the preliminary seismic microzonation maps drawn up by the PFG of CNR (Finalized Geodynamic Project (PFG) of the National Research Council (CNR), the first major national project for seismic risk assessment and reduction), which provided useful indications for reconstruction [29]. Moreover, two others towns, Bisaccia (Figures 105–108) and Calitri (Figures 109–114), although characterized by a lower intensity (I = VIII MCS), have been added to our study due to the different pathways for reconstruction, despite having both serious hydrogeological problems. Bisaccia has been often affected by landslides, due to its geological formations, mostly made by conglomerates resting on varicolored clays; indeed it was among the villages admitted to consolidation as early as 1917. Nowadays there are two Bisaccia villages: the old, ancient village recovered around the ducal castle and the new Bisaccia of the 'Piano di Zona', the latter almost completely rebuilt according to the urban plan drawn up by the architect Aldo Loris Rossi. Regards Calitri, it was built on the top of a hill made of sandstone and conglomerate rocks, with the middle-lower slopes made of intensely tectonized clay-rich units; therefore, it was often affected by landslides. A great landslide occurred due to the 1980 earthquake, approximately 850 m long and up to 100 m deep; it had terrible consequences on the urban and road structure of the village [19]. Moreover, other significant coseismic environmental effects occurred such as fracturing and liquefaction phenomena. Further landslides occurred in Calitri due to the 1694, 1805, 1910 and 1930 earthquakes [1,7,11,21,22,29–31]. We also added the Avellino city

(Figures 115–119), because it is one of the largest cities in Campania, with level of damage I = VIII [32].

In this photographic reportage, we have deliberately chosen not to reproduce the tragic images of the catastrophe, destruction, death of the places visited, that characterized all the front pages of the newspapers of the time (e.g., Figure 3) [33]. We have chosen to show the current, rebuilt urban centers.

Figure 3. The front page of "Il Mattino" newspaper (26 November 1980) transformed by Andy Warhol into a pop art manifesto [33].

We analyzed the reconstruction through the representative buildings as the churches, the town halls, the sports centers. In some cases, there are pictures of temporary villages, generally made of wooden houses, where people lived for many years waiting for the final accommodation, now used for local tourism. Almost all the villages were rebuilt in the same place despite some of them were completely razed to the ground not only by the earthquake but also by bulldozers that destroyed everything, even more than necessary.

The buildings of the old urban centers were mostly made in natural stone or baked bricks with poor mortar and wooden floors while the rebuilding involved reinforced concrete buildings, earthquake-proof. The new urban centers were rebuilt with wide roads and outsized areas for new housing compared to the current number of inhabitants, that over the last forty years has decreased dramatically, especially for the most internal areas.

As mentioned above, all of the villages affected by the earthquake were rebuilt in situ, with the exception of Conza della Campania, Romagnano al Monte and Bisaccia. For Conza della Campania, a town almost completely destroyed by the earthquake with a high number of deaths (184 victims), the political choice of the urban center relocation prevailed, supported also by the results of the numerous geological surveys some of which also emphasized local amplification phenomena due to morphology. Indeed, Conza was built on two small hills made by clay and sandy clay in the lower part, conglomerates with sands and sandstones in the middle part, and conglomerates of middle-low resistance in the upper part. The 1980 earthquake induced in this village several different environmental effects such as landslides, ground cracks and ground settlement. Moreover, this choice was also due to the historical memory of the destruction suffered by the community in past earthquakes (1466, 1517,1694, 1732 and 1930 seismic events) [7,18,29–31,34]. At present two Conza villages f coexist: the ancient village recovered and enhanced with the creation of an archaeological park that preserves the remains of the ancient Roman 'Compsa'; and the new Conza, built in Piano delle Briglie, 4 km far from the original nucleus where the topography of this flat area ensured safer conditions. This is a modern village characterized by earthquake-proof houses and wide roads, designed by the architect Beguinot of the University of Naples, [35–38].

Even for Romagnano al Monte, a small village in the Salerno province with only 370 inhabitants, located 650 m a.s.l., overlooking the gorges of the Platano River a few kilometres from the epicentre of the 1980 earthquake, the political decision to relocate prevailed. The main reason was due to the declared inhabitability of 446 residential units, after the earthquake, that also caused the collapse of some churches and heavy damage to the town hall. The geomorphological and geological assessment, that accentuated the seismic shaking, also contributed to the choice in the reconstruction process. The old town, in fact, is located at the highest point on the ridge and along the slopes are frequent phenomena of rock fall caused by the high degree of fracturing bedrock.

The village was evacuated and abandoned becoming a "ghost town" [39]. The new town was located in Ariola, 2 km from the old center, in a less panoramic position, but providing more convenient access for the inhabitants.

More complex and longer is the history of Bisaccia's relocation. The village located at 860 m a.s.l., was affected by landslides due to the geological conditions on which it stands and was already destroyed by historical earthquakes (1694 1732, 1930 and 1980 seismic events). These aspects have heavily conditioned its rebuilding. The 1980 earthquake, once again highlighted the territory's extremely unstable conditions, so the Municipal Administrations opted for the reconstruction of the village in another site, more stable from a geological point of view, called "Piano di Zona", which was already identified in a previous urban planning, following the 1930 earthquake [7,40,41].

As a matter of fact, there are currently two Bisaccia towns: the old one, an ancient village recovered around the ducal castle; and the new Bisaccia of the Piano di Zona, the latter almost completely rebuilt according to the urban planning drawn up by architect Aldo Loris Rossi [42].

About the other villages we can say that among those among those which have decided to rebuild in situ, some of them have chosen the recovery or rebuilding of the old urban center, respecting the original architectural design, combined with new buildings in the expansion areas.

Among these we mention certainly Sant'Angelo dei Lombardi, with a careful reconstruction of the historical centre and the Abbey of Goleto (Figure 36) [7,29,31,43]; Calitri despite the historical large landslide triggered by the earthquakes (Figure 110) [7,29]; Guardia dei Lombardi (Figure 73) [7,29]; Torella dei Lombardi [7,29];(Figure 75); Caposele [29,34]; Lioni (Figure 28); Balvano (Figure 41); Pescopagano (Figure 66); and Quaglietta (Figure 53), whose medieval village has been recovered to transform it into an "albergo diffuso" ("scattered hotel") for tourism purposes [34,44].

Even Senerchia, which has been currently rebuilt, tries to recover the remaining houses of the ancient village built on the solid calcareous substratum (Figure 98) [7,29,34]. In all the other villages the new buildings are predominant with some valuable innovative edifices such as the town hall of Lioni (Figure 25), a modern and functional building realized by the architect Verderosa, or in Balvano where the artists Boffo and Eibl designed the houses, or architectural structures often in contrast with the original planning of the village (Figures 42 and 43).

Unfortunately, in many cases during the rebuilding process the identity of Apennine villages such as Laviano (Figures 19 and 22), San Michele di Serino (Figure 62), Castelnuovo di Conza (Figure 5), Santomenna (Figure 30), where the new buildings prevail over the old ones, was lost, becoming only "rebuilt villages".

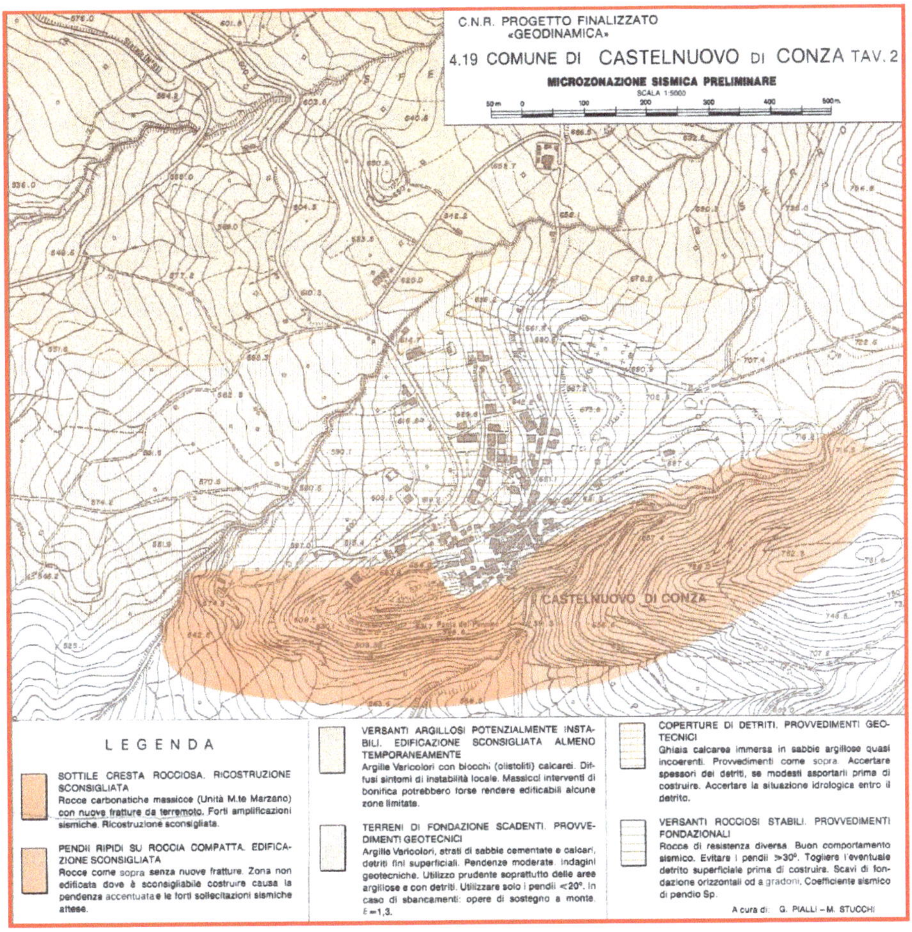

Figure 4. Original map of the seismic microzonation of Castelnuovo di Conza according to [29].

Figure 5. Castelnuovo di Conza: overview of the new village completely rebuilt after the1980 earthquake (photos by [35,36]).

Figure 6. Castelnuovo di Conza: detail of the new village completely rebuilt after the1980 earthquake (photos by [35,36]).

Figure 7. Castelnuovo di Conza: the new sports hall (photos by [35,36]).

Figure 8. Castelnuovo di Conza: Temporary village, consisting of wooden chalets, built immediately after the 1980 earthquake and still used today (photos by [35,36]).

Figure 9. Castelnuovo di Conza: detail of new reinforced concrete buildings (photos by [35,36]).

Figure 10. Original map of the seismic microzonation of Conza della Campania according to [29].

Figure 11. Conza della Campania: overview of the old village destroyed by the 1980 earthquake (photos by [35,36]).

Figure 12. Conza della Campania: old village with ruins of houses after the 1980 earthquake (photos by [35,36]).

Figure 13. Conza della Campania: old village with houses rebuilt after the 1980 earthquake (photos by [35,36]).

Figure 14. Conza della Campania: old village with the Archaeological Park realized a few years ago for the tourists (photos by [35,36]).

Figure 15. Conza della Campania: detail of the Archaeological Park (photos by [35,36]).

Figure 16. Conza della Campania: temporary village, consisting of wooden housing, built immediately after the earthquake of 1980 and now almost completely abandoned (photos by [35,36]).

Figure 17. Conza della Campania: new village with the new Cathedral "Concattedrale di Santa Maria Assunta" in the center of village built after the 1980 earthquake (photos by [35,36]).

Figure 18. Conza della Campania: panoramic view of the new Conza built in Piano delle Briglie locality, 4 km far from the original nucleus of the old Conza (photos by [35,36]).

Figure 19. Laviano: the village completely rebuilt after the 1980 earthquake (photos by [35,36]).

Figure 20. Laviano: the new "Chiesa Madre" (photos by [35,36]).

Figure 21. Laviano: the new Town Hall of Laviano (photos by [35,36]).

Figure 22. Laviano: Piazza della Repubblica (photos by [35,36]).

Figure 23. Lioni: one of the first buildings built in 1985, the "Bergamo condominium" is in reinforced concrete, built thanks to the solidarity of the inhabitants of Bergamo (Northern Italy) (photos by [35,36]).

Figure 24. Lioni: a detail of the "Bergamo condominium", the recent murales was done to remember the solidarity of the people forty years after the catastrophic earthquake (photos by [35,36]).

Figure 25. Lioni: part of modern Town Hall built by the architect A. Verderosa (1984–1994) (photos by [35,36]).

Figure 26. Lioni: modern bus station (photos by [35,36]).

Figure 27. Lioni: the new sports facilities (photos by [35,36]).

Figure 28. Lioni: restored Church and bell tower of S. Maria Assunta (photos by [35,36]).

Figure 29. Santomenna: overview of the village completely rebuilt (photos by [35,36]).

Figure 30. Santomenna: the center of the village completely rebuilt (photos by [35,36]).

Figure 31. Santomenna: a detail of the new buildings (photos by [35,36]).

Figure 32. Santomenna: the new Town Hall in reinforced concrete (photos by [35,36]).

Figure 33. Santomenna: S. Maria delle Grazie church (photos by [35,36]).

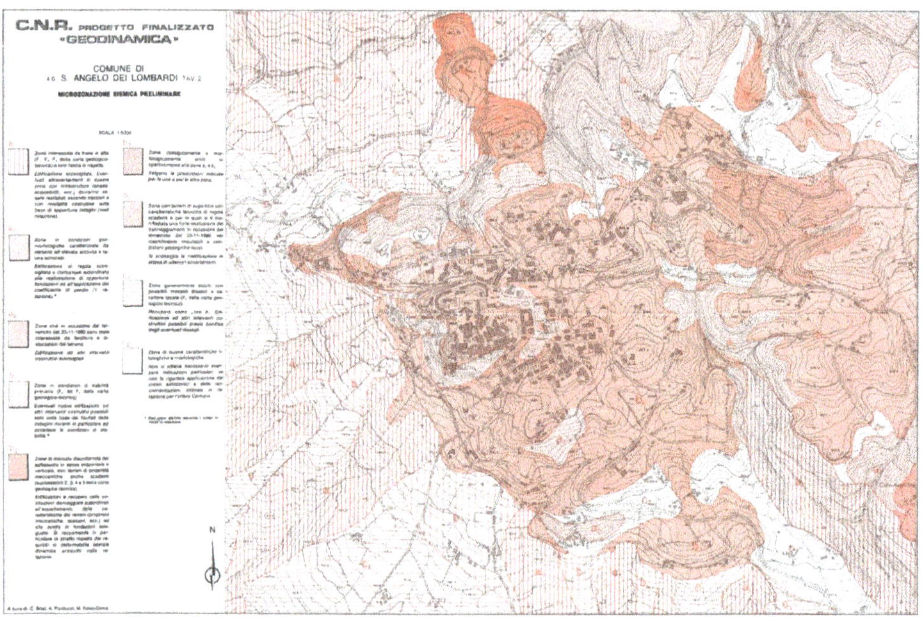

Figure 34. Sant'Angelo dei Lombardi: original map of the seismic microzonation according to [29].

Figure 35. Sant'Angelo dei Lombardi: the historical center with the Town Hall, reconstructed taking into account the original urban design (photos by [35,36]).

Figure 36. Sant'Angelo dei Lombardi: the remains of the Goleto Abbey (photos by [35,36]).

Figure 37. Sant'Angelo dei Lombardi: Archdiocese of Sant'Angelo dei Lombardi in the historical center: (photos by [35,36]).

Figure 38. Sant'Angelo dei Lombardi: Castle of the Imperials of Sant'Angelo dei Lombardi in the Historical center (photos by [35,36]).

Figure 39. Sant'Angelo dei Lombardi: overview of new buildings and the football field (photos by [35,36]).

Figure 40. Balvano: Panoramic view with the Castle in the foreground (photos by [35,36]).

Figure 41. Balvano: overview of new buildings (photos by [35,36]).

Figure 42. Balvano: the new buildings details of the architecture by R. Boffo and K. Eibl (photos by [35,36]).

Figure 43. Balvano: the new buildings details of the architecture by R. Boffo and K. Eibl (photos by [35,36]).

Figure 44. Caposele: the Town Hall in the historical center (photos by [35,36]).

Figure 45. Caposele: the historical center with the new buildings reconstructed taking into account the original urban design (photos by [35,36]).

Figure 46. Caposele: the historical center with the new buildings reconstructed taking into account the original urban design (photos by [35,36]).

Figure 47. Caposele: the historical center with artistic murals (photos by [35,36]).

Figure 48. Calabritto: panoramic view of the village completely rebuilt (photos by [35,36]).

Figure 49. Calabritto: The new Church of Santissima Trinità (photos by [35,36]).

Figure 50. Calabritto: "Largo 23 November 1980"—Memorial dedicated to the earthquake victims (photos by [35,36]).

Figure 51. Quaglietta: Panoramic view of the castle and its medieval village completely restored after the 1980 hearthquake. Currently the medieval village is a tourist attraction as "albergo diffuso" (photos by [35,36]).

Figure 52. Quaglietta: detail of the medieval village (photos by [35,36]).

Figure 53. Quaglietta: detail of the medieval village (photos by [35,36]).

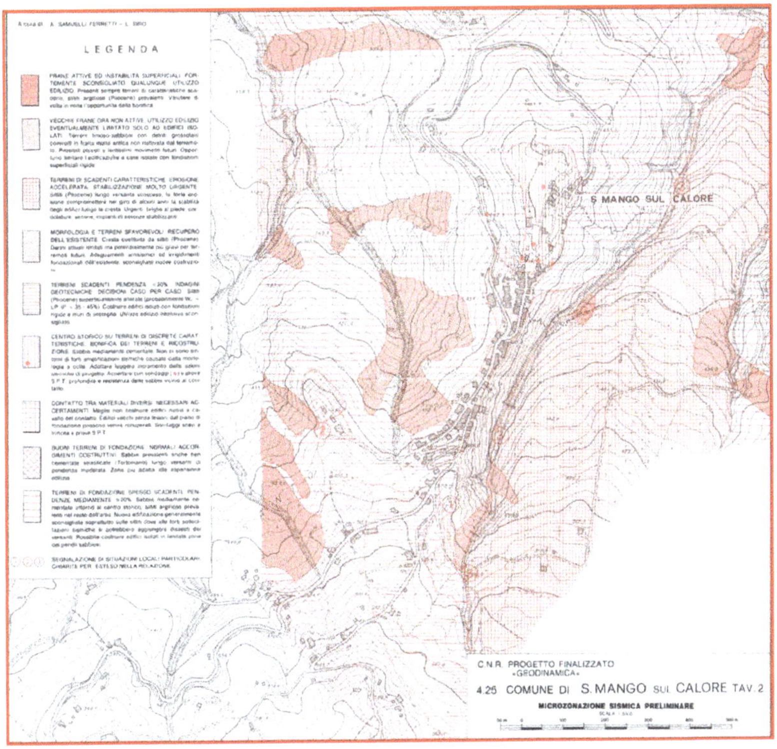

Figure 54. Original map of the seismic microzonation of San Mango sul Calore according to [29].

Figure 55. San Mango sul Calore: Panoramic view of the village completely rebuilt after the 1980 earthquake (photos by [35,36]).

Figure 56. San Mango sul Calore: the new Church of Santa Maria degli Angeli (photos by [35,36]).

Figure 57. San Mango sul Calore: "villaggio italo-canadese" new homes built with the help of Canadians (photos by [35,36]).

Figure 58. San Mango sul Calore: "Villaggio S. Stefano". Temporary village, consisting of wooden chalets, built immediately after the 1980 earthquake (photos by [35,36]).

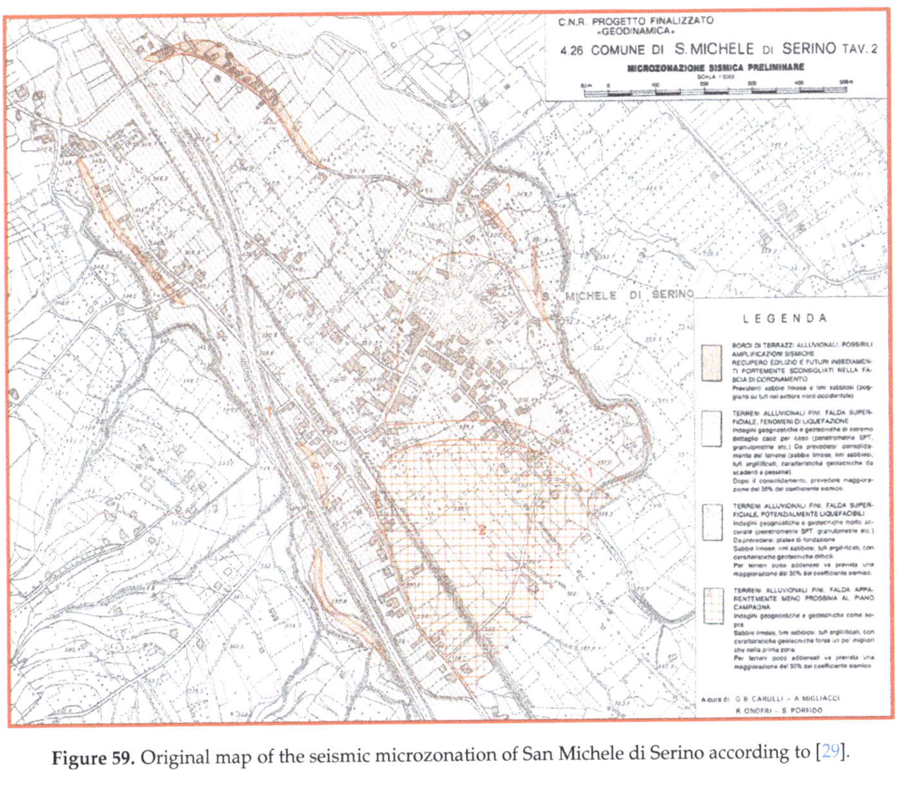

Figure 59. Original map of the seismic microzonation of San Michele di Serino according to [29].

Figure 60. San Michele di Serino: "Mariconda" palace facade, one of the few historical facades not destroyed by the 1980 earthquake (photos by [35,36]).

Figure 61. San Michele di Serino: the new municipal building (photos by [35,36]).

Figure 62. San Michele di Serino: new reinforced concrete buildings (photos by [35,36]).

Figure 63. San Michele di Serino: the new Church of San Michele Arcangelo (photos by [35,36]).

Figure 64. San Michele di Serino: the monument to the victims of the 1980 earthquake (photos by [35,36]).

Figure 65. San Michele di Serino: detail of Cotone Street in which there were phenomena liquefaction triggered by the earthquake [22], (photos by [35,36]).

Figure 66. Pescopagano: a panoramic view of the new village (photos by [35,36]).

Figure 67. Pescopagano: panoramic view of a part of village (photos by [35,36]).

Figure 68. Pescopagano: the new Town Hall of village (photos by [35,36]).

Figure 69. Pescopagano: "Porta della Sibilla" into the historical centre (photos by [35,36]).

Figure 70. Pescopagano: ruins of the medieval castle (photos by [35,36]).

Figure 71. Guardia dei Lombardi: panoramic view of the new village (photos by [35,36]).

Figure 72. Guardia dei Lombardi: the restored Church of Santa Maria delle Grazie (photos by [35,36]).

Figure 73. Guardia dei Lombardi: the restored historical centre (photos by [35,36]).

Figure 74. Guardia dei Lombardi: the new Town Hall of village (photos by [35,36]).

Figure 75. Torella dei Lombardi: the restored castle of Ruspoli di Candriano (photos by [35,36]).

Figure 76. Torella dei Lombardi: the new Church of S. Maria del Popolo (photos by [35,36]).

Figure 77. Torella dei Lombardi: stone plaque to remember the victims of the 1980 earthquake (photos by [35,36]).

Figure 78. Torella dei Lombardi: details of the new buildings (photos by [35,36]).

Figure 79. Torella dei Lombardi: details of the new buildings (photos by [35,36]).

Figure 80. Colliano: the new Town Hall of the village (photos by [35,36]).

Figure 81. Colliano: modern access structure with lift to Piazza Epifani (photos by [35,36]).

Figure 82. Colliano: New parking area of the Epifani Square (photos by [35,36]).

Figure 83. Colliano: New Epifani Square (photos by [35,36]).

Figure 84. Colliano: detail of the restored historical centre (Church of Santa Maria del Borgo) (photos by [35,36]).

Figure 85. Colliano: new residential settlement known as "Piano di zona" (photos by [35,36]).

Figure 86. Romagnano al Monte: panoramic view of the village, completely abandoned after 1980 earthquake, now ghost town (photos by [35,36]).

Figure 87. Romagnano al Monte: details of the old village completely abandoned after 1980 earthquake, (photos by [35,36]).

Figure 88. Romagnano al Monte: the Church of Maria SS. del Rosario in the new town, located in Ariola, 2 km from the old center (photos by [35,36]).

Figure 89. Romagnano al Monte: details of new buildings in the new town (photos by [35,36]).

Figure 90. Salvitelle: Panoramic view of the new village (photos by [35,36]).

Figure 91. Salvitelle: the new Town Hall of the village (photos by [35,36]).

Figure 92. Salvitelle: new buildings mainly in reinforced concrete (photos by [35,36]).

Figure 93. Salvitelle: details of new buildings (photos by [35,36]).

Figure 94. Salvitelle: plaque of a new street that remember the earthquake of 23 November 1980 (photos by [35,36]).

Figure 95. Senerchia: Overview of the remains of the old village (photos by [35,36]).

Figure 96. Senerchia: a small street of the old village (photos by [35,36]).

Figure 97. Senerchia: the new village, "Piazza 23 Novembre 1980" and monument dedicated to the earthquake victims (photos by [35,36]).

Figure 98. Senerchia: details of the new buildings (photos by [35,36]).

Figure 99. Senerchia: new school complex (photos by [35,36]).

Figure 100. Teora: panorama of the reconstructed village (photos by [35,36]).

Figure 101. Teora: the new Church of S. Nicola di Mira (photos by [35,36]).

Figure 102. Teora: new school complex and Town Hall (photos by [35,36]).

Figure 103. Teora: part of the restored historical centre (photos by [35,36]).

Figure 104. Teora: new reinforced concrete houses (photos by [35,36].

Figure 105. Bisaccia: the new village (Piano di Zona) with new Church of Sacro Cuore built by the architect A. L. Rossi, 1998 (photos by [41]).

Figure 106. Bisaccia: the new village (Piano di Zona) with a details of some reinforced concrete buildings realized by architect A. L. Rossi, (photos by [41]).

Figure 107. Bisaccia: the new village (Piano di Zona) with a detail of the new reinforced concrete homes designed by the architect A. L. Rossi, (photos by [41]).

Figure 108. Bisaccia: the new village (Piano di Zona) with the remains of the new social housing still to be completed (Istituto Autonomo Case Popolari-IACP). A part of them has been demolished in 2020 (photos by [41]).

Figure 109. Calitri: panoramic view of the village (photos by [35,36]).

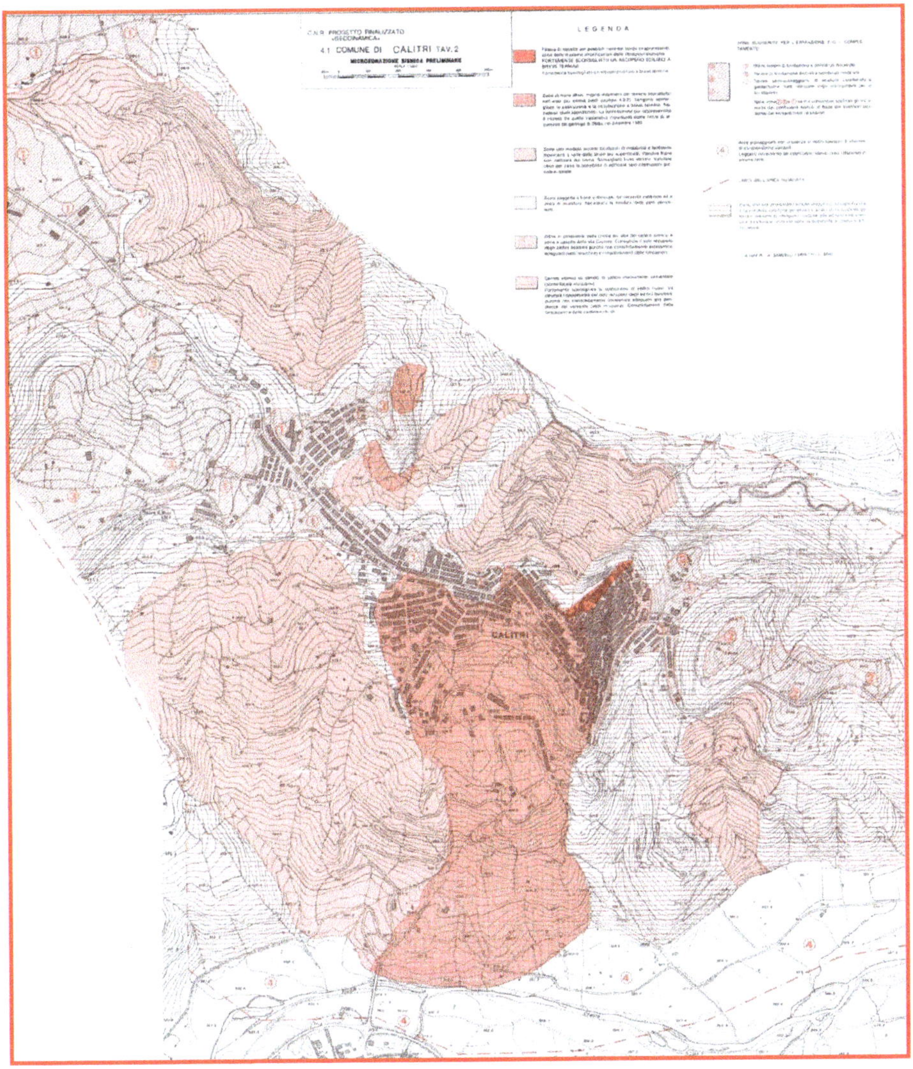

Figure 110. Original map of the seismic microzonation of Calitri according to [29].

Figure 111. Calitri: historical centre, remains of some buildings that were recovered after the 1980 earthquake (photos by [35,36]).

Figure 112. Calitri: historical centre, remains of some buildings that were recovered after the 1980 earthquake (photos by [35,36]).

Figure 113. Calitri: panoramic view of new buildings (photos by [35,36]).

Figure 114. Calitri: panoramic view of the new constructions, the historical centre and the remains of the castle on the hill (photos by [35,36]).

Figure 115. Avellino: panoramic view of the town (photos by [35,36]).

Figure 116. Avellino: Rifugio Street after 1980 earthquake (cortesy of F. Capossela).

Figure 117. Avellino: Rifugio Street today completely reconstructed (photos by [35,36]).

Figure 118. Avellino: "Piazza del popolo" after the 1980 earthquake (courtesy of F. Capossela).

Figure 119. Avellino: "Piazza del Popolo" today completely reconstructed (photos by [35,36]).

3. Conclusions

The rebuilding process, in many localities, has been driven by microzonation maps drawn up as part of the Finalized Geodynamic Project (PFG) of the National Research Council (CNR), created immediately after the 1980 earthquake [29].

The seismic microzonation maps show the areas with different seismic characteristics, illustrating the morphological structure of the territory, the distribution of the building heritage existing in 1980, and in some cases, the level of damage suffered due the 1980 event and provide suggestions for reconstruction. We reported some examples of these maps: Castelnuovo di Conza (Figure 4), Conza della Campania (Figure 11), S. Angelo dei Lombardi (Figure 34), San Mango sul Calore (Figure 55), San Michele di Serino (Figure 60), Calitri (Figure 110).

However, the choices made for the reconstruction did not depend only on the geological assessment of the territory and therefore on the indications given by the microzoning maps, but also on the urban and socio-economic context. Nevertheless, there has not been a socio-economic redevelopment policy for the entire territory which, despite the settlement of some important factories, is currently suffering from unemployment and depopulation [45–49].

Our reportage is certainly not exhaustive, in fact there are still many other places that have undergone considerable transformations during the reconstruction process. The images can guide the future urban development of ancient villages after a major destructive seismic event, with a view to safeguarding the territory and cultural heritage. Even more than this, documenting the rebuilding process in a large epicentral area reveals the human legacy to the natural landscape, and our ability, or failure, to properly interpret the environmental fate of a site. Often in the post-earthquake reconstruction process of 1980, instead of taking into accounts the socio-economic, historical and geological local conditions, different case by case for each affected village, the policy of rebuilding at any cost has prevailed, even with buildings unsuitable for the context of the Apennine inland areas.

Author Contributions: Conceptualization, S.P. and E.S.; methodology, S.P.; software, R.N.; validation, R.N., G.A., G.G., E.S., A.M.M. and S.P.; formal analysis, S.P., E.S.; investigation, G.G., G.A.; resources, R.N.; data curation, R.N., G.G., G.A.; writing—original draft preparation, S.P., G.G., R.N., A.M.M. and G.A.; writing—review and editing, R.N., G.G., G.A., A.M.M.; visualization, A.M.M.; supervision,

S.P.; project administration, E.S., S.P. All authors have read and agreed to the published version of the manuscript.

Funding: This research received no external funding.

Acknowledgments: This manuscript is dedicated to all the inhabitants of these villages who experienced the terrible earthquake of 1980, but nevertheless have been able to recover with great dignity.

Conflicts of Interest: The authors declare no conflict of interest. The funders had no role in the design of the study; in the collection, analyses, or interpretation of data; in the writing of the manuscript, or in the decision to publish the results.

References

1. Postpischl, D.; Branno, A.; Esposito, E.; Ferrari, G.; Marturano, A.; Porfido, S.; Rinaldis, V.; Stucchi, M. The Irpinia earthquake of November 23, 1980. In *Atlas of Isoseismal Maps of Italian Earthquakes*; CNR-PFG: Bologna, Italy, 1985; Volume 114, pp. 152–157.
2. Rovida, A.; Locati, M.; Camassi, R.; Lolli, B.; Gasperini, P. CPTI15, the 2015 Version of the Parametric Catalogue of Italian Earthquakes. INGV 2016. [CrossRef]
3. Serva, L.; Esposito, E.; Guerrieri, L.; Porfido, S.; Vittori, E.; Comerci, V. Environmental Effects from some historical earthquakes in Southern Apennines (Italy) and macroseismic intensity assessment. Contribution to INQUA EEE scale project. *Quat. Int.* **2007**, *173–174*, 30–44. [CrossRef]
4. Gizzi, F.T.; Potenza, M.R. The scientific landscape of November 23rd, 1980 Irpinia Basilicata earthquake: Taking stock of (almost) forty years of studies. *Geosciences* **2020**, *10*, 482. [CrossRef]
5. Commissione Parlamentare d'Inchiesta Sulla Attuazione Degli Interventi per la Ricostruzione e lo Sviluppo dei Territori della Basilicata e della Campania Colpiti dai Terremoti del Novembre 1980 e Febbraio 1981. Relazione Conclusiva e Propositiva. 1991, Volume I. Available online: https://storia.camera.it/lavori/sedute/27-novembre-1990-bf-10-19901127-00-01 (accessed on 24 December 2020).
6. Proietti, G. *"Dopo la Polvere". Rilevazione Degli Interventi di Recupero Post-Sismico del Patrimonio Archeologico, Architettonico ed Artistico delle Regioni Campania e Basilicata Danneggiato dal Terremoto del 23 Novembre 1980 e del 14 Febbraio 1981*; Ministero per i Beni Culturali e Ambientali: Rome, Italy, 1994; Volume 5.
7. Mazzoleni, D.; Sepe, M. (Eds.) *Rischio Sismico, Paesaggio, Architettura: L'Irpinia Contributi per un Progetto*; CRdC AMRA: Naples, Italy, 2005.
8. Bollettinari, G.; Panizza, M. Una faglia di superficie presso San Gregorio Magno in occasione del sisma del 23/11/1980 in Irpinia. *Rend. Soc. Geol. Ital.* **1981**, *4*, 135–136.
9. Cinque, A.; Lambiase, S.; Sgrosso, I. Su due faglie nell'alta valle del Sele legate al terremoto del 23.11.1980. *Rend. Soc. Geol. Ital.* **1981**, *4*, 127–129.
10. Carmignani, L.; Cello, G.; Cerrina Feroni, A.; Funiciello, R.; Klin, O.; Meccheri, M.; Patacca, E.; Pertusati, P.; Plesi, G.; Salvini, F.; et al. Analisi del campo di fratturazione superficiale indotto dal terremoto Campano-Lucano del 23/11/1980. *Rend. Soc. Geol. Ital.* **1981**, *4*, 451–465.
11. Cotecchia, V. Ground deformations and slope instability produced by the earthquake of 23 November 1980 in Campania and Basilicata. *Geol. Appl. Idrogeol.* **1986**, *21*, 31–100.
12. Pantosti, D.; Valensise, G. Faulting mechanism and complexity of the November 23, 1980, Campania-Lucania earthquake, inferred from surface observations. *J. Geophys. Res.* **1990**, *95*, 319–341. [CrossRef]
13. Westaway, R.; Jackson, J. Surface faulting in the southern Italian Campania-Basilicata earthquake of 23 November 1980. *Nature* **1984**, *312*, 436–438. [CrossRef]
14. Blumetti, A.M.; Esposito, E.; Ferreli, L.; Michetti, A.M.; Porfido, S.; Serva, L.; Vittori, E. Ground effects of the 1980 Irpinia earthquake revisited: Evidence for surface faulting near Muro Lucano. In *Large-Scale Vertical Movements and Related Gravitational Processes*; Dramis, F., Farabollini, P., Molin, P., Eds.; Studi Geol. Camerti: Camerino, Italy, 2002; pp. 19–27.
15. Bernard, P.; Zollo, A. The Irpinia (Italy) 1980 earthquake: Detailed analysis of a complex normal faulting. *J. Geophys. Res.* **1989**, *94*, 1631–1648. [CrossRef]
16. Ascione, A.; Mazzoli, S.; Petrosino, P.; Valente, E. A decoupled kinematic model for active normal faults: Insights from the 1980, MS = 6.9 Irpinia earthquake, southern Italy. *Geol. Soc. Am. Bull.* **2013**, *125*, 1239–1259. [CrossRef]
17. Galli, P. Roman to Middle Age Earthquakes Sourced by the 1980 Irpinia Fault: Historical, Archaeoseismological, and Paleoseismological Hints. *Geosciences* **2020**, *10*, 286. [CrossRef]
18. Matano, F.; Di Nocera, S.; Criniti, S.; Critelli, S. Geology of the Epicentral Area of the November 23, 1980 Earthquake (Irpinia, Italy): New Stratigraphical, Structural and Petrological Constrains. *Geosciences* **2020**, *10*, 247. [CrossRef]
19. Del Prete, M.; Trisorio Liuzzi, G. Risultati dello studio preliminare della frana di Calitri (AV) mobilitata dal terremoto del 23/11/1980. *Geol. Appl. Idrogeol.* **1981**, *16*, 153–165.
20. Esposito, E.; Gargiulo, A.; Iaccarino, G.M.; Porfido, S. Distribuzione dei fenomeni franosi riattivati dai terremoti dell'Appennino meridionale. Censimento delle frane del terremoto del 1980. In *Proceeding Internal Convention on Prevention of Hydrogeological Hazards*; CNR(IRPI): Torino, Italy, 1998; pp. 409–429.

21. Porfido, S.; Esposito, E.; Michetti, A.M.; Blumetti, A.M.; Vittori, E.; Tranfaglia, G.; Guerrieri, L.; Ferreli, L.; Serva, L. Areal distribution of ground effects induced by strong earthquakes in the Southern Apennines (Italy). *Surv. Geophys.* **2002**, *23*, 529–562. [CrossRef]
22. Porfido, S.; Esposito, E.; Guerrieri, L.; Vittori, E.; Tranfaglia, G.; Pece, R. Seismically induced ground effects of the 1805, 1930 and 1980 earthquakes in the Southern Apennines. *Boll. Soc. Geol. Ital.* **2007**, *126*, 333–346.
23. Wasowski, J.; Pierri, V.; Pierri, P.; Capolongo, D. Factors controlling seismic susceptibility of the Sele valley slopes: The Case of the 1980 Irpinia Earthquake Re-examined. *Surv. Geophys.* **2002**, *23*, 563–593. [CrossRef]
24. Carulli, G.B.; Migliacci, A.; Onofri, R.; Porfido, S. Indagini geologiche ed in prospettiva sismica a S. Michele di Serino. *Rend. Soc. Geol. Ital.* **1981**, 161–164.
25. Cotecchia, V.; Salvemini, A. Correlation between seismic events and discharge variations at Caposele and Cassano Irpino springs, with particular reference to the 23 November 1980 earthquake. [Correlazione fra eventi sismici e variazione di portata alle sorgenti di Caposele e Cassano Irpino, con particolare riferimento al sisma del 23 Novembre 1980]. *Geol. Appl. Idrogeol.* **1981**, *16*, 167–192.
26. Galli, P. New empirical relationships between magnitude and distance for liquefaction. *Tectonophysics* **2000**, *324*, 169–187. [CrossRef]
27. Esposito, E.; Pece, R.; Porfido, S.; Tranfaglia, G. Hydrological anomalies connected to earthquakes in Southern Apennines, Italy. *Nat. Hazards Earth Syst. Sci. EGS* **2001**, *1*, 137–144. [CrossRef]
28. Osservatorio Permanente sul Doposisma. *Edizioni MIdA-2010.* p. 114. Available online: http://www.laquilaemotion.it/idee/idee/osservatorio-permanente-sul-doposisma.html (accessed on 24 December 2020).
29. Siro, L.; Bigi, G.; Testoni, P.; Luongo, G. Indagini di microzonazione sismica. In *Intervento Urgente in 39 Comuni della Campania e della Basilicata Colpiti dal Terremoto del 23 Novembre 1980*; Pubbl. n. 492; CNR-PFG: Bologna, Italy, 1983; 221p.
30. Guarino, M. Verso una Nuova Qualità dell'abitare: La Riqualificazione dell'edilizia Residenziale Pubblica. Tesi di Dottorato di Ricerca in Ingegneria delle Strutture e del Recupero Edilizio ed Urbano (2007–2010). Available online: http://elea.unisa.it/jspui/handle/10556/156 (accessed on 24 December 2020).
31. Moscaritolo, G.I. Reconstruction as a Long-Term Process. Memory, Experiences and Cultural Heritage in the Irpinia Post-Earthquake (November 23, 1980). *Geosciences* **2020**, *10*, 316. [CrossRef]
32. Maresca, R.; Nardone, L.; Gizzi, F.T.; Potenza, M.R. Ambient noise HVSR measurements in the Avellino historical centre and surrounding area (southern Italy). Correlation with surface geology and damage caused by the 1980 Irpinia-Basilicata earthquake. *Measurement* **2018**, *130*, 211–222. [CrossRef]
33. Artibrune. Dal 2011 Arte Eccetera Eccetera. Available online: https://www.artribune.com/attualita/2011/11/%E2%80%9Cfate-presto%E2%80%9D-stratigrafia-di-un-titolo/ (accessed on 5 October 2020).
34. Scirè, E.; Siro, L.; Stucchi, M.; Gaiazzi, M. Geo-seismic investigations and urban planning after the Irpinia- 971 Basilicata 1980 earthquake: Part 1, the case of Caposele and Conza della Campania. *Geol. Appl. Idrogeol.* **1986**, *21*, 441–450.
35. Porfido, S.; Spiga, E. *Ricostruzione 1980–2020 Ed*; Blurb: San Francisco, CA, USA, 2020; Volume I, ISBN 978-1-71-571504-5.
36. Porfido, S.; Spiga, E. *Ricostruzione 1980–2020 Ed*; Blurb: San Francisco, CA, USA, 2020; Volume II, ISBN 978-1-71-572142-8.
37. Coletta, T. La Conservazione dei Centri Storici Minori Abbandonati. Il Caso Della Campania. Ph.D. Thesis, Università "Federico II" Napoli, Naples, Italy, 2005.
38. Ventura, S. I ragazzi dell'Ufficio del Piano la ricostruzione urbanistica in Irpinia. In *I frutti di Demetra, Bollettino di Storia Ambientale*; Donzelli: Roma, Italy, 2010; Volume 22.
39. Porfido, S.; Alessio, G.; Gaudiosi, G.; Nappi, R.; Spiga, E. The resilience of some villages 36 years after the Irpinia-Basilicata (Southern Italy) 1980 earthquake. In *Workshop on World Landslide Forum*; Springer: Cham, Switzerland, 2017. [CrossRef]
40. Porfido, S.; Spiga, E. Il terremoto del 23 novembre 1980: Ricostruzioni e abbandoni di alcuni paesi nell'Appennino meridionale. The earthquake of November 23rd, 1980: Reconstructions and abandonments of some villages in the southern Apennines. In *La Città Altra Storia e Immagine della Diversità Urbana: Luoghi e Paesaggi dei Privilegi e del Benessere, Dell'isolamento, del Disagio, della Multiculturalità*; Capano, F., Pascariello, M.I., Visone, M., Eds.; CIRICE: Napoli, Italy, 2018; ISBN 978-88-99930-03-5. [CrossRef]
41. Spiga, E.; Porfido, S. *Bisaccia Piano di Zona Ed*; Blurb: San Francisco, CA, USA, 2020; ISBN 978-1-71-555296-1.
42. Locci, M.; Loris Rossi, A. *La Concretezza Dell'utopia*; Testo & Immagine: Torino, Italy, 1997.
43. Scirè, E.; Siro, L.; Stucchi, M.; Gaiazzi, M. Geo-seismic investigations and urban planning after the Irpinia-Basilicata 1980 earthquake: Part 2, the case of S. Angelo dei Lombardi (Italy). *Geol. Appl. Idrogeol.* **1986**, *21*, 387–397.
44. Ventura, F. The long-term effects of the 1980 earthquake on the villages of southern Italy. *Disasters* **1984**, *894*, 11. [CrossRef]
45. Alexander, D. Housing crisis after natural disasters: The aftermath of the November 1980 southern Italian earthquake. In *Geoforum*; Elsevier: Amsterdam, The Netherlands, 1984; pp. 489–516.
46. Guadagno, L. Disastri Naturali e Vulnerabilità Sociale. Un'analisi del Terremoto in Campania. Ph.D. Thesis, Università degli Studi del Sannio, Benevento, Italy, 2010.
47. Wyss, M.; Rosset, P. Near-Real-Time Loss Estimates for Future Italian Earthquakes Based on the M6.9 Irpinia Example. *Geosciences* **2020**, *10*, 165. [CrossRef]

48. Colucci, M.; Gallo, S. Campania in movimento-Rapporto sulle migrazioni interne in Italia. In *Collana "Fuori Collana"*; Società editrice il Mulino: Bologna, Italy, 2020; p. 200. ISBN 978-88-15-28720-5.
49. Ricciardi, T. Avellino. L'Irpinia 40 anni dopo il terremoto, tra cambiamenti e amare continuità. In *Rapporto Italiani nel Mondo 2020*; Fondazione Migrantes: Roma, Italy, 2020; pp. 95–106.

MDPI
St. Alban-Anlage 66
4052 Basel
Switzerland
Tel. +41 61 683 77 34
Fax +41 61 302 89 18
www.mdpi.com

Geosciences Editorial Office
E-mail: geosciences@mdpi.com
www.mdpi.com/journal/geosciences

www.ingramcontent.com/pod-product-compliance
Lightning Source LLC
Chambersburg PA
CBHW061617220926
51850CB00005B/526

* 9 7 8 3 0 3 6 5 6 8 1 6 4 *